Modern Nanomaterials and Nanotechnology for Cancer Therapy

Maoquan Chu (储茂泉)
Editor

Modern Nanomaterials and Nanotechnology for Cancer Therapy

Springer

Editor
Maoquan Chu (储茂泉)
Tongji University
Shanghai, China

ISBN 978-981-95-6848-2 ISBN 978-981-95-6849-9 (eBook)
https://doi.org/10.1007/978-981-95-6849-9

This Springer imprint is published by the registered company Springer Nature Singapore Pte Ltd.
The registered company address is: 152 Beach Road, #21-01/04 Gateway East, Singapore 189721, Singapore

Preface

Although modern science and technology are advancing rapidly, cancer treatment still faces great challenges. Due to the tumor heterogeneity and the complexity of pathogenesis, the development of novel technologies to treat tumors as effectively as we treat common colds remains impossible now. Therefore, further and deep explorations of new strategies and novel approaches are necessary and urgent.

Research over the past several decades shows that cancer treatment requires cross-disciplinary integration of physics, chemistry, pharmacology, biotechnology, and bioinformatics with medical science. New conceptual thinking and innovative approaches need to be introduced into the design of cancer therapy.

Nanomaterials and nanotechnology play important roles in cancer therapy and may represent some of the most promising materials and approaches for overcoming current challenges in cancer treatment. Nanomaterials possess exceptional capabilities in targeted drug delivery in vivo, disrupting the tumor microenvironment (TME), enhancing oxidative damage to tumors, activating antitumor immune responses, and improving the efficacy of conventional therapies such as radiotherapy and chemotherapy. For example, biomimetic nanoparticles coated with cell membranes can evade clearance by the immune system during systemic delivery, functioning as a "Trojan Horse" strategy to enhance tumor-targeting efficiency. Hedgehog-inspired spiky nanoparticles loaded with drugs may facilitate cancer cell uptake due to their hedgehog-like structure, potentially reversing drug resistance in cancer cells. Certain nanomaterials (e.g., manganese dioxide nanoparticles, platinum nanoparticles) can react with the TME to generate oxygen, alleviating tumor hypoxia, which may suppress tumor growth and metastasis and enhance the effectiveness of radiotherapy, chemotherapy, and other traditional cancer therapies. Additionally, some nanoparticles (e.g., Fe_3O_4) can produce hydroxyl radicals through the Fenton reaction within tumor tissue, a process known as chemodynamic therapy.

Importantly, some nanoparticles exhibit enhanced cancer therapeutic efficacy when activated by external stimuli—including light, heat, sound, magnetic fields, electricity, microwaves, X-rays, and others—producing novel treatment modalities. For instance, dark-colored inorganic nanoparticles (e.g., gold nanocages) exposed to a near-infrared (NIR) laser can convert laser light energy into heat due to the intense surface plasmon resonance effect, thereby suppressing tumor growth via localized hyperthermia, a process known as photothermal therapy. Photothermal effects or other heat sources can also trigger the thermal decomposition of specific compounds, generating alkyl radicals for tumor ablation, a technique referred to as thermodynamic therapy. Additionally, photosensitizers exposed to NIR laser radiation can release energy that excites ground-state oxygen ($^3O_2$) into cytotoxic singlet oxygen ($^1O_2$), as well as generate other reactive oxygen species, such as superoxide anions ($•O_2^-$). This oxidative damage to cancer cells and tissues is called photodynamic therapy.

Given the limited penetration depth of external light in biological tissues, alternative stimuli—such as X-rays, ultrasound, microwaves, or electric fields—can activate specific nanomaterials to generate reactive oxygen species. For instance, some nanoscintillators (e.g., copper-cysteamine) can be directly excited by X-ray irradiation to produce reactive oxygen species. Alternatively, X-ray-irradiated nanoscintillators emit fluorescence that excites nearby photosensitizers via fluorescence resonance energy transfer, triggering the photosensitizers to generate reactive oxygen species; this process is called X-ray-induced photodynamic therapy.

Sonoluminescence (light emission resulting from acoustic cavity collapse during ultrasonic cavitation) can activate photosensitizers to produce reactive oxygen species. Additionally, mechanical stress waves and thermal energy generated by cavitation contribute to tumor cell destruction, enabling sonodynamic therapy.

Certain nanomaterials—such as metal-organic frameworks—can generate reactive oxygen species when stimulated by microwaves. This reactive oxygen species production may be further enhanced by microwave-induced hyperthermia, while the thermal effect simultaneously causes direct damage to cancer cells. This combined approach is known as microwave dynamic therapy.

During exposure to electric fields (e.g., high-voltage pulsed electric fields), certain nanomaterials (such as platinum nanoparticles) generate significant reactive oxygen species while simultaneously enhancing the electric field effect. This combination enables electrodynamic therapy for tumors. When magnetic nanoparticles are exposed to high-frequency alternating magnetic fields, they produce thermal effects that facilitate magnetic hyperthermia therapy. In contrast, low-frequency alternating magnetic field or other low-frequency dynamic magnetic fields (e.g., oscillating or rotating magnetic fields) induce negligible heating. However, magneto-mechanical actuation caused by these low-frequency magnetic fields can cause nanoparticles to physically damage cancer cells or tissues through direct impact, inducing irreversible mechanical disruption and concurrent reactive oxygen species production. This approach is known as magneto-mechanical force therapy.

The common feature of these nanomaterials-mediated cancer therapies is that the nanoparticles damage cancer cells and tumor tissues through toxic reactive oxygen species. These effects may enhance the host's antitumor immune response to varying degrees. Furthermore, integrating microneedle technology, traditional Chinese medicine, and stem cell technology with micro/nanotechnology holds significant potential for advancing cancer treatment efficacy.

To capture recent developments in this research field, this book focuses on the fundamental theories, methodologies, and applications at the intersection of nanomaterials/nanotechnology and tumor therapy. Thirteen technologies, including nanomaterials-mediated cancer photodynamic therapy, photothermal therapy, electrodynamic therapy, sonodynamic therapy, chemodynamic therapy, microwave dynamic therapy, thermodynamic therapy, magneto-mechanical force therapy, X-ray-induced photodynamic therapy, magnetic hyperthermia therapy, microneedle-mediated cancer therapy, stem cell therapy, and nanotechnology in traditional Chinese medicine for cancer treatment, are summarized and extensively reviewed. This book includes numerous color figures and experimental examples for the convenience of the reader. All contents have been written by outstanding researchers whose fields include biomaterials, nanobiology, nanomedicine and pharmacy. I sincerely thank Professors Chen Bo (陈波), Ding Yue (丁越), Feng Nianping (冯年平), Hu Yihui (胡益辉), Le Wenjun (乐文俊), Liu Jinliang (刘金亮), Meng Xianwei (孟宪伟), Sun Tiedong (孙铁东), Zhang Jun (张君), Zhang Yongtai (张永太), Zhang Yu (张宇), and Xie Xin (谢鑫), along with their coworkers and students, as well as my own students, for their contributions. This publication is supported by the Fundamental Research Funds for the Central Universities.

This book is likely to be of interest to undergraduate and graduate students, researchers, R&D engineers, and clinicians who wish to understand the core principles, methods, and applications of nanomaterials and nanotechnology-based cancer therapies.

Shanghai, China
July 5, 2025

Maoquan Chu (储茂泉)

Contents

About the Editor in Chief

Maoquan Chu (储茂泉) was born in Anhui Province in China. He received his PhD in Chemical Engineering from East China University of Science and Technology in 2001. He then did his postdoctoral work in the School of Life Science and Technology in Shanghai Jiaotong University, where he delved into biomaterials and nanotechnology. In 2004, he attended the School of Life Science and Technology at Tongji University. In 2007, he won the China Education Ministry's "New Century Excellent Talents Supporting Plan." In 2008, he was hired as a Professor and PhD supervisor. He is now a principal investigator at the School of Life Science and Technology, Tongji University, Shanghai, China. His main research area is nanobiomedicine, with a current focus on cancer and diabetes therapies using nanobiomaterials. His research work has been published in many important journals such as *Small*, *Biomaterials*, *Acta Biomaterialia*, *Nano Research*, *Carbon*, *Theranostics*, *ACS Nano*, etc. As sole author and editor in chief, he has published six books on biomaterials and nanobiology besides this book.

Nanomaterial-Mediated Cancer Photodynamic Therapy

1

Jinliang Liu, Rongjie Gao, and Xingyao Shen

Abstract

The high morbidity and mortality rates associated with cancer necessitate innovative strategies, as traditional therapies often demonstrate limited efficacy in advanced stages primarily due to drug resistance and collateral damage to healthy tissues. Nanomaterial-mediated photodynamic therapy (PDT) has emerged as a promising anticancer approach effectively addressing the limitations of traditional modalities such as surgery, radiotherapy, and chemotherapy. PDT employs photosensitizer (PS) that, upon light activation, generates reactive oxygen species (ROS) to selectively target and destroy malignant cells while preserving the surrounding healthy tissue.

This chapter systematically reviews advancements in nanobiotechnology that have revolutionized efficacy of PDT through the use of nanomaterials. These materials can serve as PS or multifunctional carriers, improving the stability and tumor-targeting capabilities of PS and enabling the combinatorial therapeutic approaches through co-delivery of additional therapeutic agents. The discussion encompasses various types of nanoparticles, including quantum dots (QDs), carbon dots (CDs), metal-organic frameworks (MOFs), organic polymers, upconversion nanoparticles (UCNPs), and X-ray-excited scintillators, highlighting their roles in optimizing PDT outcomes.

Furthermore, this chapter delves into the synergistic potential of combining PDT with other therapeutic strategies, including photothermal therapy (PTT), radiotherapy (RT), chemotherapy, chemodynamic therapy (CDT), and immunotherapy, to address the intrinsic limitations of monotherapy PDT. Despite promising advancements, challenges remain, including the influence of the tumor

J. Liu (✉) · R. Gao · X. Shen
School of Environmental and Chemical Engineering, Shanghai University, Shanghai, People's Republic of China
e-mail: liujl@shu.edu.cn

M. Chu (ed.), *Modern Nanomaterials and Nanotechnology for Cancer Therapy*,
https://doi.org/10.1007/978-981-95-6849-9_1

microenvironment (TME) on ROS generation, the photostability of PS, and the biocompatibility of nanomaterials. Addressing these challenges is crucial for the successful clinical translation of nanomaterial-based PDT. Finally, this chapter aims to provide insights into the future of PDT, emphasizing the necessity for personalized treatment approaches and the development of intelligent, biodegradable nanomaterials that can enhance therapeutic efficacy while minimizing side effects.

Keywords

Photodynamic therapy · Reactive oxygen species · Nanomaterial-mediated · Combination therapy · Photosensitizer

1 Introduction

Cancer is characterized by the abnormal proliferation of cells, seriously impacting public health. With high morbidity and mortality rates, it has become one of the deadliest and most persistent diseases in the world. In 2022, global cancer statistics reported approximately 20 million new cancer cases globally, with an estimated 9.7 million cancer-related deaths [1]. The mortality rate continues to rise annually [2]. Presently, surgical treatment, radiation therapy (RT), and chemotherapy [3, 4] are among the conventional methods applied in clinical practice. Surgical treatment is highly effective for early-stage solid tumors. Nevertheless, its efficacy diminishes for advanced or metastatic tumors. Chemotherapy uses chemotherapeutic agents to inhibit tumor cell growth. Prolonged use may induce drug resistance in tumor cells, affecting the efficacy of chemotherapy. RT delivers targeted ionizing radiation directly to tumor tissue, prompting tumor cell death [5]. While radiation therapy allows for precise targeting and control of radiation doses to reduce systemic toxicity, it still poses collateral damage to surrounding normal tissues [6]. These limitations seriously hinder the therapeutic efficacy of these treatment modalities [7].

In recent years, photodynamic therapy (PDT) has become increasingly recognized in the medical field as an effective modality for treating various cancers [8]. The origins of PDT can be traced back to the 1970s when porphyrin derivatives were first used as photosensitizers to treat bladder cancer. This groundbreaking application marked the inception of PDT in cancer treatment [9]. PDT leverages photosensitizers to trigger the generation of cytotoxic reactive oxygen species (ROS). These ROS, including singlet oxygen ($^1O_2$), hydrogen peroxide (H_2O_2), hydroxyl radicals (·OH), and superoxide anion radicals ($\cdot O_2^-$), play a crucial role in targeting and destroying malignant tumors [10]. PDT has some advantages in cancer treatment, for example, PDT can enable precise targeting of cancerous tissues, minimizing damage to surrounding healthy cells. By selectively targeting cancer cells, PDT can help preserve organ structure and function, enhancing patient outcomes. As a minimally invasive procedure, it significantly reduces the risks associated with conventional cancer treatments. The ability of PDT to effectively treat tumors without harming normal tissues has garnered widespread recognition in the medical community.

Over the past few decades, nanobiotechnology has gradually matured and has become a promising treatment modality in cancer therapy [11, 12]. Nanomaterials can be used as carriers or agents in PDT, serving as a multifunctional delivery system. This system can enhance the stability and tumor-targeting ability of photosensitizers and enable the co-delivery of other anticancer drugs for cancer combination therapy.

This chapter summarizes the research progress of PDT based on nanomaterials in the field of tumor therapy, aiming to provide optimized application strategies for PDT. The discussion begins by systematically and concisely introducing the fundamental principles and commonly used photosensitizers of PDT. Subsequently, the development of PDT in combination applications with other therapeutic strategies is discussed to overcome various limitations of PDT, as well as to integrate diagnostic and therapeutic platforms. Finally, the potential of nanomaterial-based photodynamic combination therapy strategies for future cancer treatments is explored.

2 Mechanisms of PDT

PDT relies on three fundamental components: photosensitizer (PS), appropriate light source, and oxygen in the tumor tissue. In the absence of light exposure, PS itself is harmless. However, when PS accumulates in the tumor area and is exposed to the light of specific wavelengths, it reacts with the oxygen in the region to generate cytotoxic ROS which can damage proteins, nucleic acids, lipids, cell membranes, and organelles within cells, ultimately leading to cell death [13]. The anticancer effects of PDT stem from three interrelated mechanisms: (1) direct cytotoxic effects on tumor cells, (2) damage to the tumor vasculature, and (3) induction of a strong inflammatory response that can lead to systemic immune activation [14].

Unexcited PS molecules are mostly in the ground state with two electrons of opposite spin. When PS is excited by the light of a specific wavelength, it will absorb photon energy, and one electron will be promoted to a higher energy orbital. When the spin direction of the excited state is reversed, the PS molecule will transition from an excited singlet state to an excited triplet state. Subsequently, as the electron returns to a lower energy state, it dissipates energy by radiation, producing phosphorescence. According to physical and chemical processes, PDT can be further classified into type I and type II reactions [15–19] (Fig. 1.1).

Type I reaction involves direct interaction of the PS molecule with organic molecules in the cellular microenvironment, acquiring hydrogen atoms or electrons to promote free radical formation. Subsequently, the reduced PS molecule oxidizes itself to generate $\cdot O_2^-$, generates H_2O_2 through disproportionation or single-electron reduction process, and then transforms into a potent oxidizing agent $\cdot OH$ [20]. This series of reactions leads to oxidative stress, ultimately damaging tumor cells. Type

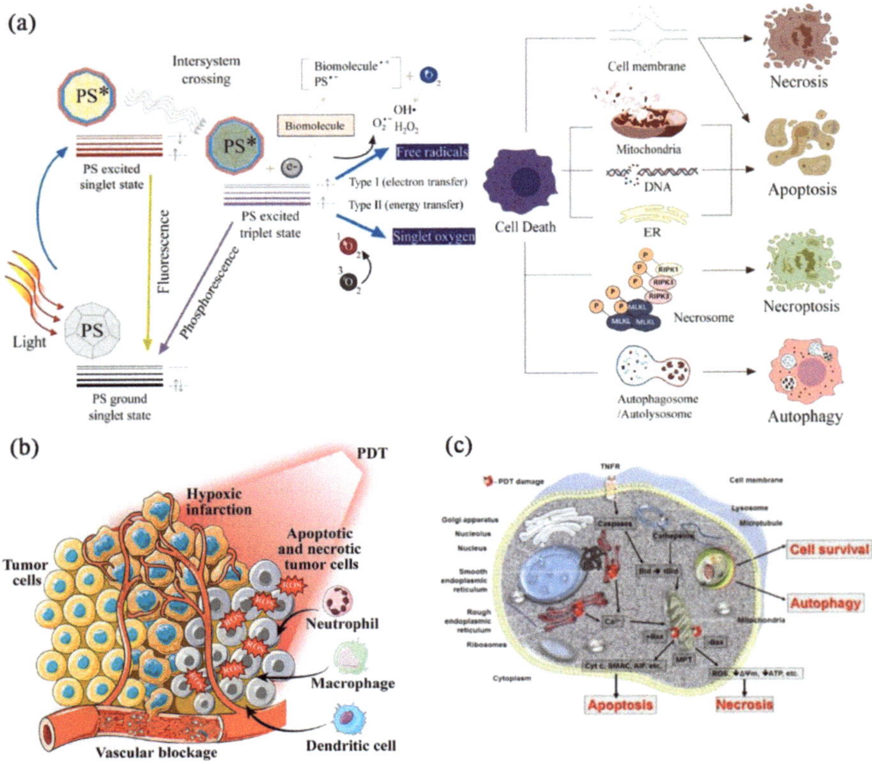

Fig. 1.1 The photophysical mechanism and the two classic types of PDT and the tumor cells damage mechanism caused by PDT. (**a**) Diagrammatic representation of the cell death pathways and photodynamic reactions (type I or type II) in PDT (Copyright © 2024 Elsevier Ltd. Reprinted with permission from Ref. [17]). (**b**) The tumor damage mechanism caused by PDT (Copyright © 2021 Frontiers in Oncology. Reprinted with permission from Ref. [19]). (**c**) Cell death pathways in PDT. The mode of cell death observed after PDT depends to some extent on the intracellular localization of the PS and PDT-related organelle damage. PDT with PS localized in mitochondria will lead to loss of membrane permeability and release of pro-apoptotic mediators, while endoplasmic reticulum damage will lead to the release of cellular calcium deposits. PS that accumulates in lysosomes will lead to the release of proteolytic enzymes upon illumination. Lysosomes may also fuse with autophagosomes to hydrolyze damaged organelles and recycle them during autophagy. In the case of excessive damage, the cell will not survive despite the initiation of autophagy. Certain PS may localize in more than one organelle, and the activation of cell death pathways may occur concurrently (Copyright © 2021 Cancers. Reprinted with permission from Ref. [18])

II reaction involves the energy transfer of molecular oxygen by the excited triplet state PS molecule, yielding $^1O_2$ [14–21]. Type II PDT is heavily dependent on molecular oxygen concentration, a characteristic that limits its practical application in treating hypoxic tumors. In contrast, type I PDT can effectively generate oxygen free radicals even under low-oxygen conditions, giving it an advantage in the treatment of hypoxic tumors. In actual therapy, both reactions usually occur simultaneously and are primarily influenced by factors such as PS concentration, substrate, and oxygen levels [22, 23]. Active species like $^1O_2$ and free radicals possess high reactivity and short half-lives; therefore, their effect is limited to the targeted area where the PS is located, ensuring lower side effects of PDT.

3 Nanomaterial-Mediated PDT

Presently, nanomaterials are gradually replacing traditional polymeric materials due to their unique advantages. These nanomaterials exhibit outstanding performances, such as high specific surface area (BET), good conductivity, light absorption capacity after spectral redshift, fluorescent properties, and degradation capability [24]. These characteristics have positioned nanomaterials as pivotal players in cancer research [25, 26], driving progress in oncology. Nanomaterials with diameters of 100 nm or less have been widely employed in the field of medical treatment. They demonstrate immense potential in drug delivery systems to control drug release, enhance drug permeability, penetrate through biological barriers, and improve biocompatibility [27]. The unique size and properties of nanoparticles have significant impacts on various therapeutic approaches and greatly enhance treatment efficacy [28].

3.1 Nanomaterials as Direct Photosensitizers

3.1.1 Quantum Dots

Semiconductor quantum dots (QDs) are nanocrystals with unique optical properties, predominantly derived from two-dimensional materials such as graphitic carbon nitride (g-C_3N_4) [29, 30], graphene [31], hexagonal boron nitride (h-BN) [32], monolayer buckled crystals (phosphorene), germanene, silicene [33], and transition metal dichalcogenides (TMDCs) [34–36]. Semiconductor quantum dots exhibit distinctive optical characteristics, including high extinction coefficients and emission quantum yields, broad absorption spectra and narrow symmetric emission wavelengths, and chemically stable properties, good biocompatibility, relative ease of surface modification, high specific surface area, and easy functionalization, which enable their wide applications in bioimaging [37, 38], sensing, optoelectronics [39], catalysis, energy storage, and so on [40, 41]. QDs, particularly as photosensitizers in PDT, can generate $^1O_2$ upon light excitation, thereby effectively destroying cancer cells.

Compared to traditional photosensitizers, QDs offer higher photostability and larger two-photon absorption cross section, allowing for effective PDT in deep tissues. QDs can also facilitate energy transfer through Förster resonance energy transfer (FRET) processes [42], providing the required light absorption wavelengths for PDT, thereby overcoming the optical weaknesses of conventional phototherapeutic agents. Crucially, QDs exhibit size-dependent luminescent property, allowing for the design of QDs tailored for specific luminescent applications [43]. Unlike the emission spectra of conventional photosensitizers in the visible range, the emission spectra of QDs can be easily modified to emit at any position between ultraviolet and near-infrared (NIR) light regions by simply adjusting the size. This enables excitation in the NIR range, enhancing tissue penetration, thereby facilitating targeting of deep-seated tumors [44]. Moreover, dual-photon emission (TPE) and FRET can enhance the $^1O_2$ quantum yield [45], making QDs promising in the field of PDT.

Traditional metal QDs with CdSe cores induce cell death by increasing ROS and inhibiting survival-related signals, and the cytotoxicity of CdSe core is related to the release of free Cd^{2+} [46]. One effective method to reduce their toxicity is coating CdSe cores with zinc sulfide (ZnS). However, extended circulation often results in the leaching of bare QDs from the shell, leading to cellular toxicity [47]. Although cadmium-based semiconductor quantum dots have advantages in optical properties, their biocompatibility and toxicity issues limit their applications [48–50]. In contrast, carbon-based quantum dots exhibit lower toxicity and higher biocompatibility, making them promising alternatives to cadmium-based quantum dots [51]. Liang et al. [52] designed a novel photodynamic nanoplatform called Mn-functionalized graphene quantum dots (Mn-FGQD), which can effectively induce autophagy-related cancer cell death by precisely damaging lysosomes (Fig. 1.2a). The introduction of Mn significantly enhances the ROS generation capacity of Mn-FGQD, and experimental results demonstrated excellent ROS generation capabilities in both type I and type II pathways. Moreover, the high structural stability of Mn-FGQDs maintains high photodynamic efficiency in lysosomal environments,

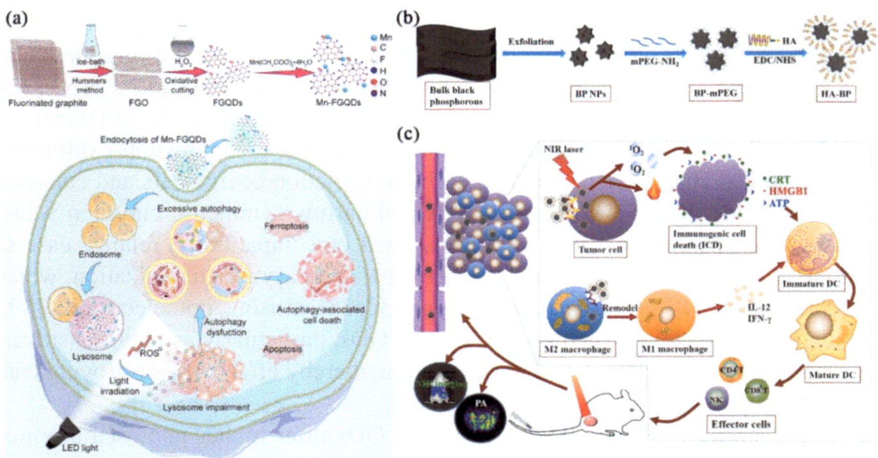

Fig. 1.2 PDT based on QDs as PSs. (**a**) The preparation principle of Mn-FGQDs and the mechanism of photodynamic lysosome damage induced autophagy-related tumor cell death (Copyright © 2023 Journal of Materials Chemistry B: Materials for Biology and Medicine. Reprinted with permission from Ref. [52]). (**b**) Synthesis scheme of HA-BP nanoparticles, and (**c**) schematic diagram of the combined treatment of PDT/PTT/immunotherapy of tumor cells with HA-BP nanoparticles (Copyright © 2020 Elsevier B.V. on behalf of KeAi Communications Co., Ltd. Reprinted with permission from Ref. [56])

enabling lysosome destruction under light irradiation while inducing severe autophagic dysfunction and further enhancing PDT efficiency.

Black phosphorus, an allotrope of phosphorus composed of phosphorus layers, can be further exfoliated into ultrathin nanomaterials [53]. Its nanosheets exhibit thickness-dependent tunable bandgaps ranging from approximately 0.3 eV in bulk to 2.0 eV in a single layer, bridging the gap between zero-bandgap graphene and wider-bandgap transition metal sulfides. The absorption range of black phosphorus nanosheets spans the ultraviolet and visible light regions, making them suitable for NIR photothermal applications. Recently, another zero-dimensional nanostructured material derived from black phosphorus, known as black phosphorus quantum dots (BPQDs) [54], has garnered significant attention due to their broad-spectrum light absorption in the ultraviolet and NIR regions, enabling their use as biocompatible PSs for photothermal therapy (PTT) and PDT [55]. A novel targeted multifunctional black phosphorus (BP) nanoparticle, PEGylated hyaluronic acid (HA)-modified, was designed by Zhang et al. [56] for PTT/PDT/photo-immunotherapy (Fig. 1.2b, c). The HA-BP nanoparticles exhibit good biocompatibility and stability, reshaping the phenotypes of tumor-associated macrophages (TAMs) in in vitro experiments. The HA-BP nanoparticles selectively accumulate at tumor sites through CD44-mediated active targeting and passive enhanced permeability and retention (EPR) effects. The combined treatment of PDT, PTT, and immunotherapy using HA-BP nanoparticles not only significantly inhibits primary tumors but also induces immunogenic cell death (ICD) and releases damage-associated molecular patterns (DAMPs), thereby inducing dendritic cell (DC) maturation and activation of effector cells, significantly enhancing anti-tumor immune responses and effectively treating cancer.

3.1.2 Carbon Dots

Carbon dots (CDs) are one of the most important and innovative materials in the new generation of carbon nanomaterials, consisting of quasi-spherical carbon nanoparticles with diameters of <10 nanometers [57]. They possess multifunctional advantages [58], including non-blinking strong fluorescent emission [59], high conductivity, good physicochemical properties, high quantum yield (QY) [60], low photobleaching, tunable photoluminescence (PL) [61], high light absorption rate, excellent biocompatibility [62], good water solubility, low toxicity, high chemical and photostability, and environmental friendliness [63, 64]. These properties have led to their wide applications in research areas such as biosensing [65, 66], bioimaging [67], and disease diagnosis and treatment [68, 69], particularly demonstrating significant potential in PDT applications.

As inorganic PSs, CDs can generate a variety of ROS due to their structural characteristics and facile electron transfer properties [70]. Additionally, CDs have the advantage of broad absorption, allowing for a wide response to stimuli from light sources of different wavelengths, thereby increasing ROS generation rates with minimal constraints from the excitation wavelength [71, 72]. The core of CDs is composed of graphitic sp^2 hybridized carbon [73], while the shell is rich in organic functional groups such as carboxyl, amino, and hydroxyl groups. This structural composition makes CDs easily amenable to various surface modifications and functionalizations, such as atomic doping (e.g., N, P, S, F, and inorganic functional groups, e.g., carbonyl, carboxyl, hydroxyl, etc.) modifications, enabling controlled tuning of their optical activities (PL, QY, energy levels, spectral range, and photosensitivity), as well as obtaining good water solubility [74]. Utilizing them as photosensitizers in PDT overcomes drawbacks of traditional photosensitizers [75], such as poor water solubility, high toxicity, low stability, and low photosensitivity, while maximizing ROS production during photoexcitation [76], thus enhancing their biological utility.

Zhang et al. [77] prepared three types of ROS-regulating carbon dots (RCDs) using persimmon leaf extract, among which the adjusted p-RCD could simultaneously generate quantum yields of 0.42 for $^1O_2$ and a certain amount of $O_2^{-}\cdot$ (Fig. 1.3a, b). In vitro experiments revealed that after treatment with h-RCDs at 200 μg/mL and exposure to 100 mW/cm^2 of 635 nm laser, the death rate of 4T1 cells reached as high as 95%. Cai et al. [78] designed a novel light-controlled NO-releasing CD platform for synergistic enhancement of tumor PDT (Fig. 1.3c). These CDs, synthesized from methylene blue and metformin through a hydrothermal strategy, can target cancer cell lysosomes. Under light irradiation, the metformin groups on the surface can supply abundant NO, which reacts with simultaneously produced ROS to form more toxic peroxynitrite ($ONOO^-$), subsequently disrupting lysosomal structure and permeability to hasten cancer cell death.

3.1.3 Metal-Organic Frameworks

Metal-organic frameworks (MOFs) are porous polymeric materials assembled from organic ligands and metal ions or metal clusters through specific functional

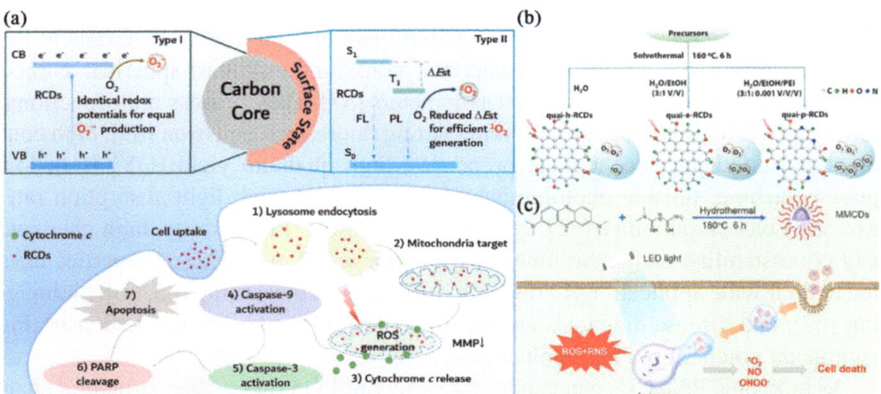

Fig. 1.3 PDT based on CDs as PSs. (**a**) Mechanism of type I and II ROS produced by the prepared RCDs and the schematic diagram of PDT in tumor cells. (**b**) Preparation process of the type I/II RCD PSs with equal O_2^- and incremental $^1O_2$ production (Copyright © 2023 Biomaterials. Reprinted with permission from Ref. [77]). (**c**) Schematic diagram of MMCD preparation principle and photodynamic lysosome damage-induced tumor cell death (Copyright © 2024 Chinese Chemical Letters. Reprinted with permission from Ref. [78])

connections. This structural design allows MOFs to exhibit high order and tunability at the molecular level, imparting them with unique chemical and physical properties. By scaling down the size of MOFs to the micro- and nanometer scale, nanoscale metal-organic frameworks (NMOFs) are formed, showcasing immense potential in substance separation, enrichment, and detection. This size adjustment significantly expands the application scope of porous MOFs in various fields such as analytical chemistry, environmental science, and life sciences.

Recently, due to their excellent photophysical and photochemical properties [79], as well as their inherent porosity, NMOFs have been explored for tumor phototherapy. Furthermore, many NMOFs exhibit favorable metabolism in the kidneys [80], indicating high biocompatibility and safety, further enhancing their potential as phototherapeutic agents [81].

Most PS-based NMOFs are prepared from organic PS complexes, such as BODIPY (4,4-difluoro-4-bora-3a,4a-diaza-s-indacene) dyes and porphyrins and their derivatives [82], serving as organic ligands and metal ions or clusters [83, 84]. The resulting PS-based NMOFs can absorb light to undergo photochemical pathways and generate cytotoxic $^1O_2$ in the presence of molecular oxygen, thereby inducing cell death in tumors [85]. In PS-based NMOFs, the metal ions/clusters play a crucial role in enhancing the efficacy of PDT by alleviating hypoxia, increasing the production of cytotoxic $^1O_2$ or ROS [82].

Lu et al. [86] successfully designed and synthesized DBP-UiO NMOF through a solvothermal method, utilizing Hf(IV) as the heavy metal center, which endowed it with PS functionality (Fig. 1.4a). Compared to conventional PSs, DBP-UiO NMOF demonstrates significant $^1O_2$ quantum yields and enhanced PDT efficacy due to its discrete organic ligands, strengthened Hf(IV) intersystem crossing, and inherent porosity of NMOFs. Chen et al. [87] developed a novel self-oxygen-enriched nanoplatform 69-L_2@F for PDT by combining fluoropolymer and BODIPY-based MOF (Fig. 1.4b). 69-L_2 is constructed from the post-synthetic exchange of BODIPY-derived ligand L_2, followed by methoxy poly(ethylene glycol) functionalization with fluorinated phosphoric acid to obtain 69-L_2@F. This platform exhibited significant inhibition of tumor growth in in vivo experiments, especially under simulated hypoxic conditions in deep-seated tumors. This research provides new insights for designing and developing novel MOF-based PSs, offering the potential to enhance therapeutic outcomes in PDT, particularly for treating hypoxic tumors.

3.1.4 Organic Polymers

The PSs used in traditional PDT exhibit diverse molecular structures. The first-generation PS includes various forms of hematoporphyrin derivatives (hematoporphyrin, HPD), which were employed in cancer treatment in the 1950s [88]. However, despite the widespread use of first-generation PSs, they come with several limitations, such as skin photosensitivity [89, 90], low efficacy at low doses [91], instability, and lack of selectivity. Another limitation of this generation is that they are primarily excited by short waves, resulting in poor tissue penetration [74]. Second-generation PSs mainly consist of porphyrins, dihydroxyphenyl, phenocyanine, anthocyanins, and curcumin, among others. They possess advantages such as higher

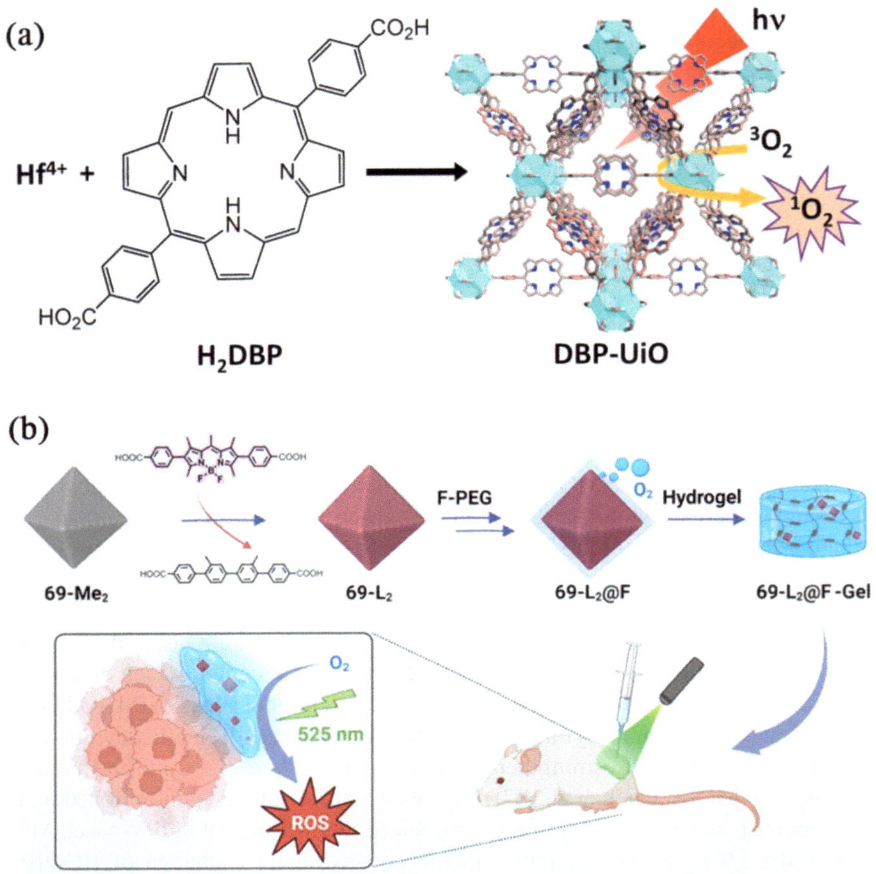

Fig. 1.4 PDT based on MOFs as PSs. (**a**) Schematic of the synthesis of Hf-DBP NMOF and the generation of singlet oxygen (Copyright © 2014 Journal of the American Chemical Society. Reprinted with permission from Ref. [86]). (**b**) Schematic of the preparation and PDT process of the oxygen self-sufficient nanoscale platform 69-L_2@F (Copyright © 2024 Journal of the American Chemical Society. Reprinted with permission from Ref. [87])

chemical purity, longer absorption wavelengths, greater tissue penetration, and higher molar absorption coefficients, enhancing their mediated tumor PDT effects [92]. However, most second-generation PSs are highly hydrophobic, limiting their administration via veins, are prone to aggregation in physiological environments, and exhibit poor tumor targeting, thus reducing the efficiency of ROS production. The development of third-generation PSs is based on the rapid advancements in nanotechnology and nanomedicine. This involves the incorporation of second-generation PSs into nanoformulations such as nanomicelles, nanoliposomes, and nanovesicles or the construction of novel nano-PSs by covalently binding them with antibodies or polymers [18]. These systems can actively or passively target diseased tissues, enabling efficient optical diagnosis and treatment of diseases [93].

Polymer nanoparticles are composed of natural polymers like sodium alginate, albumin, chitosan, or synthetic polymers such as polyacrylamide, polyvinylpyrrolidone, and polylactic acid-glycolic acid copolymers [94]. Polymer nanoparticles offer advantages such as good biocompatibility, biodegradability, ease of surface modification, and excellent stability, enhancing the solubility of traditional PSs and increasing their accumulation at tumor sites [95]. Zheng et al. [96] reported the development of polymer nanoparticles (poly(TPP)NPs) constructed by loading tetraphenylporphyrin (TPP) onto a biocompatible copolymer mPEG-PLA (Fig. 1.5a). This drug delivery system exhibited high drug loading capacity for TPP, enhanced TPP accumulation at tumor sites, and improved in vivo anti-tumor efficacy. Yu et al. [97] developed a series of novel nanoparticles formed by the electrostatic interaction between negatively charged 1–2 generation aryl benzyl ether dendrimer zinc (II) phthalocyanines and positively charged triblock copolymer poly(L-lysine)-b-poly(ethylene glycol)-b-poly(L-lysine) for effective PSs in PDT, targeting mitochondria and efficiently inducing apoptotic cell death in tumor cells.

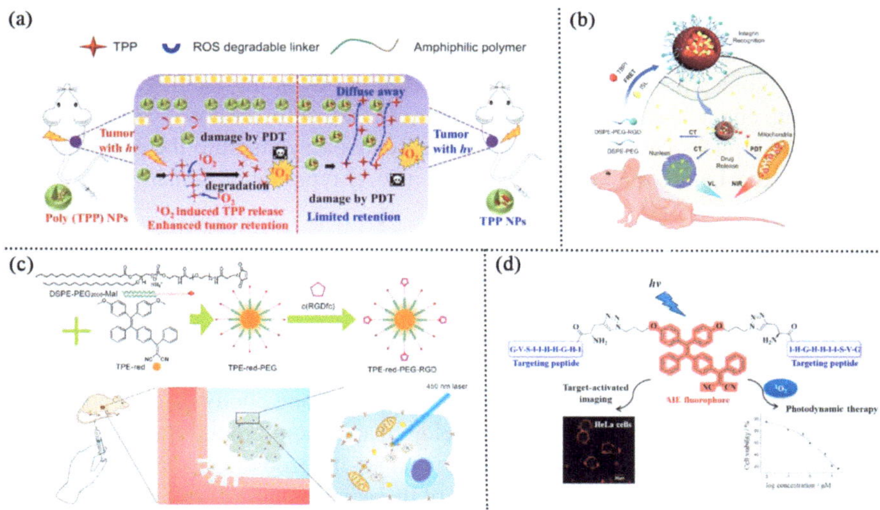

Fig. 1.5 PDT based on organic polymers as PSs. (**a**) Enrichment and PDT schematic diagram of poly(TPP) NP drug delivery system at tumor site (Copyright © 2019 ACS Applied Materials and Interfaces. Reprinted with permission from Ref. [96]). (**b**) Composition of IT-PEG-RGD nanoparticles and schematic diagram of PDT and chemotherapy combination therapy for tumor cells (Copyright © 2023 ACS Nano. Reprinted with permission from Ref. [106]). (**c**) The preparation process of TPE-red-PEG-RGD nanoparticles and the schematic diagram of in vivo PDT via intravenous injection (Copyright © 2018 Nano Research. Reprinted with permission from Ref. [107]). (**d**) A novel red light-emitting light source TPE-red with typical AIE characteristics, as well as the chemical structure and synthesis route of TPE-red-2AP2H that can be used for PDT (Copyright © 2014 Analytical Chemistry. Reprinted with permission from Ref. [108])

However, when traditional PSs are formulated into nanoparticles, their fluorescence and generation efficiency of active oxygen are compromised due to a phenomenon known as aggregation-induced quenching [98, 99]. This effect has been addressed by the aggregation-induced emission (AIE) phenomenon, which provides a new perspective [100–102]. Organic propeller-shaped AIE luminogens, such as tetraphenylethene (TPE), exhibit no or weak emission in good solvents (often organic) but become brighter when aggregated in poor solvents (like water) [103]. This characteristic has allowed for the development of new PSs by combining AIE luminogens with traditional PSs, further benefiting PDT. AIE luminogens have been combined with popular PSs like protoporphyrin IX to serve as enhancers of ROS generation [104, 105]. Sun et al. [106] synthesized a D-A asymmetric small organic molecule TBPI with strong intramolecular charge transfer (ICT) effects and AIE properties for mitochondria-targeted photosensitizers in tumor photodynamic therapy (Fig. 1.5b). They conjugated TBPI to ISL (a natural flavonoid with anti-tumor activity) and constructed IT-PEG-RGD nanoparticles using DSPE-PEG-RGD. TBPI exhibited higher intersystem crossing (ISC) efficiency and longer triplet diffusion lengths inside the nanoparticles. This provides direct evidence and theoretical support that the AIE effect significantly enhances ROS generation capabilities of PS.

Sun et al. [107] designed an easily prepared TPE-red-PEG-arginine-glycine-aspartic acid (RGD) PS nanoparticle for targeted and image-guided in vivo PDT (Fig. 1.5c). The TPE-red-PEG-RGD nanoparticles harness the AIE properties of the TPE-red molecule, enhancing ROS generation, avoiding aggregation-induced quenching, and achieving higher anti-tumor efficacy at lower doses. Hu et al. [108] introduced a push-pull electron-withdrawing cyanoacrylate onto the AIE molecule TPE, synthesizing the donor-acceptor (D-A)-type photosensitizer TPE-red (Fig. 1.5d).

3.1.5 Metal-Based Nanomaterials

Metal nanomaterials such as gold (Au), silver (Ag), manganese (Mn), copper (Cu), and other transition metals are renowned for their outstanding optical properties, particularly the localized surface plasmon resonance (LSPR) effect they exhibited [109, 110]. These metal nanoparticles, due to their strong photon energy absorption capability, create a significant electric field enhancement around them through the LSPR effect. This effect, along with various light-induced effects and intermolecular interactions, has led to the application of these nanomaterials in targeted PDT for tumors. The high cost and non-biodegradability in biological systems of nanomaterials based on Au or Ag have limited their widespread clinical application [111]. In contrast, Cu as an abundantly available metal possesses unique biological activity and can effectively eliminate tumor cells by impacting multiple cell death pathways [112].

Copper nanoparticles have photothermal performance and relatively high NIR absorption capacity; thus, during NIR irradiation, Cu-containing PSs located on the tumor cell membrane interact with surrounding oxygen molecules, leading to an increase in energy levels and the formation of $^1O_2$ through oxygen energy transfer.

Hosseini et al. [113] reported interactions between Cu^{2+} and respiratory complexes (I, II, and IV); the addition of Cu^{2+} to isolated mitochondria leads to damage in electron transfer chains of complexes I, II, and IV, resulting in increased ROS production. Yang et al. [114] designed AIE-active NS-STPA and CuCy-assembled NSCuCy NPs; under 660 nm irradiation, the generation of ·OH from NSCuCy NPs at the same concentration far exceeds that of NS-STPA NPs (Fig. 1.6a). CuCy, predominantly based on Cu, undergoes a Fenton-like reaction with O_2^- to generate ·OH, which is more effective in combating the glutathione (GSH)-induced oxidative stress in the tumor environment and disrupting the redox balance to induce tumor cell disintegration.

Metal coordination self-assembly has recently emerged as a new method for creating therapeutic diagnostic nanosystems utilizing the coordination force between metal ions and organic ligands [115–117]. Ru(II) complexes exhibit superior photostability, satisfactory TPA cross section, significant Stokes shift, and high quantum yield of $^1O_2$ generation compared to clinically used PSs [118]. Liu et al. [119] designed and constructed a novel AIE-active Ir(III) complex, IrDPTP, with high ROS generation capacity, efficient hydrogen production, and NIR-II emission capabilities due to the presence of iridium atoms (Fig. 1.6b). IrDPTP was then

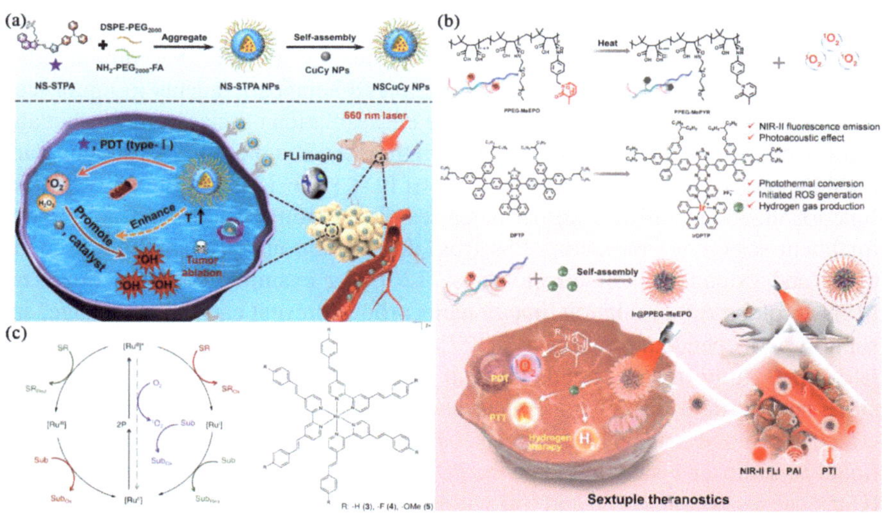

Fig. 1.6 PDT based on metal nanomaterials as PSs. (**a**) Schematic diagram of the preparation of NSCuCy NPs and the mechanism of PDT and promotion of ·OH generation in tumor cells (Copyright © 2023 ACS Applied Nano Materials. Reprinted with permission from Ref. [114]). (**b**) Design principles, preparation, and application of Ir@PPEG-MeEPO on NIR-II FLI-PAI-PTI trimodal imaging-guided photodynamic-photothermal-hydrogen therapy synergistic cancer therapy (Copyright © 2024 Science China: Chemistry. Reprinted with permission from Ref. [119]). (**c**) Possible reaction pathways for two-photon-absorbing ruthenium PSs following both energy- and electron-transfer processes ([Ru] ruthenium PS, SR sacrificial reagent, Sub substrate, Ox oxidation, Red reduction) and molecular structures of Ru TPA complexes 3, 4, and 5 (Copyright © 2022 Nature Communications. Reprinted with permission from Ref. [120])

encapsulated in peroxide-functionalized amphiphilic polymers to form Ir@PPEG-MeEPO and has proven effective for ROS generation, especially under hypoxic conditions. Han et al. [120] reported a series of ruthenium-based PSs coordinated with 4,4′-bis(ethylene)-2,2′-bipyridine (bpyvp), which, under 740 nm light, achieve excellent rates of energy transfer reactions (e.g., reactions involving singlet oxygen) and redox reactions (hydrogenation dehalogenation, C-H cyanation reactions, and nickel-catalyzed aldehyde olefin isopropylation) (Fig. 1.6c). Similar to Ru(II) complexes, Ir(III) complexes exhibit excellent $^1O_2$ generation capability due to their longer T_1 lifetime.

3.2 Nanomaterials as Carriers for Delivery of Photosensitizers

3.2.1 Mesoporous Silica

Mesoporous silica nanoparticles (MSNs) are a type of porous material with pore sizes typically ranging from 2 nm to 50 nm. Featuring highly ordered pore structures and tunable pore sizes [121], MSNs possess characteristics such as biocompatibility, high loading capacity, stability, adjustable pore sizes, and ease of functionalization, which have garnered increasing attention in the field of cancer phototherapy [122]. Reports indicate that MSN can address some challenges associated with the use of PSs in PDT, such as poor selectivity for tumor tissues, hydrophobicity, and prolonged accumulation in healthy tissues. Serving as ideal nanocarriers, MSNs can effectively deliver phototherapeutic agents to tumor sites for cancer PDT. They not only enhance the encapsulation efficiency of PSs but also achieve controlled release of the PS through their ordered mesoporous structure and tunable pore sizes, optimizing the efficacy of PDT [123, 124]. Additionally, surface functionalization of MSNs further enhances the targeting of PSs [125], improving treatment selectivity, protecting PSs from degradation under physiological conditions, and extending their effectiveness [126]. For example, PEG-functionalized MSNs prolong the circulation time of nanocarriers, prevent early drug leakage, and support pH-dependent drug release [127]. PEG-functionalized MSNs can accumulate successfully in cancer cells through EPR. These attributes make mesoporous silica nanoparticles promising PS carriers in PDT, aiding in enhancing treatment effectiveness and reducing side effects. Güleryüz et al. [126] developed a multifunctional nanocarrier system based on the PS IR780 and the anticancer agent curcumin (Cur), encapsulating IR780 and Cur within polyethylene glycol-coated mesoporous silica nanoparticles (Cur&IR780@MSN) (Fig. 1.7a). Characterization revealed that this nanocarrier system exhibits excellent hydrophobic drug loading, temperature-suppressive properties, high biocompatibility, and enhanced generation of reactive oxygen species. Yu et al. [128] prepared a pH-responsive biodegradable Mg-doped MSN by dissolution-regeneration method, where Mg^{2+} can replace certain Si in Si-O-Si to form Mg-O-Si (Fig. 1.7b). The resulting Mg-O bond is acid-responsive, accelerating the biodegradation of silica particles.

Tian et al. [129] conducted a gram-scale synthesis of nonmetal mesoporous N-doped carbon dots (NCDs)/MSN I-type PSs with POD-like activity for hypoxic

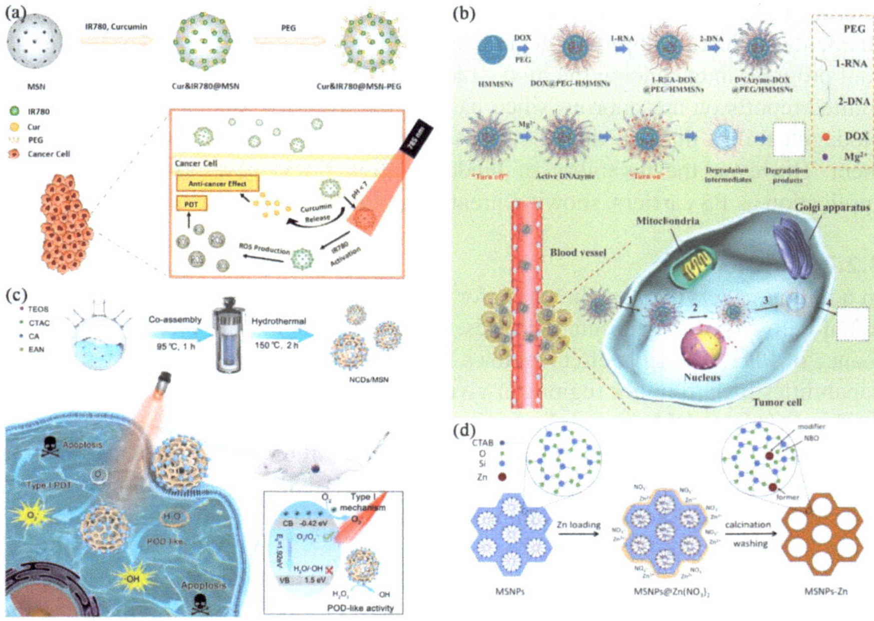

Fig. 1.7 PDT with mesoporous silica as the carrier. (**a**) Schematic illustration of the nanocarrier synthesis and combinational-therapeutic applications of Cur&IR780@MSN toward deep-seated tumors with NIR light (Copyright © 2024 Lasers in Medical Science. Reprinted with permission from Ref. [126]). (**b**) Schematic illustration of the design, fabrication, controllable drug delivery/release, and intracellular biodegradation of DNAzyme-doxorubicin (DOX)@PEG/HMMSNs (Copyright © 2018 Small. Reprinted with permission from Ref. [128]). (**c**) The synthesis process of nonmetallic mesoporous NCD/MSN type I PS and its schematic diagram for PDT of hypoxic tumors (Copyright © 2023 Advanced Functional Materials. Reprinted with permission from Ref. [129]). (**d**) Schematic illustrating the procedure used to fabricate the MSNPs-Zn (Copyright © 2020 Materials Today Advances. Reprinted with permission from Ref. [134])

tumor synergistic therapy (Fig. 1.7c). The NCDs within the NCDs/MSN nanohybrids, synthesized via a one-pot co-assembly route, were covalently anchored to the MSN framework through strong Si-C bonds, preventing NCDs from escaping the nanocarrier and maintaining the mesoporous structure of MSN. Both in vitro and in vivo experimental results demonstrate that NCDs/MSN can serve as ROS-enhancing generators for synergistic eradication of hypoxic tumors, exhibiting good hydrophilicity, excellent photostability, and superior biocompatibility.

Concerns about using mesoporous silica as a delivery vehicle involve the possibility of long-term retention in the human body, which could pose health risks [130]. However, it has been reported that the physicochemical properties of mesoporous silica can be altered at the atomic level by introducing modified cations into the Si-O network to enhance its biodegradability [131–133]. Chen et al. [134] designed a monodisperse biodegradable platform of zinc-containing mesoporous silica nanoparticles (MSNPs-Zn) for therapeutic cationic delivery that degrades inside cells upon internalization (Fig. 1.7d). MSNPs-Zn exhibit pH-responsive cation release, with zinc release rates in acidic conditions much faster than in neutral environments, indicating improved degradability of zinc-doped MSNPs. Despite significant progress in enhancing the biodegradability of MSNs in recent years, there are limited reports on mesoporous silica-based PS delivery nanoplatforms that modify biodegradability through doped cations, indicating the need for further development. However, the tunability of biodegradation offers new prospects for their application as PS carriers, showing great potential for future exploration.

3.2.2 MOFs

In addition to serving as a direct PS for PDT, NMOFs can also facilitate the functional delivery of PSs in tumor tissues, thereby enhancing the therapeutic performance of PDT. The EPR effect endows NMOFs with precise and passive targeting capabilities toward tumor regions in vivo. Moreover, the various functional groups on the surface of NMOFs can be conjugated with a range of target molecules. The high specific surface area and flexible structure of NMOFs enable the encapsulation of PS molecules within the framework, effectively reducing self-aggregation and self-quenching of common PSs, improving their solubility, and maximizing the utilization efficiency of the PSs [135].

For the first time, Negi et al. [136] coupled a novel dual-photon-active PS, CDs, which exhibit a redshifted absorption effect, onto the surface of NMOFs using the EDC/NHS chemical method. This conjugation renders the PS less likely to escape from cells, thereby enhancing the therapeutic efficacy. Consequently, the conjugation of CDs on the surface of NMOFs leads to an enhanced photodynamic effect. Importantly, the resulting NMOF/CD composite material exhibits photodynamic activity under 980 nm laser irradiation.

3.3 Nanomaterials as Light Transducer for PDT

3.3.1 Upconversion Nanomaterials

Upconversion nanoparticles (UCNPs) are optical nanocrystals that feature a crystal structure doped with lanthanide elements. Upon NIR excitation, these nanoparticles can convert multiple low-energy photons into a single high-energy photon. This unique property of UCNPs allows for deeper penetration into biological tissues, while the emitted ultraviolet or visible light can effectively activate PSs, generating highly ROS that aid in directly destroying tumor cells or activating the immune system. Due to their low toxicity, high chemical stability, and excellent photostability, UCNPs show great potential for combating tumors in PDT [137].

The composition of UCNPs includes sensitizer ions, activator ions, and the matrix material serving as the carrier. Sensitizing ions, such as Yb^{3+}, are responsible for absorbing NIR light and transferring the energy to emitting lanthanide ions [138], which are typically Er^{3+}, Ho^{3+}, or Tm^{3+}. The matrix material provides a stable crystal framework for these activator ions, ensuring they are correctly placed within the lattice and can function under the appropriate light conditions [139].

The NIR light wavelength range spans from 780 nm to 2526 nm, with lower absorption rates and stronger scattering effects in biological tissues, allowing for deeper tissue penetration [140]. This characteristic, known as the "optical window" in medicine, provides strong support for cancer treatment research [141]. In the field of PDT research, typical NIR light sources include wavelengths of 808 nm, 980 nm, and 1550 nm. The second NIR window (NIR-II, 1000–1700 nm) offers deeper tissue penetration compared to the first NIR window (NIR-I, 700–950 nm) [142, 143], effectively enhancing the treatment of deep-seated tumors [144], which is also considered a promising research area due to wavelength redshift reducing light attenuation [145, 146].

UCNPs excited by NIR-II exhibit significant potential in PDT. UCNPs can convert NIR-II light, which has stronger penetration capabilities in deep tissues, into high-energy light, effectively stimulating photosensitizers to produce reactive oxygen species for selectively killing tumor cells. Compared to the traditional NIR-I region, NIR-II light reduces heat effects during treatment due to lower absorption in biological tissues and, through mitochondrial targeting strategies, enhances the efficacy of PDT.

3.3.2 X-Ray-Excited Nanoscintillator

Compared to NIR, X-ray possesses a greater tissue penetration depth. X-ray-mediated photodynamic therapy (X-PDT) can be achieved by converting high-energy X-rays into ultraviolet or visible light using nanoscintillators [147]. When nanoscintillators interact with X-rays, photon excitation and Compton scattering occur, generating a large amount of electron-holes and hot electrons in situ, which transfer between the valence and conduction bands and recombine at luminescent centers, converting ionizing radiation (such as X-rays) into (ultraviolet/visible/infrared) light to activate coupled PSs for ROS production [148]. Various X-ray luminescent nanoparticle (XLNP) types have been developed, including lanthanide-doped fluoride nanoparticles (LaF_3, $NaYF_4$), inorganic persistent luminescent nanoparticles ($ZnGa_2O_4:Cr^{3+}$), metal nanoparticles (Au, etc.), quantum dots (CdTe, CdSe), and chalcogenide compounds ($CsPbBr_3$, Rb_2CuBr_3) [149].

For instance, Sun et al. [150] designed aggregation-induced emission gold nanoclusters conjugated with PS RB (AIE-Au) (Fig. 1.8a). Under low-dose X-ray irradiation, AIE-Au efficiently absorbed X-rays, generated hydroxyl radicals, and enhanced the radiative therapy effect by disrupting double-stranded DNA. Additionally, AIE-Au effectively converted X-rays into optical luminescence and excited conjugated PS to trigger ROS generation, thereby reducing X-ray dosage. Liu et al. [151] designed a silica-based nanoscintillator (PEG/$LiLuF_4:Ce$@DMSN-SNO-g-C_3N_4), which can achieve X-ray-controlled release of $ONOO^-$ to enhance postoperative radiotherapy for colon cancer (Fig. 1.8b). Under 1.0 Gy X-ray irradiation, NO and O_2^- were synchronously released, forming highly cytotoxic $ONOO^-$; meanwhile, SCNPs acted as radiosensitizers, absorbing X-rays and depositing radiation energy to enhance RT. Sun et al. [152] synthesized uniformly sized monodisperse silicate nanoscintillators $Zn_2SiO_4:Mn$ using a silica template method and conjugated them with PSs rose bengal (RB) and RGD peptide to

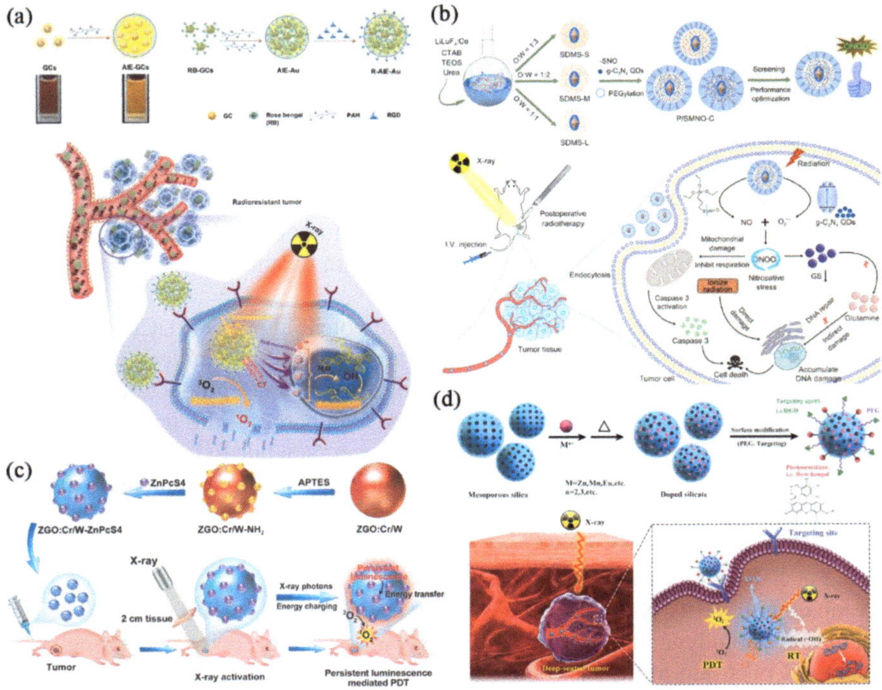

Fig. 1.8 NPs for X-ray-mediated PDT. (**a**) Schematics of the preparation of AIE-GCs and R-AIE-Au nanosensitizers and the working model of R-AIE-Au for fluorescence- and CT imaging-guided X-ray-induced enhanced RT and PDT (Copyright © 2019 Angewandte Chemie International Edition. Reprinted with permission from Ref. [150]). (**b**) Schematic illustration of P/SMNO-C with tunable structural and X-ray-induced nitrosative stress to accumulate DNA damage within tumors via an open source and reduce expenditure route for postoperative RT of colon cancer (Copyright © 2022 Nano Letters. Reprinted with permission from Ref. [151]). (**c**) Schematic illustration for the design of X-ray-activated PLNP-mediated PDT nanoplatform (Copyright © 2018 Advanced Functional Materials. Reprinted with permission from Ref. [153]). (**d**) Schematic illustrations of the preparation process of RGD-ZSM-RB nanosensitizers and the mechanism of RGD-ZSM-RB-mediated X-PDT (Copyright © 2019 Advanced Materials. Reprinted with permission from Ref. [152])

prepare RB-ZSM-RGD nanosensitizers for integrin-targeted tumor delivery (Fig. 1.8d). Under low-dose X-ray irradiation (1.0 Gy), the silicate nanosensitizers effectively generated ROS and significantly inhibited tumor growth.

Nevertheless, the luminescence of X-rays immediately disappears after X-ray excitation ceases; thus, continuous X-ray exposure may cause additional damage to healthy tissues. Hence, persistent luminescent nanoparticle (PLNP) that maintains long afterglow luminescence after removing the light source is considered to reduce X-ray irradiation dosage. Song et al. [153] synthesized W(VI)-doped $ZnGa_2O_4$:Cr (ZGO:Cr/W) PLNPs and prepared a PDT nanoplatform (ZGO:Cr/W-$ZnPcS_4$) by coupling with PS Zn(II) phthalocyanine tetrasulfonic acid ($ZnPcS_4$) (Fig. 1.8c). After X-ray irradiation removal, the sustained luminescence continued to excite coupled PSs, significantly reducing X-ray dosage (≈0.18 Gy). Furthermore, lanthanide ion-doped $NaLnF_4$ (Ln = Y, Gd, or Lu) nanoparticles have emerged as a new X-ray-activatable PLNP. Ou et al. [154] discovered that the afterglow emission of $NaLuF_4$:Tb@$NaYF_4$ nanoscintillators lasted for more than 30 days after X-ray cessation, far exceeding the reported afterglow lifetime of $ZnGa_2O_4$:Cr^{3+} phosphors. Pei et al. [155] stored pre-irradiated $NaYF_4$:3%Er@$NaYF_4$-PLNPs at −20 °C for 0.5 h, then heated them to room temperature, and intravenously injected them into live mice, achieving high-contrast and high-resolution NIR-II vascular and tumor afterglow PL imaging with significantly higher signal-to-noise ratio compared to fluorescence FL imaging.

4 Single PDT and Combination Therapy

Although PDT has great potential in clinical treatment, the traditional PDT treatment still has many serious defects. For example, traditional PSs can only be stimulated by short wavelengths of light; as a result, PDT is limited to the treatment of superficial tumors and cannot penetrate the center of tumor lesions [156]. In addition, the low bioavailability of traditional PSs, the low targeting efficiency of tumor cells, and the potential toxicity caused by off-target also become major obstacles to the clinical application of PDT [157]. Moreover, the tumor microenvironment also has a great impact on the efficiency of PDT. For example, hypoxia inhibits the production of $^1O_2$ by PDT [158], endogenous antioxidant defense system (such as GSH, SOD, etc.) can realize the rapid removal of free radicals and reduce the escape of cancer cells from oxidative stress damage [159], and the immunosuppressive microenvironment also greatly limits the anti-tumor immune response [160].

For deep tumor therapy, the use of photoconverting nanomaterials that can be activated by NIR light or X-rays facilitates light transport in deep tissues and activates PSs coupled with them through FRET [161]. In order to improve the targeting ability of PDT, PDT nanoplatforms with tumor-specific parts can be designed to accurately target cancer cells through cell surface receptors, tumor antigens, and tumor vascular system [162]. However, due to the small diffusion radius (10 nm) and short lifetime (<40 ns) of ROS generated in PDT, the location of ROS generation is crucial to the effect of PDT. By modifying organelle-specific targeting ligands on the surface of the PDT nanoplatform, the accumulation of nanoparticles in the organelles can be increased [163]. At present, nanomaterials targeting the plasma membrane, endoplasmic reticulum, mitochondria, lysosome, and nucleus have been reported successively [164]. Bai et al. [165] coupled the PS 5-ALA with lipophilic cationic triphenylphosphine targeting mitochondria and packaged it in biocompatible BSA nanoparticles and finally modified folic acid targeting receptors on its surface to enhance tumor cell delivery. Cell experiments and experiments on BALB/c mice carrying MCF-7 xenografts have shown that the nanoparticle can reduce cell respiration and increase the anti-tumor effect of PDT.

To solve the problem of hypoxia, it can design type I PSs with low O_2 dependence [166], which mainly include metal oxides/sulfides, organic dye molecules,

MOFs, and transition metal complexes [167]. Tian et al. [168] developed a type I PS agent (TPFN-AzoCF_3) that can be switched on demand, which consists of two parts: anoxic-normoxic cyclic response fragment (AzoCF_3) and type I PS (TPFN). Only under the condition of anoxic environment and light, the aryl azo group can be converted into hydrazine group to restore the generation of ROS. However, the photosensitivity of PS will be quenched under normal oxygen or no light conditions, thus minimizing the side effects. Although type I PS can achieve partial oxygen cycling to reduce O_2 dependence through Haber-Weiss reaction or superoxide dismutase (SOD)-mediated disproportionation reaction, tumor endogenous O_2 is still required [169]. Li et al. [170] synthesized an organic polymer PS with NIR photoexcitation and used glycol oligomers as side chains of the organic skeleton, which alleviated the rapid recombination of electron-hole pairs, promoted substrate H_2O sensitization to $\bullet O_2^-$ and •OH, and fundamentally realized O_2_independent PDT. Moreover, the hydrophilic glycol oligomer H_2O molecules are brought into the vicinity of the electron-hole pair, improving the photosensitivity reaction efficiency and successfully reducing the laser power to 15 mW/cm^2. In addition, overcoming hypoxia by delivering O_2 vectors or reagents that catalyze H_2O_2 to produce O_2 in situ (e.g., perfluorocarbons, artificial red blood cells, and metal oxide catalysts) is also a promising innovative strategy [171]. Wei et al. [172] synthesized PFC-SICy-PEG-Biotin (FCy) by fluorinating indocyanine-based PSs and linking them with PEG-modified biotin and self-assembled nano-PS (FCy NPs). The introduction of perfluoroalkyl promoted the self-assembly process of FCy, endowed FCy NP with better endoplasmic reticulum targeting ability, and increased the yield of $^1O_2$ to 32.6%, which greatly improved the therapeutic effect of hypoxic tumors. Other hypoxic mitigation strategies include inhibiting tumor cell respiration to reduce endogenous oxygen consumption [173], using solid peroxides (such as CaO_2 and MgO_2) to react with H_2O to generate O_2 [174], using materials with high electron-hole separation efficiency to catalyze H_2O to form O_2 [175], reversing the tumor microenvironment (such as destroying ECM, reducing oxygen consumption). Normalization of tumor blood vessels and inhibition of HIF-1 signaling pathway [158] also have good application prospects [176].

In addition, well-designed nanomaterials can achieve multiple functions. Combining PDT with RT, chemotherapy, CDT, immunotherapy, gas therapy, starvation therapy, and other therapies can reduce the side effects of single PDT treatment, increase the sensitivity of tumor cells, and improve the therapeutic effect [177].

4.1 The Combination of PDT and PTT

Photothermal therapy (PTT), utilizing photothermal transducers (PTAs) under NIR light irradiation to convert light energy into heat, is a method to increase the local temperature and ablation efficiency of tumor tissues, thereby killing cancer cells [178]. Due to its noninvasiveness, high selectivity, and minimal side effects, PTT is a promising cancer treatment modality. Similar to PSs in PDT, PTA molecules undergo transitions from the ground state to the singlet excited state, followed by non-radiative relaxation processed back to the ground state and releasing heat [179]. Presently, all PTAs used in research have not been tested in large-scale clinical trials, while laser ablation procedures without PTT drugs have been applied in clinical settings [180]. This is mainly due to limitations that restrict the efficacy of PTT in many cases, hindering further clinical applications, including limited light penetration depth, relatively low accumulation efficiency of PTAs in tumors, damage to normal tissues, and resistance to PTT due to overexpression of heat shock proteins (HSP) [181].

Generally speaking, both PDT and PTT have specific limitations. For example, PTT is superior to PDT in terms of being oxygen-independent, but PTT offers weaker protection to surrounding tissues compared to PDT, and it involves higher costs [182]. The cell death pathway of PTT is also different from PDT, as it can induce denaturation of proteins involved in DNA repair, increasing the sensitivity of tumors to PDT [183]. At the same time, PDT and PTT can activate adaptive immune responses, control primary and metastatic tumor lesions, and enhance immune responses [184]. Therefore, the combination therapy of PDT and PTT can leverage the advantages of each modality and potentially overcome their respective shortcomings [185, 186].

One strategy is to use photoactive molecules that can serve both PDT and PTT simultaneously, such as porphyrins, indocyanine green (ICG) [187], heptamethine carbocyanines, BODIPY, AIE molecules, semiconductor polymers, etc. [188], and encapsulate them into liposomes, nanovesicles, lipoproteins, nanoemulsions, proteins, silica, or polymer nanoparticles to improve the biological distribution of photoactive molecules. For example, Tian et al. constructed supramolecular nanovesicles (NV) with GSH-triggered PTT/PDT conversion property, which connected hydrophobic Gd-chelated porphyrin molecules and hydrophilic PEG molecules via disulfide bonds (CCP), self-assembling into NV (Fig. 1.9a). After GSH triggers disulfide bond cleavage, the NV undergoes self-disassembly polymerization, thereby achieving the transition from PTT to PDT [189]. Xiao et al. [190] developed light-emitting nanoparticles with AIE properties composed of the PS BODTPE and the targeting ligand DBCO, which can target cell membranes via bioorthogonal copper-free click chemistry and exhibit PDT/PTT effects under 808 nm laser irradiation, inducing mitochondrial and DNA damage as well as immune responses (Fig. 1.9b). Li et al. [191] designed a heavy metal-free organic semiconductor PS, BTAC6, with a donor-acceptor alternative structure and conjugated molecular skeleton. Under 808 nm laser irradiation, it demonstrates type I and type II PDT as well as PTT effects, generating cytotoxic $^1O_2$, •OH, and $O_2^{\bullet -}$, as well as local heat, which is not limited by the anoxic microenvironment of the tumor.

Another strategy is to integrate multiple components such as light-sensitive materials, photothermal conversion materials, and nanocarriers (including gold nanoparticles [192], carbon-based nanomaterials like graphene and carbon nanotubes [193], MoS_2, CuS, black phosphorus, upconversion nanoparticles, polydopamine, etc.) into the same nanoparticle to achieve synergistic PTT and PDT.

Shen et al. [194] prepared a gold nanorod-manganese porphyrin system (GMCM) encapsulated in a 4T1 cell membrane, demonstrating specific targeting capabilities

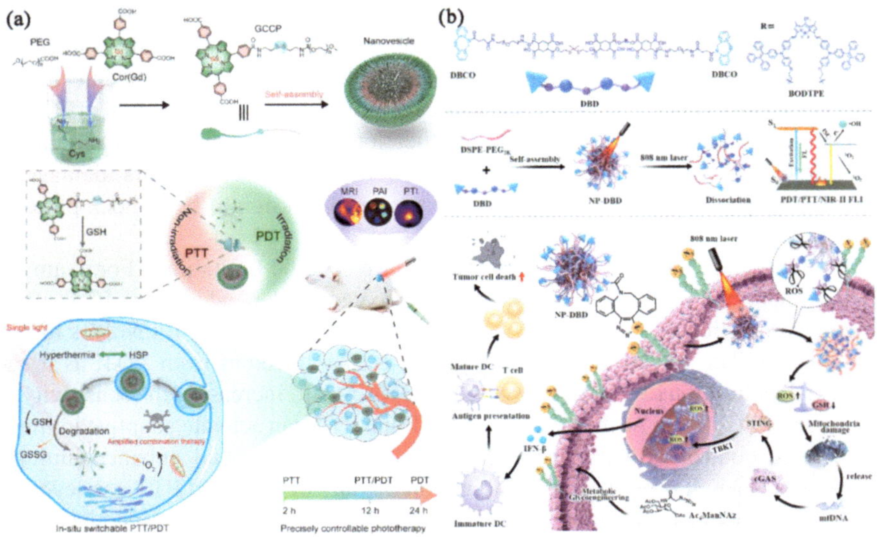

Fig. 1.9 Photoactive molecular monomers capable of both PDT and PTT. (**a**) Schematic illustration of switchable PTT/PDT triggered by GSH and switchable light therapy guided by multimodal imaging using a single-laser source in unbalanced redox homeostasis (Copyright © 2023 Chemical Engineering Journal. Reprinted with permission from Ref. [189]). (**b**) Schematic illustration of NP-DBD anchored to the cell membrane through biological orthogonal copper-free tapping reaction to enhance the PDT/PTT effect and activate the cGAS-STING pathway, promote DC maturation, and activate effective anti-tumor immune response (Copyright © 2023 Advanced Materials. Reprinted with permission from Ref. [190])

towards 4T1 cells. Under 808 nm laser irradiation, GMCM generates $^1O_2$ and releases heat, while Mn^{2+} can consume endogenous GSH to further enhance the efficacy of PTT/PDT treatment. Liu et al. [195] coated mesoporous silica with $NaGdF_4$:Yb, Er, Tm upconversion nanocrystals and loaded them with photothermal agent AuNPs and photosensitizer Cys molecules, achieving efficient photothermal conversion and $^1O_2$ generation under NIR light triggering, as well as autonomous motion to improve tumor cell permeability. Lin et al. [196] constructed multifunctional nanoparticles by self-assembling PS Ce6-modified carbon dots (CDs-Ce6) with Cu^{2+}, where TME stimulation triggers Cu/CCNP decomposition and restoration of PDT and PTT performance, allowing selective activation in tumors and reducing adverse reactions in normal tissues. Cu^{2+} reacts with endogenous H_2O_2 to generate highly toxic •OH and significantly depletes GSH.

Qian et al. [197] fabricated unique $NaYF_4$:Yb/Er/Nd@$NaYF_4$:Nd-UCNPs@$AgBiS_2$ core-shell nanoparticles, where the light-to-heat conversion efficiency of $AgBiS_2$ was enhanced to 45% due to the cross-relaxation between the continuous energy band of $AgBiS_2$ and the ladder-like energy levels of Nd^{3+}, and the simultaneous generation of ROS and heat was achieved under 808 nm laser irradiation, which has a good therapeutic effect on 4T1 tumors. Li et al. [198] anchored spherical manganese ferrite nanoparticles (SMFN) onto black phosphorus nanosheets (BPNS) and then modified with RGD peptide to design and prepare a uniformly dispersed water-soluble composite nanoplatform. Compared to single material, this nanoplatform has precise tumor cell delivery capabilities and enhanced photon responsiveness. Under laser irradiation, it can induce ROS production and temperature elevation in HeLa cells, while black phosphorus's catalase-like activity can reshape the tumor's hypoxic microenvironment, further enhancing the efficiency of PDT/PTT synergistic therapy. Wang et al. [199] loaded hemoglobin Hb and PS Ce6 into polydopamine PDA and encapsulated them in micelles formed by aldehyde-modified PEG and polyethyleneimine (PEI). This nanoparticle could release oxygen rapidly in highly acidic tumor regions, effectively boosting the efficiency of PTT/PDT synergistic treatment.

4.2 The Combination of PDT and RT

Since the discovery of X-rays by Roentgen, RT has been widely used in cancer treatment [200]. Compared to traditional PDT, the X-rays used in RT have greater tissue penetration depth, which can overcome the challenges faced by traditional PDT in treating deep-seated tumors [201]. When high-energy photon beams interact with biological media, they will induce photoelectric effect and Compton scattering, leading to the cascade generation of numerous secondary electrons, the effects of which depend on the incident photon energy and atomic number (Z). On the one hand, these secondary electrons disrupt DNA molecular structures, impeding cell division and proliferation, thereby activating metabolic pathways that lead to programmed cell death, directly killing tumor cells. On the other hand, in the cellular environment, ionizing radiation induces the radiation decomposition of water, generating entities such as e^{-}_{aq} (hydrated electrons), •OH (hydroxyl radicals), H• (hydrogen radicals), H_2 (radiolysis of water), H_2O_2, and HO_2• (hydroperoxyl radicals), which can produce additional ROS to promote oxidative stress through further reactions, indirectly causing cancer cell destruction and inhibiting tumor growth [202].

Although RT can treat various types of cancers, the resistance of certain cancer cells to RT remains a serious issue, often leading to treatment failure and tumor recurrence [203]. Moreover, the side effects of high-dose RT cannot be ignored [204], which can increase the risk of systemic toxic effects, induce local inflammation and fibrosis in normal tissues, and lower the quality of life for patients [205]. Combining RT with PDT is an attractive approach. Although both treatments involve the generation of cytotoxic ROS, the mechanisms of which they induce cancer cell death differ. RT mainly induces cell apoptosis by damaging cell DNA, but certain gene mutations can render cancer cells incapable of initiating apoptosis, leading to resistance to RT. In PDT, traditional PSs typically accumulate on cell and mitochondrial membranes, and ROS generated after light exposure causes damage to these membranes, resulting in cell necrosis or apoptosis. Due to the unique mechanisms of action of PDT and RT, there is a good complementarity between the two treatments [206].

One method to combine RT and PDT is to use X-rays as the excitation source for PDT, known as X-PDT, as detailed in Sect. 3. However, this scintillator-mediated strategy commonly faces challenges such as insufficient energy transfer efficiency and tumor hypoxia in the microenvironment. Designing scintillators with good performance and matching them with suitable PSs poses many difficulties, severely hindering the effectiveness of X-PDT [207, 208]. Additionally, a simpler approach involves the separate use of laser and X-ray to combine PDT with RT. For example, Chen et al. [209] designed novel covalent polymer nanodots based on RB and Gd (referred to as GRDs) as clinical PSs for both MRI and radiation sensitization (Fig. 1.10a). Under laser and X-ray exposure, the combination of PDT effects with radiation sensitization leads to efficient cancer cell death. However, this approach requiring two light sources adds complexity to treatment.

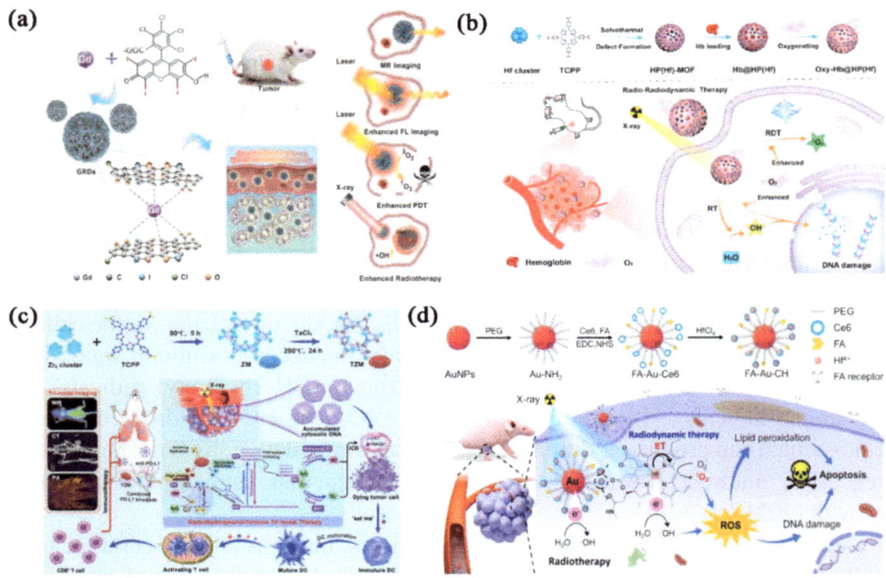

Fig. 1.10 Schematic diagram of the combined treatment strategy of RT and PDT. (**a**) Schematic diagram of GRD preparation and in vivo fluorescence/MR image-guided PDT and RT of 4T1 tumor cells (Copyright © 2020 Advanced Materials. Reprinted with permission from Ref. [209]). (**b**) The synthesis scheme of oxygen-rich Hf-MOF nanosensitizer loaded with Hb and its application in RT-RDT cancer therapy (Copyright © 2023 ACS Materials Letters. Reprinted with permission from Ref. [210]). (**c**) Schematic diagram of TZM preparation and mechanism of multimodal image-guided RT/RDT/immunotherapy for metastatic osteosarcoma with TZM. Under X-ray irradiation, Ta-Zr co-doping promotes TZM to absorb X-ray energy and produce •OH for RT, and TZM also efficiently transfers energy to TCPP to produce •OH and $^1O_2$ to enhance RT and RDT effects, respectively. Finally, the combination action of RT, RDT, and anti-PD-L1 successfully triggered a powerful anti-tumor immune response (Copyright © 2023 Advanced Science. Reprinted with permission from Ref. [211]). (**d**) Preparation diagram of FA-Au-CH nanosensitizer and mechanism of RT-RDT in tumor therapy (Copyright © 2023 ACS Nano. Reprinted with permission from Ref. [208])

Radiation dynamic therapy (RDT) based on metal complexes or metal-based hybrid structures does not require fluorescence to activate PSs. Instead, it mainly generates $^1O_2$, O_2^{-}•, and •OH through energy transfer within nanomaterials [201]. Typically, high Z elements are introduced to enhance their ability to attenuate high-energy X-rays and absorb energy, thereby enhancing the radiation sensitization effect and reducing radiation doses [177].

Introducing high Z elements into NMOFs can achieve radiosensitizing functionality, while using PSs as organic ligands can endow NMOFs with RDT capabilities [210] (Fig. 1.10b). Li et al. [211] developed a NMOF named TZM by co-doping Ta and Zr as metal nodes and using porphyrin molecules (tetra(4-carboxyphenyl) porphyrin (TCPP)) as the photosensitive ligand. The co-doping of Ta-Zr in TZM promotes energy transfer between TZM and the PS TCPP, facilitating the generation of $^1O_2$ (Fig. 1.10c). In the study of Li et al., the Hf-chelated PS Ce6 and folic acid are connected to the surface of gold nanoparticles (Fig. 1.10d). Under radiation exposure, Hf^{4+} transfers radiation energy to the PS to produce ROS, while Hf and Au together play a radiosensitizing role [208]. Lin et al. [212] coordinated $Ir[bpy(ppy)_2]^+$ or $[Ru(bpy)_3]^{2+}$ derivatives of tricarboxylic ligands with the heavy element Hf in the $[Hf_6O_4(OH)_4(HCO_2)_6]$SBU to construct a two-dimensional metal-organic layer (MOL), which can directly excite the PS under X-ray to generate $^1O_2$.

4.3 The Combination of PDT and Chemotherapy

Chemotherapy, as one of the most commonly used tumor treatment modalities in clinical research, can be applied to various types of malignant tumors such as breast cancer, lung cancer, lymphoma, multiple myeloma, sarcoma, acute leukemia, etc. [213]. Common chemotherapy drugs in research include DOX [214], paclitaxel (PTX) [215], methotrexate (MTX) [216], and platinum-based drugs [217]. However, due to the lack of tumor tissue selectivity, chemotherapy can lead to severe side effects [218, 219]. Moreover, the multidrug resistance (MDR) of tumor cells to chemotherapy drugs can make it difficult to completely eradicate tumors through a single mode of chemical treatment, which may lead to recurrence [220, 221], thus failing to meet the requirements of cancer therapy. Many studies have shown that PDT can reduce the activity of proteins related to drug efflux, overcome chemotherapy resistance, and inhibit the activation of compensatory signaling pathways [222, 223]. Moreover, at the tumor microenvironment level, PDT can alter the permeability of tumor blood vessels and extracellular matrix, enhance drug delivery, and improve the efficacy of tumor perfusion chemotherapy. Conversely, chemotherapy can increase the sensitivity of tumor cells to PDT-induced ROS. Therefore, integrating chemotherapy and PDT into the same nanoplatform has shown a great potential to enhance cancer treatment efficacy and reduce the risk of recurrence [224, 225].

The key to achieving synergistic effects between the two treatment strategies is to design safe and effective nanodelivery systems [162]. One approach is to passively target drug delivery to the target tissue based on the unique structure of tumor tissue vasculature [226, 227]. Common carriers include micelles, dendritic polymers, liposomes, and various inorganic, organic, or hybrid materials. These systems can encapsulate, covalently link, or physically adsorb drug molecules to improve the pharmacokinetics of drugs [228]. For example, Mao et al. [229] used MSN as a delivery system for the PS Ce6 and surface-modified beta-cyclodextrin grafted polyethyleneimine (CD-PEI) for the coupling of cisplatin prodrug, thereby achieving the chemotherapy/PDT combination therapy.

Furthermore, by modifying specific active ligands/molecules (such as peptides, antibodies, proteins, polysaccharides, nucleic acids, and receptors) on the surface of nanocarriers, active targeted drug delivery can be achieved, contributing to precise optimization of drug administration parameters [230, 231] and enhancing therapeutic effects at lower doses [232]. Liu et al. [233] developed Yb/Er upconversion nanoparticles modified with angiopoietin-2 (ANG2) peptide, jointly loaded with Ce6 and DOX (Fig. 1.11a). These nanoparticles can cross the blood-brain barrier and be internalized into lysosomes of glioblastoma cells, thereby directly killing glioblastoma cells. Yang et al. [234] used silk fibroin (SF) as a biotemplate to synthesize nano-sized ZIF-8 and loaded the drug DOX and PS ICG (Fig. 1.11b). Through modification with PEI, they successfully connected MCF-7 breast tumor-targeting peptide AR with ZIF-8, achieving PDT/PTT/chemotherapy three-mode therapies with targeted properties.

Acidity [235], hypoxia [236], excessive ROS [237], high GSH concentration [238], and several highly expressed enzymatic activities [239] are unique features of the tumor microenvironment (TME) that distinguish it from the normal tissue environment. Targeting these characteristics to release active ingredients at the tumor site provides new opportunities for cancer treatment. Sun et al. [240] constructed a polyethylene glycol-coated PDT/chemotherapy nanoplatform, comprising a light-activated PPa-PEG_{2k} shell and a ROS-reactive PTX prodrug core. The ROS generated by PPa-PEG_{2k} is not only used for releasing chemically toxic PTX but also synergistically promotes tumor cell apoptosis. He et al. [241] developed a polymer micelle consisting of hydrophobic segment of poly-ε-caprolactone (PCL) linked to Ce6, H_2O_2-unstable group, quinone methide (QM) for GSH depletion, and anticancer drug DOX loaded in the hydrophobic cavity (Fig. 1.11c). This nanoplatform can realize a triple treatment strategy of ROS generation, GSH depletion, and chemotherapy. Ding et al. [242] designed a nanoplatform based on albumin, co-delivering the PS IR780, hypoxia-activated prodrug TPZ, and the NLG919 dimer as an immune checkpoint inhibitor (ICI) (Fig. 1.11d). In this study, NIR is used to trigger PDT, ROS in tumor cells trigger TPZ release and activate chemotherapy, and GSH triggers activation of NLG919 and enhance the immune response.

However, these stimulus-responsive delivery systems still exhibit off-target release, as these endogenous reaction processes sometimes also occur in healthy tissues. Utilizing photoresponsive nano-components can achieve precise spatiotemporal control of drug release [243]. Zhu's group [244] designed UCNPs with switchable emission properties, encapsulating the photosensitizer ZnPc and drug BBM in $mSiO_2$. A 980 nm excitation triggers azobenzene conformational change and drug release, while 808 nm excitation triggers PDT to generate ROS, achieving superior therapeutic effects compared to single treatments.

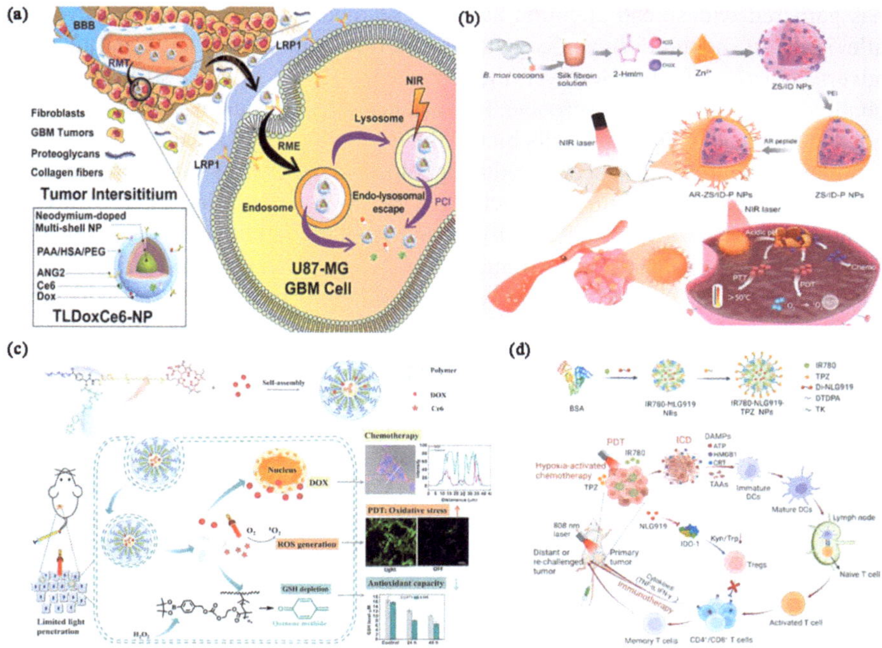

Fig. 1.11 Schematic diagram of the combined treatment strategy of chemotherapy and PDT. (**a**) TLDoxCe6-NPs consist of Nd-based upconversion nanoparticles, DOX and PS Ce6, surface-modified with poly(acrylic acid), human serum albumin layer, and poly(ethylene glycol) and coupled with BBB and U87-MG-glioma targeting ANG2 peptide. TLDoxCe6-NPs entered U87-MG-glioma cells through receptor-mediated endocytosis and endocytosis and stimulated PDT with 808 nm light source (Copyright © 2022 Advanced Materials. Reprinted with permission from Ref. [233]). (**b**) Preparation of AR-ZS/ID-P NPs and schematic diagram of MCF-7 breast tumor-targeting peptide guiding breast cancer therapy. ZIF-8 nanoparticles containing DOX and PS ICG were prepared by a one-pot method and coated with PEI to enhance stability. Then, MCF-7 breast tumor-targeting peptide AR was linked to the nanoparticles to selectively accumulate in the tumor (Copyright © 2023 Advanced Science. Reprinted with permission from Ref. [234]). (**c**) Structural diagram of the star polymer (PEG(-b-PCLCe6)-b-PBEMA) and its mechanism of action for triple cancer therapy with chemotherapy, PDT, and GSH depletion (Copyright © 2021 Chemical Engineering Journal. Reprinted with permission from Ref. [241]). (**d**) Schematic diagram of the preparation of IR780-NLG919-TPZ NPs and its mechanism of action in the synergistic therapy of chemotherapy and photodynamic immunotherapy (Copyright © 2023 Biomaterials. Reprinted with permission from Ref. [242])

4.4 The Combination of PDT and CDT

According to previous studies, solid tumors, due to the abnormal metabolism of tumor cells, are capable of generating abundant H_2O_2 (10–50 μM). Therefore, chemodynamic therapy (CDT) utilizes Fenton or Fenton-like reactions to convert endogenous H_2O_2 into highly toxic ·OH, inducing apoptosis in tumor cells [245]. Recent studies have shown that not only Fe ions but also many metal ions such as Cu^{+}, Mn^{2+}, Co^{2+}, and Ti^{3+} possess Fenton-like reaction mechanism [246–248]. CDT has garnered widespread attention among researchers due to its noninvasiveness, alleviation of the tumor hypoxic environment, high tumor specificity, and minimal adverse reactions [249]. However, the clinical translation of CDT still faces many challenges that need to be overcome. For instance, the insufficient endogenous H_2O_2 content and acidity in tumor cells hinder the efficient catalytic reaction [250, 251]. Simultaneously, the presence of reducing substances (such as GSH) in the TME also diminishes the anti-tumor effects of ·OH [252]. Combining PDT with CDT helps increase the levels and activity of ROS, amplify oxidative stress levels in tumor cells, and maximize the advantages of each treatment method [253].

Chen et al. [254] coated mesoporous silica on the surface of iron oxide nanoparticles and encapsulated the PS prodrug δ-ALA to construct the FMMSH nanoplatform (Fig. 1.12a). They performed TPP modification and tumor cell membrane (TCM) coating for respectively enhancing mitochondrial targeting and tumor cell targeting capabilities. Iron oxide and δ-ALA were used for CDT and PDT, enhancing oxidative stress and inducing cell apoptosis.

Dong et al. [255] loaded calcium peroxide (CaO_2) and ICG into MSN and further surface-coated them with phase-change material lauric acid (LA) as a protective layer to build a synergistic PDT/CDT therapeutic platform (Fig. 1.12b). Under 808 nm laser irradiation, the photothermal effect generated by ICG melts LA, exposing CaO_2 which, upon decomposition, produces O_2 to improve PDT. The generated H_2O_2 can further promote Mn^{2+}-mediated CDT. Lin et al. [256] developed lanthanide-doped nanoparticles coated with copper/manganese silicate (LDNPs@CMSNs). CMSNs act as PSs for PDT under short-wavelength light excitation. The excessive endogenous GSH triggers CMSN degradation, releasing Cu^+ and Mn^{2+} which can catalyze H_2O_2 to produce ·OH and O_2 and facilitate the consumption of GSH.

Nanoenzymes with peroxidase-like activity have been applied in the biomedical field and shown promising therapeutic effects. Similar to Fenton reagents, they can catalyze the disproportionation of H_2O_2 to generate ·OH [257]. Fan et al. [258] constructed nanorods based on CeO_2-MnO_2, surface-modified with folate and PEG to enhance targeting and water solubility, and loaded Ce6 through coordination interactions (Fig. 1.12c). Excessive H_2O_2 in tumors can competitively coordinate with Ce6 and trigger CDT, and the released Ce6 can restore the generation of $^1O_2$. MnO_2 can consume high levels of GSH, and the released Mn^{2+} catalyzes H_2O_2 to produce ·OH. Xu et al. [259] designed nanoliposomes VC@Lipo based on VOx nanoenzymes and Ce6. Ce6 was covalently linked to DSPE-mPEG-NH and self-assembled with phospholipids to obtain C@Lipo, and then VC@Lipo is obtained by loading VOx into the hydrophilic cavity. In the TME, Ce6 can generate $^1O_2$ under light irradiation, and VOx can hydrolyze to generate V^{4+} and V^{5+} compounds. V^{4+} catalyzes H_2O_2 to produce ·OH, while V^{5+} oxidizes GSH to GSSG.

NMOFs are excellent candidates for PDT/CDT therapy due to their high porosity and biocompatibility [260]. Zeng et al. [261] used a porphyrin MOF (PCN-224) as PS, doping Cu^{2+} in the core of the porphyrin ring and growing ultrasmall AuNPs in situ to construct PCN-224(Cu)-Au. The coordination of Cu^{2+} can slow down electron-hole recombination and enhance PDT performance. The ultrasmall AuNPs

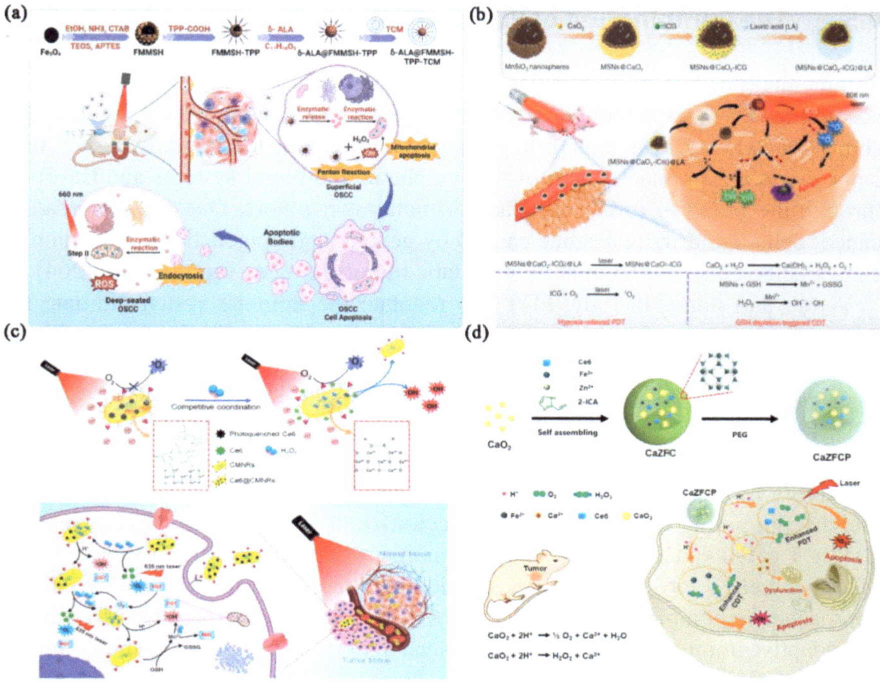

Fig. 1.12 Schematic diagram of the combined treatment strategy of chemotherapy and PDT. (**a**) Schematic diagram of magnetic mesoporous silica hybrid nanoparticle (FMMSH) drug delivery system δ-ALA@FMMSH-TPP-TCM. Iron oxide and δ-ALA were used for CDT and PDT, respectively, and triphenylphosphine (TPP) and tumor cell membrane (TCM) were modified for mitochondria and tumor cell dual-targeting (Copyright © 2023 Advanced Healthcare Materials. Reprinted with permission from Ref. [254]). (**b**) Schematic diagram of the preparation of (MSNs@ CaO_2-ICG)@LA nanoparticles and the synergistic treatment mechanism of PDT/CDT that supplies both O_2 and H_2O_2 and depletes GSH (Copyright © 2020 Nature Communications. Reprinted with permission from Ref. [255]). (**c**) Schematic diagram of competitive coordination of CeO_2-MnO_2 nanoenzyme and the MRI-guided programmed activation of PDT/CDT combination therapy (Copyright © 2023 Advanced Functional Materials. Reprinted with permission from Ref. [258]). (**d**) Schematic diagram of the preparation of CaO_2@ZIF-Fe/Ce6@PEG nanoparticles. Self-supplied H_2O_2 and O_2 can alleviate endogenous H_2O_2 and O_2 deficiency, and Ca^{2+} overload can amplify intracellular oxidative stress and enhance the synergistic therapeutic effect of CDT/PDT (Copyright © 2021 Advanced Functional Materials. Reprinted with permission from Ref. [262])

with catalase-like activity can provide H_2O_2. PCN-224(Cu) undergoes degradation in the TME, and the released Cu^{2+} can deplete overexpressed GSH, while the released Cu^+ can induce ·OH generation.

Chen et al. [262] constructed a pH-responsive degradable ZIF-90 nanoplatform surface-modified by PEG, encapsulating CaO_2 NPs, Fenton reagent Fe^{2+}, and Ce6 (Fig. 1.12d). In the slightly acidic TME, ZIF-90 degrades, and CaO_2 hydrolyzes to produce H_2O_2 and O_2, which increased endogenous H_2O_2 levels and alleviated hypoxia. Additionally, Ca^{2+} overload can further amplify oxidative stress, induce mitochondrial dysfunction, and increase tumor cell sensitivity to Fe^{2+}-mediated CDT and Ce6-mediated PDT.

4.5 The Combination of PDT and Immunotherapy

Tumor metastasis is considered to be the leading cause of cancer-related death, accounting for approximately 90% of all cancer-related deaths [263]. Studies have shown that when cancer cells from primary tumors invade surrounding tissues, they enter the microvascular system of the lymphatic and blood systems and travel far through the blood to form secondary or metastatic tumors. Due to the spread of cancer cells and drug resistance caused by genetic and epigenetic modification of cancer cells, clinical treatment of metastatic tumors remains very difficult [264].

As an anti-tumor therapy, PDT can regulate the immune system, initiate the immune response against tumor cells, and provide patients with long-term protection against tumor recurrence and metastasis [265]. ROS produced by PDT can trigger endoplasmic reticulum stress through direct or indirect action, resulting in ICD. Dying tumor cells can release DAMPs (such as CRT, HSP70/90, ATP, and HMGB1), which can bind to receptors of immature DCs. Mature DCs have a stronger ability to present tumor-associated antigen (TAA) to T cells in lymph nodes, thereby promoting the proliferation of cytotoxic T lymphocytes (CTLs, such as CD8 + T cells) and activating the host immune system, thus killing a large number of tumor cells [266–270]. ICD is considered to be one of the most promising approaches to completely eliminate tumor cells because of its ability to activate T cell adaptive immune responses and even lead to the formation of long-term immune memory [271]. In addition, studies have shown that PDT can promote the reprogramming of TAMs from the pro-tumor M2 phenotype to the anti-tumor M1 phenotype, thus overcoming the immunosuppressive effect of the TME [272, 273]. However, PDT still has many challenges in inducing ICD. For example, PDT is inefficient in generating ROS in hypoxic tumors [274], and PDT can also induce PD-L1 expression in tumor cells and promote the establishment of adaptive immune tolerance in tumor cells [275]. These factors will severely limit the efficacy of PDT and further reduce its ability to induce immune response [276–278]. Therefore, it is necessary for us to combine ICD-based PDT with other immunotherapies to achieve multimode photodynamic immunotherapy.

As a nonspecific immunotherapy, cytokine-based therapy combined with PDT can also enhance the therapeutic effect of PDT [279]. In a study by Liu's group, UCNPs-YS@ZnPc@TPZ was obtained by constructing upconversion nanoparticle UCNFS-YS coated with yolk shell and loading PS ZnPc and hypoxia-activating drug TPZ into it (Fig. 1.13). To achieve controlled release of the immune molecule IFN-γ, Tm-doped UCNP was combined with engineered bacteria, which can produce large amounts of interferon IFN-γ within 30 min under 808 nm laser irradiation of 400 mW/cm^2. Experimental results have shown that the programmed combination therapy combined with PDT and immunity can effectively inhibit both primary and distant metastatic tumors [280].

In solid tumors, in order to evade immune surveillance and prevent immune cell attack, tumor cells usually overexpress various inhibitory checkpoint receptors and recruit inhibitory cells, such as regulatory T cells (Tregs) and myeloid suppressor cells, to establish an immunosuppressive microenvironment [272]. Programmed

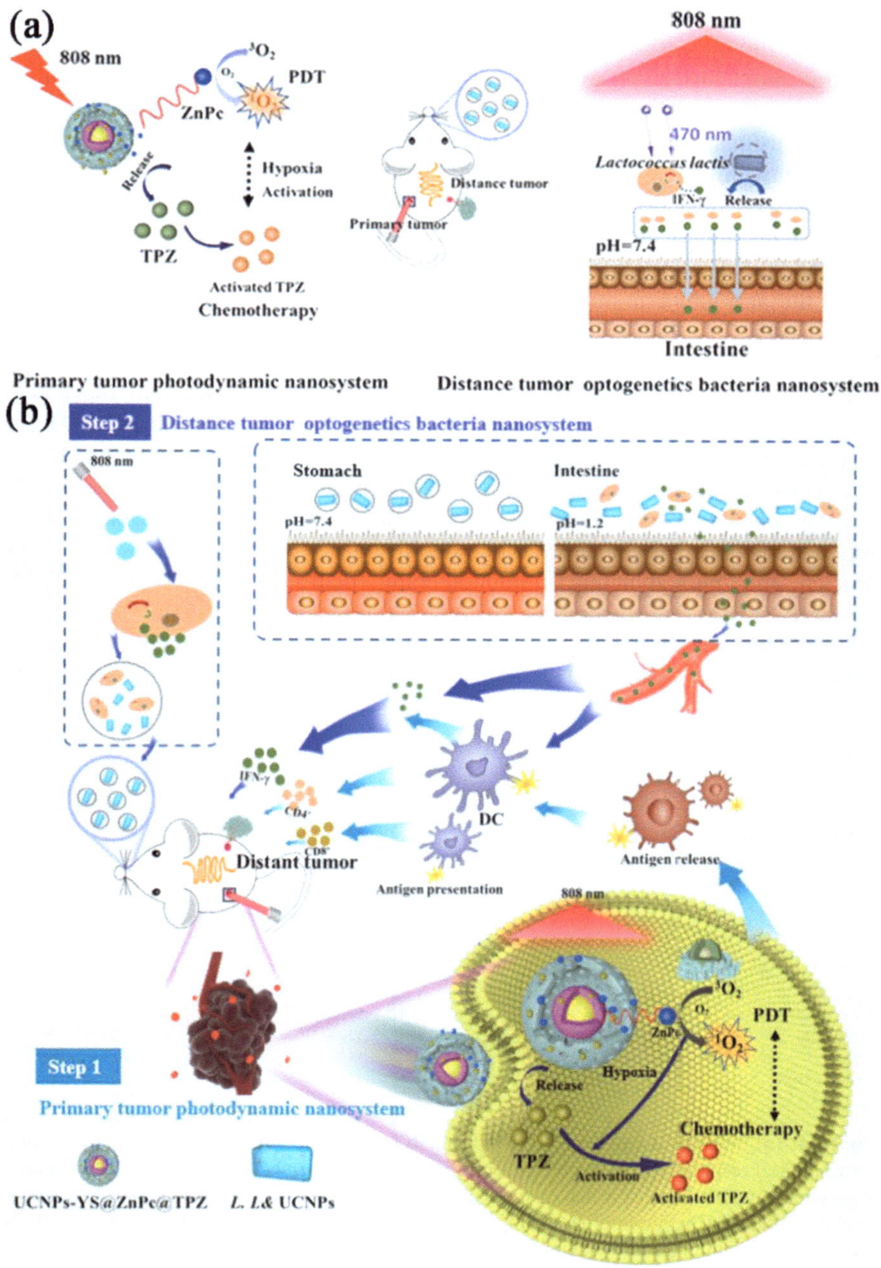

Fig. 1.13 (a) UCNPs-YS@ZnPc@TPZ therapeutic mechanism for in situ tumor PDT and optogenetic immune platform for suppression of distal tumors. (b) Programmed collaborative platform for treating primary and distant tumors by combining UCNP-based PDT nanosystem and engineered bacteria-based immunotherapy system (Copyright © 2022 ACS Applied Materials & Interfaces. Reprinted with permission from Ref. [280])

death protein 1(PD-1)/PD-L1 signal transduction pathway is one of the main mechanisms of immune escape, and immune checkpoint blocking therapy (ICB) targeting this mechanism can regulate the immunosuppressive microenvironment of tumors and restore the function of CTLs [281]. Lin et al. [282] reported a novel two-dimensional nano-PS BrP@MOL, which uses carboxylic acid on BrP to combine with two-dimensional MOL Hf_{12}-SBUs to realize efficient loading of BrP and PS (Fig. 1.14(a)). BrP@MOL can reprogram cell metabolism by inhibiting mitochondrial respiration and glycolysis, increases tumor oxygen content, and promotes PDT-induced ROS production. When combined with anti-PD-L1 immune checkpoint, the anti-tumor immune response was further enhanced, and the cure rate reached 80%. In addition, the ICB therapy targeting CTLA-4 and CD47 is also the focus of present research [283].

Immune adjuvant is a kind of chemical substance that enhances the body's immune response to antigens. The combination of immune adjuvant and PDT can enhance the anti-tumor immune response. CpG oligodeoxynucleotide (CpG ODN) is a synthetic short DNA fragment that can activate DC through interaction with TLR9 expressed by DC, which can trigger innate immunity, induce cytokine secretion, and enhance antigen-specific immune response [19]. Xiong et al. [284] constructed the photodynamic immunotherapy platform (PEG2000-SiNcTI-Ph/CpG-ZIF-8@CM) by loading the PS PEG2000-SiNcTI-Ph and the immunoadjuvant CpG into the MOF ZIF-8 NPs and then coating them with fusion membranes expressing PD-1 and CD47. This nanoplatform can evade macrophage

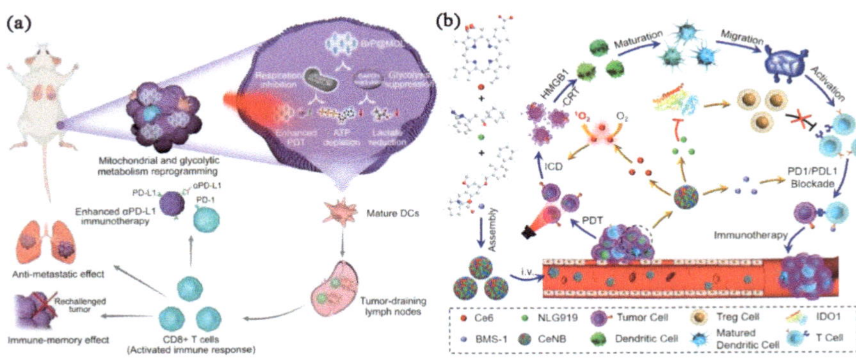

Fig. 1.14 (**a**) BrP@MOL promotes PDT-induced ROS generation by inhibiting the mechanism of mitochondrial respiration and glycolysis reprogramming cell metabolism, which further enhances anti-tumor immune response when combined with anti-PD-L1 immune checkpoint (Copyright © 2024 Angewandte Chemie International Edition. Reprinted with permission from Ref. [282]). (**b**) Schematic diagram of photodynamic immunotherapy of metastatic tumors by CeNB via cascade immunoactivation. Ce6 can promote the self-assembly of NLG919 and BMS-1 through non-covalent intermolecular interactions. The PDT effect of CeNB can inhibit the primary tumor and induce ICD, followed by activation of an immune cascade through in situ tumor vaccination. CeNB can block the immune checkpoint of IDO and PD-1/PD-L1 at the same time, reverse the immunosuppression and immune evasion behavior of metastatic tumor cells, and achieve effective immunotherapy (Copyright © 2023 Small. Reprinted with permission from Ref. [285])

phagocytosis by activating the CD47/SIRRPα pathway and target CT26 tumor cells using PD-1/PD-L1 interactions. The synergistic therapy of ICD triggered by PDT, CpG, and ICB enhances the anti-tumor immune response and effectively inhibits the recurrence and metastasis of CT26 tumors.

Immunosuppressive agents can weaken immunosuppressive cells and alter immunosuppressive TME, thereby promoting PDT-induced immune responses. Indoleamine 2,3-dioxygenase (IDO) overexpressed in tumor cells can catalyze tryptophan (Trp) metabolism to kynurenine (Kyn), which can not only inhibit the proliferation and activity of CTLs but also can abnormally accumulate and recruit Tregs cells. Zheng et al. [285] constructed a photodynamic immune nanoplatform (CeNB) by combining the PS Ce6, IDO inhibitor (NLG919), and PD-1/PD-L1 blocker (BMS-1) through a simple self-assembly method (Fig. 1.14(b)). CeNB can induce ICD and form tumor vaccine in situ, inhibit the activity of IDO and Tregs, cut off the immune escape behavior of tumor cells, and greatly inhibit the growth and metastasis of tumors.

5 Conclusion and Future Perspectives

In recent decades, significant progress has been made in the development of functional nanoparticles for targeted PDT in oncology. These nanomaterials have demonstrated superior performance in addressing existing challenges associated with PDT, including enhancing tumor accumulation, improving cellular uptake, and maximizing therapeutic efficacy. This chapter systematically summarizes the principles of PDT and the applications of nanomaterials in this context and details various categories of nanoparticles such as quantum dots, carbon dots, metal-organic frameworks, and upconversion nanoparticles. Each of these materials plays a crucial role as photosensitizers, carriers, or light converters within PDT.

Furthermore, the synergistic application of PDT with other therapeutic strategies, such as PTT, RT, chemotherapy, CDT, and immunotherapy, has been explored. These combination therapies not only enhance therapeutic outcomes but also address the inherent limitations in standalone PDT approaches. The multifunctionality of these nanomaterials facilitates the potential for synergistic effects between various treatment modalities, thereby enabling precise tumor therapy tailored to individual patient needs.

Despite the promising potential of nanomaterials in PDT and combination therapies, several challenges remain to be addressed. The efficacy of PDT is significantly influenced by the tumor microenvironment, where hypoxic conditions are known to diminish the generation and activity of ROS. Additionally, the photostability of PSs and the tissue penetration capabilities of the light source are still limited, restricting the applicability of PDT in deep-seated tumors. Furthermore, the clinical translation of PDT is hampered by concerns regarding the biocompatibility and safety of nanomaterials, and the biodegradability has been emerged as a critical issue. The persistence of nanoparticles in the body can lead to chronic inflammatory responses or pose unknown toxicological risks. Thus, the development of nanomaterials that can rapidly degrade and be cleared from the body after treatment is essential not only to mitigate potential long-term side effects but also to enhance the overall safety of PDT.

Moreover, further investigation into the synergistic mechanisms between PDT and other therapeutic modalities is warranted to achieve more efficient tumor treatment outcomes. Strengthening preclinical studies and clinical trials is imperative to validate the safety and efficacy of novel PDT strategies, thereby providing a robust scientific foundation for their clinical application. Personalized treatment approaches also represent a vital direction for the future of PDT, with the potential to develop tailored PDT regimens based on the unique characteristics of each patient's tumor, optimizing therapeutic efficacy while minimizing side effects.

In summary, PDT represents a promising modality for cancer treatment, with its advances reliant on the progress of nanotechnology and interdisciplinary collaboration. By continually optimizing the design of nanomaterials, enhancing the synergistic effects of treatment strategies, addressing biodegradability concerns, and bolstering clinical translation research, it is potential to develop intelligent, biodegradable nanomaterials capable of treating deep tumors and responding multifaceted responses within the tumor microenvironment. The integration of nanomaterial-based PDT is poised to play an increasingly significant role in the future of cancer therapy, offering patients safer and more effective treatment options.

References

1. Bray F, Laversanne M, Sung H, Ferlay J, Siegel RL, Soerjomataram I, Jemal A. Global cancer statistics 2022: GLOBOCAN estimates of incidence and mortality worldwide for 36 cancers in 185 countries. CA Cancer J Clin. 2024;74(3):229–63.
2. Gao L, Zhang C, Gao D, Liu H, Yu X, Lai J, Wang F, Lin J, Liu Z. Enhanced anti-tumor efficacy through a combination of integrin αvβ6-targeted photodynamic therapy and immune checkpoint inhibition. Theranostics. 2016;6(5):627.
3. Mun EJ, Babiker HM, Weinberg U, Kirson ED, Von Hoff DD. Tumor-treating fields: a fourth modality in cancer treatment. Clin Cancer Res. 2018;24(2):266–75.
4. Radvanyi LG. Targeting the cancer mutanome of breast cancer. Nat Med. 2018;24(6):703–4.
5. Liu Y, Wang X, He D, Cheng Y. Protection against chemotherapy-and radiotherapy-induced side effects: a review based on the mechanisms and therapeutic opportunities of phytochemicals. Phytomedicine. 2021;80:153402.
6. Li H, Wang M, Huang B, Zhu S, Zhou J, Chen D, Cui R, Zhang M, Sun Z. Theranostic near-infrared-IIb emitting nanoprobes for promoting immunogenic radiotherapy and abscopal effects against cancer metastasis. Nat Commun. 2021;12(1):7149.
7. Zhong Y, Li X, Chen J, Wang X, Wei L, Fang L, Kumar A, Zhuang S, Liu J. Recent advances in MOF-based nanoplatforms generating reactive species for chemodynamic therapy. Dalton Trans. 2020;49(32):11045–58.
8. Jiang W, Lin L, Wu P, Lin H, Sui J. Near-infrared-II nanomaterials for activatable photodiagnosis and phototherapy. Chem Eur J. 2024;30(33):e202400816.
9. Lismont M, Dreesen L, Wuttke S. Metal-organic framework nanoparticles in photodynamic therapy: current status and perspectives. Adv Funct Mater. 2017;27(14):1606314.
10. Lucky SS, Soo KC, Zhang Y. Nanoparticles in photodynamic therapy. Chem Rev. 2015;115(4):1990–2042.

11. Al-Ali AA, Alsalami KA, Athbi AM. Cytotoxic effects of CeO_2 NPs and β-Carotene and their ability to induce apoptosis in human breast normal and cancer cell lines. Iraqi J Sci. 2022;63(3):923–37.
12. Ghdeeb NJ, AbdulMajeed AM, Mohammed AH. Role of extracted Nano-metal oxides from factory wastes in medical applications. Iraqi J Sci. 2023;64(4):1704–16.
13. Halliwell B. Free radicals and antioxidants–quo vadis? Trends Pharmacol Sci. 2011;32(3):125–30.
14. Agostinis P, Berg K, Cengel KA, Foster TH, Girotti AW, Gollnick SO, Hahn SM, Hamblin MR, Juzeniene A, Kessel D. Photodynamic therapy of cancer: an update. CA Cancer J Clin. 2011;61(4):250–81.
15. Wang D, Wu H, Zhou J, Xu P, Wang C, Shi R, Wang H, Wang H, Guo Z, Chen Q. In situ one-pot synthesis of MOF–polydopamine hybrid nanogels with enhanced photothermal effect for targeted cancer therapy. Adv Sci. 2018;5(6):1800287.
16. Knoll JD, Turro C. Control and utilization of ruthenium and rhodium metal complex excited states for photoactivated cancer therapy. Coord Chem Rev. 2015;282:110–26.
17. Singh PP, Sinha S, Gahtori P, Mishra D, Pandey G, Srivastava V. Recent advancement in photosensitizers for photodynamic therapy. Dyes Pigments. 2024;229:112262.
18. Algorri JF, Ochoa M, Roldan-Varona P, Rodriguez-Cobo L, Lopez-Higuera JM. Photodynamic therapy: a compendium of latest reviews. Cancer. 2021;13(17):4447.
19. Hua J, Wu P, Gan L, Zhang Z, He J, Zhong L, Zhao Y, Huang Y. Current strategies for tumor photodynamic therapy combined with immunotherapy. Front Oncol. 2021;11:738323.
20. Zhong S, Chen C, Yang G, Zhu Y, Cao H, Xu B, Luo Y, Gao Y, Zhang W. Acid-triggered nanoexpansion polymeric micelles for enhanced photodynamic therapy. ACS Appl Mater Interfaces. 2019;11(37):33697–705.
21. Chen W, Wang Z, Tian M, Hong G, Wu Y, Sui M, Chen M, An J, Song F, Peng X. Integration of TADF photosensitizer as "Electron Pump" and BSA as "Electron Reservoir" for boosting type I photodynamic therapy. J Am Chem Soc. 2023;145(14):8130–40.
22. Castano AP, Demidova TN, Hamblin MR. Mechanisms in photodynamic therapy: part two—cellular signaling, cell metabolism and modes of cell death. Photodiagn Photodyn Ther. 2005;2(1):1–23.
23. Plaetzer K, Krammer B, Berlanda J, Berr F, Kiesslich T. Photophysics and photochemistry of photodynamic therapy: fundamental aspects. Lasers Med Sci. 2009;24:259–68.
24. Zanganeh N, Guda VK, Toghiani H, Keith JM. Sinter-resistant and highly active sub-5 nm bimetallic Au–Cu nanoparticle catalysts encapsulated in silica for high-temperature carbon monoxide oxidation. ACS Appl Mater Interfaces. 2018;10(5):4776–85.
25. Cheng Z, Li M, Dey R, Chen Y. Nanomaterials for cancer therapy: current progress and perspectives. J Hematol Oncol. 2021;14:1–27.
26. Giodini L, Re FL, Campagnol D, Marangon E, Posocco B, Dreussi E, Toffoli G. Nanocarriers in cancer clinical practice: a pharmacokinetic issue. Nanomed Nanotechnol Biol Med. 2017;13(2):583–99.
27. Xu M, Zhou L, Zheng L, Zhou Q, Liu K, Mao Y, Song S. Sonodynamic therapy-derived multimodal synergistic cancer therapy. Cancer Lett. 2021;497:229–42.
28. Ali ES, Sharker SM, Islam MT, Khan IN, Shaw S, Rahman MA, Uddin SJ, Shill MC, Rehman S, Das N. Targeting cancer cells with nanotherapeutics and nanodiagnostics: current status and future perspectives. In: Seminars in cancer biology. Elsevier; 2021. p. 52–68.
29. Noda Y, Merschjann C, Tarábek J, Amsalem P, Koch N, Bojdys MJ. Directional charge transport in layered two-dimensional triazine-based graphitic carbon nitride. Angew Chem Int Ed. 2019;58(28):9394–8.
30. Liu J, Zhang Y, Zhang L, Xie F, Vasileff A, Qiao SZ. Graphitic carbon nitride (g-C_3N_4)-derived N-rich graphene with tuneable interlayer distance as a high-rate anode for sodium-ion batteries. Adv Mater. 2019;31(24):1901261.
31. Zhu H, Zhang Y, Zhang L, Yu T, Zhang K, Jiang H, Wu L, Wang S. Highly photostable and biocompatible graphene oxides with amino acid functionalities. J Mater Chem C. 2014;2(34):7126–32.

32. Li H, Tay RY, Tsang SH, Zhen X, Teo EHT. Controllable synthesis of highly luminescent boron nitride quantum dots. Small. 2015;11(48):6491–9.
33. Tokmachev AM, Averyanov DV, Parfenov OE, Taldenkov AN, Karateev IA, Sokolov IS, Kondratev OA, Storchak VG. Emerging two-dimensional ferromagnetism in silicene materials. Nat Commun. 2018;9(1):1672.
34. Zhang H, He H, Jiang X, Xia Z, Wei W. Preparation and characterization of chiral transition-metal dichalcogenide quantum dots and their enantioselective catalysis. ACS Appl Mater Interfaces. 2018;10(36):30680–8.
35. Zhang X, Lai Z, Ma Q, Zhang H. Novel structured transition metal dichalcogenide nanosheets. Chem Soc Rev. 2018;47(9):3301–38.
36. Zhu H, Lai Z, Fang Y, Zhen X, Tan C, Qi X, Ding D, Chen P, Zhang H, Pu K. Ternary chalcogenide nanosheets with ultrahigh photothermal conversion efficiency for photoacoustic theranostics. Small. 2017;13(16):1604139.
37. Guan Q, Ma J, Yang W, Zhang R, Zhang X, Dong X, Fan Y, Cai L, Cao Y, Zhang Y. Highly fluorescent Ti_3C_2 MXene quantum dots for macrophage labeling and Cu^{2+} ion sensing. Nanoscale. 2019;11(30):14123–33.
38. Long L, Niu X, Yan K, Zhou G, Wang J, Wu X, Chu PK. Highly fluorescent and stable black phosphorus quantum dots in water. Small. 2018;14(48):1803132.
39. Wang K, Chen Y, Zheng J, Ge Y, Ji J, Song Y, Zhang H. Black phosphorus quantum dot based all-optical signal processing: ultrafast optical switching and wavelength converting. Nanotechnology. 2019;30(41):415202.
40. Wang X, Sun G, Li N, Chen P. Quantum dots derived from two-dimensional materials and their applications for catalysis and energy. Chem Soc Rev. 2016;45(8):2239–62.
41. Jaramillo TF, Jørgensen KP, Bonde J, Nielsen JH, Horch S, Chorkendorff I. Identification of active edge sites for electrochemical H_2 evolution from MoS_2 nanocatalysts. Science. 2007;317(5834):100–2.
42. Ranjbari F, Fathi F. Recent advances in chemistry, mechanism, and applications of quantum dots in photodynamic and photothermal therapy. Anti Cancer Agents Med Chem. 2024;24(10):733–44.
43. Ji X, Wang W, Mattoussi H. Controlling the spectroscopic properties of quantum dots via energy transfer and charge transfer interactions: concepts and applications. Nano Today. 2016;11(1):98–121.
44. Samia AC, Chen X, Burda C. Semiconductor quantum dots for photodynamic therapy. J Am Chem Soc. 2003;125(51):15736–7.
45. Shen Y, Shuhendler AJ, Ye D, Xu J-J, Chen H-Y. Two-photon excitation nanoparticles for photodynamic therapy. Chem Soc Rev. 2016;45(24):6725–41.
46. Mo D, Hu L, Zeng G, Chen G, Wan J, Yu Z, Huang Z, He K, Zhang C, Cheng M. Cadmium-containing quantum dots: properties, applications, and toxicity. Appl Microbiol Biotechnol. 2017;101(7):2713–33.
47. Liu Y, Chang Q, Sun Z, Liu J, Deng X, Liu Y, Cao A, Wang H. Fate of CdSe/ZnS quantum dots in cells: endocytosis, translocation and exocytosis. Colloids Surf B Biointerfaces. 2021;208:112140.
48. Nguyen KC, Rippstein P, Tayabali AF, Willmore WG. Mitochondrial toxicity of cadmium telluride quantum dot nanoparticles in mammalian hepatocytes. Toxicol Sci. 2015;146(1):31–42.
49. Liang Y, Zhang T, Tang M. Toxicity of quantum dots on target organs and immune system. J Appl Toxicol. 2022;42(1):17–40.
50. Sobhanan J, Jones P, Kohara R, Sugino S, Vacha M, Subrahmanyam C, Takano Y, Lacy F, Biju V. Toxicity of nanomaterials due to photochemical degradation and the release of heavy metal ions. Nanoscale. 2020;12(43):22049–58.
51. Ge J, Lan M, Zhou B, Liu W, Guo L, Wang H, Jia Q, Niu G, Huang X, Zhou H. A graphene quantum dot photodynamic therapy agent with high singlet oxygen generation. Nat Commun. 2014;5(1):4596.

52. Liang Q, Yu F, Cai H, Wu X, Ma M, Li Z, Tedesco AC, Zhu J, Xu Q, Bi H. Photo-activated autophagy-associated tumour cell death by lysosome impairment based on manganese-doped graphene quantum dots. J Mater Chem B. 2023;11(11):2466–77.
53. Li L, Yu Y, Ye GJ, Ge Q, Ou X, Wu H, Feng D, Chen XH, Zhang Y. Black phosphorus field-effect transistors. Nat Nanotechnol. 2014;9(5):372–7.
54. Guo T, Tang Q, Guo Y, Qiu H, Dai J, Xing C, Zhuang S, Huang G. Boron quantum dots for photoacoustic imaging-guided photothermal therapy. ACS Appl Mater Interfaces. 2020;13(1):306–11.
55. Zhou Q, Chen Q, Tong Y, Wang J. Light-induced ambient degradation of few-layer black phosphorus: mechanism and protection. Angew Chem Int Ed. 2016;55(38):11437–41.
56. Zhang X, Tang J, Li C, Lu Y, Cheng L, Liu J. A targeting black phosphorus nanoparticle based immune cells nano-regulator for photodynamic/photothermal and photo-immunotherapy. Bioact Mater. 2021;6(2):472–89.
57. Li H, Kang Z, Liu Y, Lee S. Carbon nanodots: synthesis, properties and applications. J Mater Chem. 2012;22(46):24230–53.
58. Wang R, Lu K, Tang Z, Xu Y. Recent progress in carbon quantum dots: synthesis, properties and applications in photocatalysis. J Mater Chem A. 2017;5(8):3717–34.
59. Tang L, Ji R, Li X, Bai G, Liu CP, Hao J, Lin J, Jiang H, Teng KS, Yang Z. Deep ultraviolet to near-infrared emission and photoresponse in layered N-doped graphene quantum dots. ACS Nano. 2014;8(6):6312–20.
60. Gao N, Huang L, Li T, Song J, Hu H, Liu Y, Ramakrishna S. Application of carbon dots in dye-sensitized solar cells: a review. J Appl Polym Sci. 2020;137(10):48443.
61. Zeng Q, Feng T, Tao S, Zhu S, Yang B. Precursor-dependent structural diversity in luminescent carbonized polymer dots (CPDs): the nomenclature. Light Sci Theory Appl. 2021;10(1):142.
62. Kang Y, Li Y, Fang Y, Xu Y, Wei X, Yin X. Carbon quantum dots for zebrafish fluorescence imaging. Sci Rep. 2015;5(1):11835.
63. Jiang B, Yu Y, Guo X, Ding Z, Zhou B, Liang H, Shen X. White-emitting carbon dots with long alkyl-chain structure: effective inhibition of aggregation caused quenching effect for label-free imaging of latent fingerprint. Carbon. 2018;128:12–20.
64. Gao J, Zhu M, Huang H, Liu Y, Kang Z. Advances, challenges and promises of carbon dots. Inorg Chem Front. 2017;4(12):1963–86.
65. Sharma V, Tiwari P, Mobin SM. Sustainable carbon-dots: recent advances in green carbon dots for sensing and bioimaging. J Mater Chem B. 2017;5(45):8904–24.
66. Zhu X, Yuan X, Han L, Liu H, Sun BJB. A smartphone-integrated optosensing platform based on red-emission carbon dots for real-time detection of pyrethroids. Biosens Bioelectron. 2021;191:113460.
67. Ali H, Ghosh S, Jana NR. Fluorescent carbon dots as intracellular imaging probes. Wiley Interdiscip Rev Nanomed Nanobiotechnol. 2020;12(4):e1617.
68. Truskewycz A, Yin H, Halberg N, Lai DT, Ball AS, Truong VK, Rybicka AM, Cole I. Carbon dot therapeutic platforms: administration, distribution, metabolism, excretion, toxicity, and therapeutic potential. Small. 2022;18(16):2106342.
69. Tejwan N, Saha SK, Das J. Multifaceted applications of green carbon dots synthesized from renewable sources. Adv Colloid Interface Sci. 2020;275:102046.
70. Wenger OS. A bright future for photosensitizers. Nat Chem. 2020;12(4):323–4.
71. Kuo W, Chen H, Chen S, Chang C, Chen P, Hou Y, Shao Y, Kao H, Hsu C, Chen Y. Graphene quantum dots with nitrogen-doped content dependence for highly efficient dual-modality photodynamic antimicrobial therapy and bioimaging. Biomaterials. 2017;120:185–94.
72. Yang K, Wang C, Liu C, Ding S, Tian F, Li F. Bioluminescence-initiated photodynamic therapy bridged on high-luminescent carbon dots-conjugated protoporphyrin IX. J Mater Sci. 2019;54(4):3383–91.
73. Georgakilas V, Perman JA, Tucek J, Zboril R. Broad family of carbon nanoallotropes: classification, chemistry, and applications of fullerenes, carbon dots, nanotubes, graphene, nanodiamonds, and combined superstructures. Chem Rev. 2015;115(11):4744–822.

74. Lin L, Song X, Dong X, Li B. Nano-photosensitizers for enhanced photodynamic therapy. Photodiagn Photodyn Ther. 2021;36:102597.
75. Kamanli AF, Yildiz MZ, Arslan H, Çetinel G, Lim NK, Lim HSJO, Technology L. Development of a new multi-mode NIR laser system for photodynamic therapy. Opt Laser Technol. 2020;128:106229.
76. Chung YJ, Kim J, Park CB. Photonic carbon dots as an emerging nanoagent for biomedical and healthcare applications. ACS Nano. 2020;14(6):6470–97.
77. Zhang Y, Jia Q, Nan F, Wang J, Liang K, Li J, Xue X, Ren H, Liu W, Ge J. Carbon dots nano-photosensitizers with tunable reactive oxygen species generation for mitochondrion-targeted type I/II photodynamic therapy. Biomaterials. 2023;293:121953.
78. Cai H, Wu X, Jiang L, Yu F, Yang Y, Li Y, Zhang X, Liu J, Li Z, Bi H. Lysosome-targeted carbon dots with a light-controlled nitric oxide releasing property for enhanced photodynamic therapy. Chin Chem Lett. 2024;35(4):108946.
79. SimonYarza T, Mielcarek A, Couvreur P, Serre C. Nanoparticles of metal-organic frameworks: on the road to in vivo efficacy in biomedicine. Adv Mater. 2018;30(37):1707365.
80. Cheng L, Wang X, Gong F, Liu T, Liu Z. 2D nanomaterials for cancer theranostic applications. Adv Mater. 2020;32(13):1902333.
81. Zhang S, Pei X, Gao H, Chen S, Wang J. Metal-organic framework-based nanomaterials for biomedical applications. Chin Chem Lett. 2020;31(5):1060–70.
82. Alves SR, Calori IR, Tedesco AC. Photosensitizer-based metal-organic frameworks for highly effective photodynamic therapy. Mater Sci Eng C. 2021;131:112514.
83. Sun Q, Yang J, Shen W, Lu H, Hou X, Liu Y, Xu Y, Wu Q, Xuan Z, Yang Y. Engineering mitochondrial uncoupler synergistic photodynamic nanoplatform to harness immunostimulatory pro-death autophagy/mitophagy. Biomaterials. 2022;289:121796.
84. Chen J, Zhu Y, Kaskel S. Porphyrin-based metal–organic frameworks for biomedical applications. Angew Chem Int Ed. 2021;60(10):5010–35.
85. Lim C, Heo J, Shin S, Jeong K, Seo YH, Jang W, Park CR, Park SY, Kim S, Kwon I. Nanophotosensitizers toward advanced photodynamic therapy of cancer. Cancer Lett. 2013;334(2):176–87.
86. Lu K, He C, Lin W. Nanoscale metal–organic framework for highly effective photodynamic therapy of resistant head and neck cancer. J Am Chem Soc. 2014;136(48):16712–5.
87. Chen X, Mendes BB, Zhuang Y, Conniot J, Mercado Argandona S, Melle F, Sousa DP, Perl D, Chivu A, Patra HK. A fluorinated BODIPY-based zirconium metal–organic framework for in vivo enhanced photodynamic therapy. J Am Chem Soc. 2024;146(2):1644–56.
88. Gunaydin G, Gedik ME, Ayan S. Photodynamic therapy for the treatment and diagnosis of cancer–a review of the current clinical status. Front Chem. 2021;9:686303.
89. Niculescu A, Grumezescu AM. Photodynamic therapy—an up-to-date review. Appl Sci. 2021;11(8):3626.
90. Fahmy SA, Azzazy HME, Schaefer J. Liposome photosensitizer formulations for effective cancer photodynamic therapy. Pharmaceutics. 2021;13(9):1345.
91. Chen J, Fan T, Xie Z, Zeng Q, Xue P, Zheng T, Chen Y, Luo X, Zhang H. Advances in nanomaterials for photodynamic therapy applications: status and challenges. Biomaterials. 2020;237:119827.
92. Chudy M, Tokarska K, Jastrzębska E, Bułka M, Drozdek S, Lamch Ł, Wilk KA, Brzózka Z. Lab-on-a-chip systems for photodynamic therapy investigations. Biosens Bioelectron. 2018;101:37–51.
93. Agel MR, Baghdan E, Pinnapireddy SR, Lehmann J, Schaefer J, Bakowsky U. Curcumin loaded nanoparticles as efficient photoactive formulations against gram-positive and gram-negative bacteria. Colloids Surf B Biointerfaces. 2019;178:460–8.
94. Tong Q, Chen M, Lin H, Cheng Q, Lu W, Xu L. Recent progress in delivery systems for photosensitizers and anti-cancer photodynamic therapy. Acta Pharm Sin B. 2023;11(8):2320–33.
95. Sarcan ET, SilindirGunay M, Ozer AY. Theranostic polymeric nanoparticles for NIR imaging and photodynamic therapy. Int J Pharm. 2018;551(1–2):329–38.

96. Zheng N, Zhang Z, Kuang J, Wang C, Zheng Y, Lu Q, Bai Y, Li Y, Wang A, Song W. Poly (photosensitizer) nanoparticles for enhanced in vivo photodynamic therapy by interrupting the π–π stacking and extending circulation time. ACS Appl Mater Interfaces. 2019;11(20):18224–32.
97. Yu H, Huang Y, Lv H, Zhang H, Ma D, Yang H, Xie S, Peng Y. A13458 Triblock copolymers encapsulated poly (aryl benzyl ether) dendrimer zinc (II) phthalocyanine nanoparticles for enhancement in vitro photodynamic efficacy. J Hypertens. 2018;36:e70.
98. Birks JB. Photophysics of aromatic molecules. Berichte der Bunsengesellschaft für physikalische Chemie. 1970;74(12):1294–5.
99. Sekkat N, van den Bergh H, Nyokong T, Lange N. Like a bolt from the blue: phthalocyanines in biomedical optics. Molecules. 2011;17(1):98–144.
100. Ding D, Li K, Liu B, Tang BZ. Bioprobes based on AIE fluorogens. Acc Chem Res. 2013;46(11):2441–53.
101. Luo J, Xie Z, Lam JW, Cheng L, Chen H, Qiu C, Kwok HS, Zhan X, Liu Y, Zhu D. Aggregation-induced emission of 1-methyl-1, 2, 3, 4, 5-pentaphenylsilole. Chem Commun (Cambridge, UK). 2001;18:1740–1.
102. Tang BZ, Zhan X, Yu G, Lee PPS, Liu Y, Zhu D. Efficient blue emission from siloles. J Mater Chem. 2001;11(12):2974–8.
103. Mei J, Leung NL, Kwok RT, Lam JW, Tang BZ. Aggregation-induced emission: together we shine, united we soar! Chem Rev. 2015;115(21):11718–940.
104. Chang C, Hsieh M, Lin J, Chang T. Selective photodynamic therapy based on aggregation-induced emission enhancement of fluorescent organic nanoparticles. Biomaterials. 2012;33(3):897–906.
105. Hsieh M, Chien C, Chang C, Chang T. Aggregation induced photodynamic therapy enhancement based on linear and nonlinear excited FRET of fluorescent organic nanoparticles. J Mater Chem B. 2013;1(18):2350–7.
106. Sun F, Shen H, Liu Q, Chen Y, Guo W, Du W, Xu C, Wang B, Xing G, Jin Z. Powerful synergy of traditional Chinese medicine and aggregation-induced emission-active photosensitizer in photodynamic therapy. ACS Nano. 2023;17(19):18952–64.
107. Sun X, Zebibula A, Dong X, Li G, Zhang G, Zhang D, Qian J, He S. Targeted and imaging-guided in vivo photodynamic therapy for tumors using dual-function, aggregation-induced emission nanoparticles. Nano Res. 2018;11:2756–70.
108. Hu F, Huang Y, Zhang G, Zhao R, Yang H, Zhang D. Targeted bioimaging and photodynamic therapy of cancer cells with an activatable red fluorescent bioprobe. Anal Chem. 2014;86(15):7987–95.
109. Lee H, Ahn H, Mun J, Lee YY, Kim M, Cho NH, Chang K, Kim WS, Rho J, Nam KT. Amino-acid-and peptide-directed synthesis of chiral plasmonic gold nanoparticles. Nature. 2018;556(7701):360–5.
110. Jian C, Zhang J, Ma X. Cu–Ag alloy for engineering properties and applications based on the LSPR of metal nanoparticles. RSC Adv. 2020;10(22):13277–85.
111. Gezgin SY, Kepceoğlu A, Gündoğdu Y, Zongo S, Zawadzka A, Kiliç HŞ, Sahraoui B. Effect of Ar gas pressure on LSPR property of au nanoparticles: comparison of experimental and theoretical studies. Nano. 2020;10(6):1071.
112. Xue Q, Kang R, Klionsky DJ, Tang D, Liu J, Chen X. Copper metabolism in cell death and autophagy. Autophagy. 2023;19(8):2175–95.
113. Hosseini M, Shaki F, GhaziKhansari M, Pourahmad J. Toxicity of copper on isolated liver mitochondria: impairment at complexes I, II, and IV leads to increased ROS production. Cell Biochem Biophys. 2014;70:367–81.
114. Yang J, Wang Y, Qin G, Tian T, Ran J, Wang H, Yang C. Photogeneration of hydroxyl radicals based on aggregation-induced emission Luminogen-assembled copper Cysteamine nanoparticles for photodynamic therapy. ACS Appl Nano Mater. 2022;6(1):533–43.
115. Xu J, Wang J, Ye J, Jiao J, Liu Z, Zhao C, Li B, Fu Y. Metal-coordinated supramolecular self-assemblies for cancer theranostics. Adv Sci. 2021;8(16):2101101.

116. Zhang S, Li Y, Liu C, Zhang Y, Sun P, Lan X, Liu C. Supramolecular amino acid-based metallo-nanozyme through multicomponent coordination self-assembly for in-site tumor synergistic catalytic-chemotherapy. Chem Eng J. 2022;437:135312.
117. Song E, Wu Q, Gao R, Lan X, Zhang Y, Geng H, Liu C, Xu F, Li Y, Liu C. Supramolecular catalytic nanomedicines based on coordination self-assembly of amino acids for cascade-activated and-amplified synergetic cancer therapy. J Mater Chem B. 2022;10(47):9838–47.
118. Lan M, Zhao S, Liu W, Lee CS, Zhang W, Wang P. Photosensitizers for photodynamic therapy. Adv Healthc Mater. 2019;8(13):1900132.
119. Liu X, Li J, Zhang F, Yan D, Wang L, Wang D, Tang BZ. Light-driven sextuple theranostics: nanomedicine involving second near-infrared AIE-active Ir (III) complex for prominent cancer treatment. Sci China Chem. 2024;67:1–11.
120. Han G, Li G, Huang J, Han C, Turro C, Sun Y. Two-photon-absorbing ruthenium complexes enable near infrared light-driven photocatalysis. Nat Commun. 2022;13(1):2288.
121. Kankala RK, Han YH, Na J, Lee CH, Sun Z, Wang SB, Kimura T, Ok YS, Yamauchi Y, Chen AZ. Nanoarchitectured structure and surface biofunctionality of mesoporous silica nanoparticles. Adv Mater. 2020;32(23):1907035.
122. Watermann A, Brieger J. Mesoporous silica nanoparticles as drug delivery vehicles in cancer. Nano. 2017;7(7):189.
123. Lei Q, Guo J, Noureddine A, Wang A, Wuttke S, Brinker CJ, Zhu W. Sol–gel-based advanced porous silica materials for biomedical applications. Adv Funct Mater. 2020;30(41):1909539.
124. Feng Y, Liao Z, Li M, Zhang H, Li T, Qin X, Li S, Wu C, You F, Liao X. Mesoporous silica nanoparticles-based nanoplatforms: basic construction, current state, and emerging applications in anticancer therapeutics. Adv Healthc Mater. 2023;12(16):2201884.
125. Ahmadi Nasab N, Hassani Kumleh H, Beygzadeh M, Teimourian S, Kazemzad M. Delivery of curcumin by a pH-responsive chitosan mesoporous silica nanoparticles for cancer treatment. Artif Cells Nanomed Biotechnol. 2018;46(1):75–81.
126. Güleryüz B, Işık A, Gülsoy M. Synergistic effect of mesoporous silica nanocarrier-assisted photodynamic therapy and anticancer agent activity on lung cancer cells. Lasers Med Sci. 2024;39(1):91.
127. Suk JS, Xu Q, Kim N, Hanes J, Ensign LM. PEGylation as a strategy for improving nanoparticle-based drug and gene delivery. Adv Drug Deliv Rev. 2016;99:28–51.
128. Yu L, Chen Y, Lin H, Gao S, Chen H, Shi J. Magnesium-engineered silica framework for pH-accelerated biodegradation and DNAzyme-triggered chemotherapy. Small. 2018;14(35):1800708.
129. Tian B, Liu S, Yu C, Liu S, Dong S, Feng L, Hu N, Yang P. A metal-free mesoporous carbon dots/silica hybrid type I photosensitizer with enzyme-activity for synergistic treatment of hypoxic tumor. Adv Funct Mater. 2023;33(25):2300818.
130. Yang G, Phua SZF, Bindra AK, Zhao Y. Degradability and clearance of inorganic nanoparticles for biomedical applications. Adv Mater. 2019;31(10):1805730.
131. Liberman A, Martinez HP, Ta CN, Barback CV, Mattrey RF, Kono Y, Blair SL, Trogler WC, Kummel AC, Wu Z. Hollow silica and silica-boron nano/microparticles for contrast-enhanced ultrasound to detect small tumors. Biomaterials. 2012;33(20):5124–9.
132. Yu L, Chen Y, Wu M, Cai X, Yao H, Zhang L, Chen H, Shi J. "Manganese extraction" strategy enables tumor-sensitive biodegradability and theranostics of nanoparticles. J Am Chem Soc. 2016;138(31):9881–94.
133. Greasley SL, Page SJ, Sirovica S, Chen S, Martin RA, Riveiro A, Hanna JV, Porter AE, Jones JR. Controlling particle size in the Stöber process and incorporation of calcium. J Colloid Interface Sci. 2016;469:213–23.
134. Chen S, Greasley S, Ong Z, Naruphontjirakul P, Page S, Hanna J, Redpath A, Tsigkou O, Rankin S, Ryan M. Biodegradable zinc-containing mesoporous silica nanoparticles for cancer therapy. Mater Today Adv. 2020;6:100066.
135. Huang L, Zhao S, Wu J, Yu L, Singh N, Yang K, Lan M, Wang P, Kim JS. Photodynamic therapy for hypoxic tumors: advances and perspectives. Coord Chem Rev. 2021;438:213888.

136. Negi K, Pathak NK, Tripathy U, Dey SK, Sahu SK. Two-photon NIR-responsive carbon dots incorporated into NMOFs for targeted photodynamic therapy. Colloids Surf A Physicochem Eng Asp. 2024;694:134163.
137. Mahata MK, De R, Lee KT. Near-infrared-triggered upconverting nanoparticles for biomedicine applications. Biomedicine. 2021;9(7):756.
138. Del Valle CA, Hirsch T, Marín MJ. Recent advances in near infrared upconverting nanomaterials for targeted photodynamic therapy of cancer. Methods Appl Fluoresc. 2022;10(3):034003.
139. Liang G, Wang H, Shi H, Wang H, Zhu M, Jing A, Li J, Li G. Recent progress in the development of upconversion nanomaterials in bioimaging and disease treatment. J Nanobiotechnol. 2020;18:1–22.
140. Lu S, Tian H, Li L, Li B, Yang M, Zhou L, Jiang H, Li Q, Wang W, Nice EC. Nanoengineering a Zeolitic Imidazolate Framework-8 capable of manipulating energy metabolism against cancer chemo-phototherapy resistance. Small. 2022;18(48):2204926.
141. Yang Y, Zhou X, Chan YK, Wang Z, Li L, Li J, Liang K, Deng Y. Photo-activated nanofibrous membrane with self-rechargeable antibacterial function for stubborn infected cutaneous regeneration. Small. 2022;18(12):2105988.
142. Zhao H, Xu J, Feng C, Ren J, Bao L, Zhao Y, Tao W, Zhao Y, Yang X. Tailoring aggregation extent of photosensitizers to boost phototherapy potency for eliciting systemic antitumor immunity. Adv Mater. 2022;34(8):2106390.
143. Juengpanich S, Li S, Yang T, Xie T, Chen J, Shan Y, Lee J, Lu Z, Chen T, Zhang B. Pre-activated nanoparticles with persistent luminescence for deep tumor photodynamic therapy in gallbladder cancer. Nat Commun. 2023;14(1):5699.
144. Wei S, Qiao Y, Wu Z, Liu X, Li Y, Cui Z, Li C, Zheng Y, Liang Y, Li Z. Na^{+} inserted metal-organic framework for rapid therapy of bacteria-infected osteomyelitis through microwave strengthened Fenton reaction and thermal effects. Nano Today. 2021;37:101090.
145. Xu Y, Li C, Lu S, Wang Z, Liu S, Yu X, Li X, Sun Y. Construction of emissive ruthenium (II) metallacycle over 1000 nm wavelength for in vivo biomedical applications. Nat Commun. 2022;13(1):2009.
146. Yang Y, Fan X, Li L, Yang Y, Nuernisha A, Xue D, He C, Qian J, Hu Q, Chen H. Semiconducting polymer nanoparticles as theranostic system for near-infrared-II fluorescence imaging and photothermal therapy under safe laser fluence. ACS Nano. 2020;14(2):2509–21.
147. Sun B, Teo JY, Wu J, Zhang Y. Light conversion nanomaterials for wireless phototherapy. Acc Chem Res. 2023;56(10):1143–55.
148. Karnkaew A, Chen F, Zhan YH, Majewski RL, Cai WB. Scintillating nanoparticles as energy mediators for enhanced photodynamic therapy. ACS Nano. 2016;10(4):3918–35.
149. Yao B, Liu XX, Zhang WL, Lu HB. X-ray excited luminescent nanoparticles for deep photodynamic therapy. RSC Adv. 2023;13(43):30133–50.
150. Sun WJ, Luo L, Feng YS, Cai YT, Zhuang YX, Xie RJ, Chen XY, Chen HM. Aggregation-induced emission gold Clustoluminogens for enhanced low-dose X-ray-induced photodynamic therapy. Angew Chem Int Ed. 2020;59(25):9914–21.
151. Liu SK, Li WT, Zhang YY, Zhou JL, Du YQ, Dong SM, Tian BS, Fang LY, Ding H, Gai SL, Yang PP. Tailoring silica-based Nanoscintillators for Peroxynitrite-potentiated Nitrosative stress in postoperative radiotherapy of colon cancer. Nano Lett. 2022;22(15):6409–17.
152. Sun WJ, Shi TH, Luo L, Chen XM, Lv P, Lv Y, Zhuang YX, Zhu JJ, Liu G, Chen XY, Chen HM. Monodisperse and uniform mesoporous silicate Nanosensitizers achieve low-dose X-ray-induced deep-penetrating photodynamic therapy. Adv Mater. 2019;31(16):e1808024.
153. Song L, Li PP, Yang W, Lin XH, Liang H, Chen XF, Liu G, Li J, Yang HH. Low-dose X-ray activation of W(VI)-doped persistent luminescence nanoparticles for deep-tissue photodynamic therapy. Adv Funct Mater. 2018;28(18):1707496.
154. Ou X, Qin X, Huang B, Zan J, Wu Q, Hong Z, Xie L, Bian H, Yi Z, Chen X, Wu Y, Song X, Li J, Chen Q, Yang H, Liu X. High-resolution X-ray luminescence extension imaging. Nature. 2021;590(7846):410–5.

155. Pei P, Chen Y, Sun C, Fan Y, Yang Y, Liu X, Lu L, Zhao M, Zhang H, Zhao D, Liu X, Zhang F. X-ray-activated persistent luminescence nanomaterials for NIR-II imaging. Nat Nanotechnol. 2021;16(9):1011–8.
156. An YB, Xu DZ, Wen XF, Chen C, Liu G, Lu ZX. Internal light sources-mediated photodynamic therapy Nanoplatforms: Hope for the resolution of the traditional penetration problem. Adv Healthc Mater. 2024;13(1):2301326.
157. Sivasubramanian M, Chuang YC, Lo LW. Evolution of nanoparticle-mediated photodynamic therapy: from superficial to deep-seated cancers. Molecules. 2019;24(3):520.
158. Wan YL, Fu LH, Li CY, Lin J, Huang P. Conquering the hypoxia limitation for photodynamic therapy. Adv Mater. 2021;33(48):2103978.
159. Li KX, Deng ZYZ, Lei CR, Ding XQ, Li J, Wang CS. The role of oxidative stress in tumorigenesis and progression. Cells. 2024;13(5):441.
160. Fu SL, Chang LL, Liu SJ, Gao T, Sang X, Zhang ZP, Mu WW, Liu XQ, Liang S, Yang H, Yang HZ, Ma QP, Liu YJ, Zhang N. Temperature sensitive liposome based cancer nanomedicine enables tumour lymph node immune microenvironment remodelling. Nat Commun. 2023;14(1):2238.
161. Sun BW, Teo JY, Wu JX, Zhang Y. Light conversion nanomaterials for wireless phototherapy. Acc Chem Res. 2023;56(10):1143–55.
162. Dutta B, Barick KC, Hassan PA. Recent advances in active targeting of nanomaterials for anticancer drug delivery. Adv Colloid Interface Sci. 2021;296:102509.
163. Wang R, Li XS, Yoon J. Organelle-targeted photosensitizers for precision photodynamic therapy. ACS Appl Mater Interfaces. 2021;13(17):19543–71.
164. Dirak M, Yenici CM, Kolemen S. Recent advances in organelle-targeted organic photosensitizers for efficient photodynamic therapy. Coord Chem Rev. 2024;506:215710.
165. Bai XS, Lin Y, Gong LY, Duan JF, Sun XD, Wang CG, Liu ZR, Jiang J, Zhou XY, Zhou ML, Zhang ZR, Liu ZB, Jing P, Zhong ZR. Nanoparticles that target the mitochondria of tumor cells to restore oxygen supply for photodynamic therapy: design and preclinical validation against breast cancer. J Control Release. 2023;362:356–70.
166. Li ML, Xiong T, Du JJ, Tian RS, Xiao M, Guo LY, Long SR, Fan JL, Sun W, Shao K, Song XZ, Foley JW, Peng XJ. Superoxide radical photogenerator with amplification effect: surmounting the Achilles' heels of photodynamic oncotherapy. J Am Chem Soc. 2019;141(6):2695–702.
167. Tian BS, Liu SH, Yu CH, Liu SK, Dong SM, Feng LL, Hu NRS, Yang PP. A metal-free mesoporous carbon dots/silica hybrid type I photosensitizer with enzyme-activity for synergistic treatment of hypoxic tumor. Adv Funct Mater. 2023;33(25):2300818.
168. Tian JW, Li BW, Zhang F, Yao Z, Song WT, Tang YF, Ping Y, Liu B. Activatable type I photosensitizer with quenched photosensitization pre and post photodynamic therapy. Angew Chem Int Ed. 2023;62(46):e202307288.
169. Li ML, Xu YJ, Peng XJ, Kim JS. From low to no O_2-dependent hypoxia photodynamic therapy (hPDT): a New perspective. Acc Chem Res. 2022;55(22):3253–64.
170. Tang YF, Li YY, Li BW, Song WT, Qi GB, Tian JW, Huang W, Fan QL, Liu B. Oxygen-independent organic photosensitizer with ultralow-power NIR photoexcitation for tumor-specific photodynamic therapy. Nat Commun. 2024;15(1):2530.
171. Huang L, Zhao SJ, Wu JS, Yu L, Singh N, Yang K, Lan MH, Wang PF, Kim JS. Photodynamic therapy for hypoxic tumors: advances and perspectives. Coord Chem Rev. 2021;438:213888.
172. Wei K, Wu YX, Zheng X, Ma GP, Ji CD, Yin MZ. Fluorination-induced multi-enhancement of a Nano-photosensitizer for deep photodynamic therapy against hypoxia tumor. Adv Funct Mater. 2023;33(46):2305187.
173. Li ML, Shao YJ, Kim JH, Pu ZJ, Zhao XZ, Huang HQ, Xiong T, Kang Y, Li GZ, Shao K, Fan JL, Foley JW, Kim JS, Peng XJ. Unimolecular photodynamic O_2-Economizer to overcome hypoxia resistance in phototherapeutics. J Am Chem Soc. 2020;142(11):5380–8.
174. Sun XR, Chen KX, Liu YY, Zhang GD, Shi M, Shi PF, Zhang SS. Metal-organic framework combined with CaO nanoparticles for enhanced and targeted photodynamic therapy. Nanoscale Adv. 2021;3(23):6669–77.

175. Liu BY, Lan YF, Yang T, Tang X, Zhang C, Zuo BM, Li XD, Yue Q, Wang Z, Zhong XY, Luo X, Zeng TC, Zhang MQ, Wang WL, Li GQ, Guo HB. InN@In_2S_3 Core-Shell nanorod for oxygen-free and penetration depth-unrestricted photodynamic therapy against glioblastoma. Chem Eng J. 2024;479:147386.
176. Zhang C, Hu XM, Jin L, Lin LS, Lin HX, Yang Z, Huang W. Strategic design of conquering hypoxia in tumor for advanced photodynamic therapy. Adv Healthc Mater. 2023;12(24):2300530.
177. Yang J, Dai DH, Zhang X, Teng LS, Ma LJ, Yang YW. Multifunctional metal-organic framework (MOF)-based nanoplatforms for cancer therapy: from single to combination therapy. Theranostics. 2023;13(1):295–323.
178. Kong CQ, Chen XC. Combined photodynamic and photothermal therapy and immunotherapy for cancer treatment: a review. Int J Nanomedicine. 2022;17:6427–46.
179. Lv LX, Fan BL, Ji XR, Liu YX, Chen T, Li YL, Gao XY, Chen P, Tang B, Chen G. From the clinical perspective of photodynamic therapy and photothermal therapy: structure-activity-practice. Coord Chem Rev. 2024;507:215733.
180. Li XS, Lovell JF, Yoon J, Chen XY. Clinical development and potential of photothermal and photodynamic therapies for cancer. Nat Rev Clin Oncol. 2020;17(11):657–74.
181. Liu YJ, Bhattarai P, Dai ZF, Chen XY. Photothermal therapy and photoacoustic imaging nanotheranostics in fighting cancer. Chem Soc Rev. 2019;48(7):2053–108.
182. Wilson BC, Patterson MS. The physics, biophysics and technology of photodynamic therapy. Phys Med Biol. 2008;53(9):R61–R109.
183. Tang ZM, Zhao PR, Ni DL, Liu YY, Zhang M, Wang H, Zhang H, Gao HB, Yao ZW, Bu WB. Pyroelectric nanoplatform for NIR-II-triggered photothermal therapy with simultaneous pyroelectric dynamic therapy. Mater Horiz. 2018;5(5):946–52.
184. Kobayashi H, Furusawa A, Rosenberg A, Choyke PL. Near-infrared photoimmunotherapy of cancer: a new approach that kills cancer cells and enhances anti-cancer host immunity. Int Immunol. 2021;33(1):7–15.
185. Overchuk M, Weersink RA, Wilson BC, Zheng G. Photodynamic and Photothermal therapies: synergy opportunities for Nanomedicine. ACS Nano. 2023;17(9):7979–8003.
186. Bienia A, WiecheeCudak O, Murzyn AA, KrzykawskaSerda M. Photodynamic therapy and hyperthermia in combination treatment-neglected forces in the fight against cancer. Pharmaceutics. 2021;13(8):1147.
187. Liang SM, Hu DH, Li GF, Gao DY, Li F, Zheng HR, Pan M, Sheng ZH. NIR-II fluorescence visualization of ultrasound-induced blood-brain barrier opening for enhanced photothermal therapy against glioblastoma using indocyanine green microbubbles. Sci Bull. 2022;67(22):2316–26.
188. Luo H, Gao S. Recent advances in fluorescence imaging-guided photothermal therapy and photodynamic therapy for cancer: from near-infrared-I to near-infrared-II. J Control Release. 2023;362:425–45.
189. Hu L, Xu Y, Zhao Y, Mei Z, Xiong C, Xiao J, Zhang J, Tian J. Supramolecular nanovesicles with in-situ switchable photothermal/photodynamic effects for precisely controllable cancer phototherapy. Chem Eng J. 2023;476:146829.
190. Cui M, Tang D, Wang B, Zhang H, Liang G, Xiao H. Bioorthogonal guided activation of cGAS-STING by AIE photosensitizer nanoparticles for targeted tumor therapy and imaging. Adv Mater. 2023;35(52):2305668.
191. Meng X, Han Y, Wang S, Wang X, Zhang Z, Yao S, Wan X, Liu Z, Ge Z, Li L. Near-infrared photosensitizers adaptive to tumor hypoxic microenvironment for synergistic photothermal-photodynamic and immunotherapy. Nano Today. 2023;53:102030.
192. Asadi S, Bianchi L, De Landro M, Korganbayev S, Schena E, Saccomandi P. Laser-induced optothermal response of gold nanoparticles: from a physical viewpoint to cancer treatment application. J Biophotonics. 2021;14(2):e202000161.
193. Chen YW, Su YL, Hu SH, Chen SY. Functionalized graphene nanocomposites for enhancing photothermal therapy in tumor treatment. Adv Drug Deliv Rev. 2016;105:190–204.

194. Liu H, Gao C, Xu P, Li Y, Yan X, Guo X, Wen C, Shen X. Biomimetic gold Nanorods-manganese porphyrins with surface-enhanced Raman scattering effect for photoacoustic imaging-guided Photothermal/photodynamic therapy. Small. 2024;20(42):2401117.
195. Liu HX, Zhang JZ, Jia YN, Liu X, Chen XQ, Zhao WB, Mao C. Theranostic nanomotors for tumor multimode imaging and photothermal/photodynamic synergistic therapy. Chem Eng J. 2022;442:135994.
196. Sun S, Chen Q, Tang Z, Liu C, Li Z, Wu A, Lin H. Tumor microenvironment stimuli-responsive fluorescence imaging and synergistic cancer therapy by carbon-dot–Cu^{2+} Nanoassemblies. Angew Chem Int Ed. 2020;59(47):21041–8.
197. Chu Z, Tian T, Tao Z, Yang J, Chen B, Chen H, Wang W, Yin P, Xia X, Wang H, Qian H. Upconversion nanoparticles@$AgBiS_2$ core-shell nanoparticles with cancer-cell-specific cytotoxicity for combined photothermal and photodynamic therapy of cancers. Bioact Mater. 2022;17:71–80.
198. Song L, Chen B, Qin Z, Liu X, Guo Z, Lou H, Liu H, Sun W, Guo C, Li C. Temperature-dependent CAT-like RGD-BPNS@SMFN Nanoplatform for PTT-PDT self-synergetic tumor phototherapy. Adv Healthc Mater. 2022;11(8):2102298.
199. Wang Y, Luo SY, Wu YS, Tang P, Liu JJ, Liu ZY, Shen SH, Ren HZ, Wu DC. Highly penetrable and on-demand oxygen release with tumor activity composite nanosystem for photothermal/photodynamic synergetic therapy. ACS Nano. 2020;14(12):17046–62.
200. Thariat J, HannounLevi JM, Myint AS, Vuong T, Gérard JP. Past, present, and future of radiotherapy for the benefit of patients. Nat Rev Clin Oncol. 2013;10(1):52–60.
201. He LR, Yu XJ, Li WW. Recent progress and trends in X-ray-induced photodynamic therapy with low radiation doses. ACS Nano. 2022;16(12):19691–721.
202. Secchi V, Monguzzi A, Villa I. Design principles of hybrid nanomaterials for radiotherapy enhanced by photodynamic therapy. Int J Mol Sci. 2022;23(15):8736.
203. Kim BM, Hong Y, Lee S, Liu P, Lim JH, Lee YH, Lee TH, Chang KT, Hong Y. Therapeutic implications for overcoming radiation resistance in cancer therapy. Int J Mol Sci. 2015;16(11):26880–913.
204. Zeng LJ, Cao YH, He L, Ding SS, Bian XW, Tian G. Metal-ligand coordination nanomaterials for radiotherapy: emerging synergistic cancer therapy. J Mater Chem B. 2021;9(2):208–27.
205. Straub JM, New J, Hamilton CD, Lominska C, Shnayder Y, Thomas SM. Radiation-induced fibrosis: mechanisms and implications for therapy. J Cancer Res Clin Oncol. 2015;141(11):1985–94.
206. Viswanath D, Won YY. Combining Radiotherapy (RT) and Photodynamic Therapy (PDT): clinical studies on conventional RT-PDT approaches and novel nanoparticle-based RT-PDT approaches under preclinical evaluation. ACS Biomater Sci Eng. 2022;8(9):3644–58.
207. Zhu SC, Yan FH, Yang LL, Li BY, Xue RX, Yu WW, Wang Y, Huang L, Wang LJ, Han RC, Jiang YQ. Low-dose X-ray radiodynamic therapy solely based on gold nanoclusters for efficient treatment of deep hypoxic solid tumors combined with enhanced antitumor immune response. Theranostics. 2023;13(3):1042–58.
208. Li JC, Lv ZZ, Guo YR, Fang J, Wang AN, Feng YL, Zhang YQ, Zhu JF, Zhao ZS, Cheng XJ, Shi HB. Hafnium (Hf)-chelating porphyrin-decorated gold Nanosensitizers for enhanced radio-Radiodynamic therapy of colon carcinoma. ACS Nano. 2023;17(24):25147–56.
209. Sun W, Luo L, Feng Y, Qiu Y, Shi C, Meng S, Chen X, Chen H. Gadolinium–Rose Bengal Coordination Polymer Nanodots for MR-/Fluorescence-Image-Guided Radiation and Photodynamic Therapy. Adv Mater. 2020;32(23):2000377.
210. Zhao Y, Liang C, Mei ZJ, Yang HX, Wang B, Xie CH, Xu Y, Tian J. Oxygen-enriched MOF-hemoglobin X-ray Nanosensitizer for enhanced cancer radio-Radiodynamic therapy. ACS Mater Lett. 2023;5(12):3237–47.
211. Li T, Gao MQ, Wu ZF, Yang JJ, Mo BH, Yu ST, Gong XY, Liu J, Wang WD, Luo SL, Li R. Tantalum-zirconium co-doped metal-organic frameworks sequentially sensitize radio-Radiodynamic-immunotherapy for metastatic osteosarcoma. Adv Sci. 2023;10(10):2206779.

212. Lan GX, Ni KY, Xu RY, Lu KD, Lin ZK, Chan C, Lin WB. Nanoscale metal-organic layers for deeply penetrating X-ray-induced photodynamic therapy. Angew Chem Int Ed. 2017;56(40):12102–6.
213. Yi JN, Liu LY, Gao WJ, Zeng J, Chen YZ, Pang E, Lan MH, Yu CZ. Advances and perspectives in phototherapy-based combination therapy for cancer treatment. J Mater Chem B. 2024;12(26):6285.
214. Yakubova AA, Mitusova KA, Darwish A, Rogova A, Ageev EI, Brodskaia A, Muslimov AR, Zyuzin MV, Timin AS. Calcium carbonate nanoparticles tumor delivery for combined chemo-photodynamic therapy: comparison of local and systemic administration. J Control Release. 2023;359:400–14.
215. Liu YJ, Zheng XH, Zhou JL, Xie ZG. Merocyanine-paclitaxel conjugates for photothermal induced chemotherapy. J Mater Chem B. 2021;9(9):2334–40.
216. Sofla SFI, Abbasian M, Mirzaei M. A novel gold nanorods-based pH-sensitive thiol-ended triblock copolymer for chemo-photothermo therapy of cancer cells. J Biomater Sci Polym Ed. 2019;30(1):12–33.
217. Krasnovskaya OO, Akasov RA, Spector DV, Pavlov KG, Bubley AA, Kuzmin VA, Kostyukov AA, Khaydukov EV, Lopatukhina EV, Semkina AS, Vlasova KY, Sypalov SA, Erofeev AS, Gorelkin PV, Vaneev AN, Nikitina VN, Skvortsov DA, Ipatova DA, Mazur DM, Zyk NV, Sakharov DA, Majouga AG, Beloglazkina EK. Photoinduced reduction of novel dual-action Riboplatin Pt(IV) prodrug. ACS Appl Mater Interfaces. 2023;15(10) 12882–94.
218. Zeng JY, Zhang MK, Peng MY, Gong D, Zhang XZ. Porphyrinic metal-organic frameworks coated gold Nanorods as a versatile Nanoplatform for combined photodynamic/Photothermal/chemotherapy of tumor. Adv Funct Mater. 2018;28(8):1705451.
219. Khan AA, Allemailem KS, Almatroudi A, Almatroodi SA, Alsahli MA, Rahmani AH. Novel strategies of third level (Organelle-specific) drug targeting: An innovative approach of modern therapeutics. J Drug Deliv Sci Technol. 2021;61:102315.
220. DeVita VT, Chu E. A history of cancer chemotherapy. Cancer Res. 2008;68(21):8643–53.
221. Xie ZJ, Fan TJ, An J, Choi W, Duo YH, Ge YQ, Zhang B, Nie GH, Xie N, Zheng TT, Chen Y, Zhang H, Kim JS. Emerging combination strategies with phototherapy in cancer nanomedicine. Chem Soc Rev. 2020;49(22):8065–87.
222. Nunes T, Hamdan D, Leboeuf C, El Bouchtaoui M, Gapihan G, Nguyen TT, Meles S, Angeli E, Ratajczak P, Lu H, Di Benedetto M, Bousquet G, Janin A. Targeting cancer stem cells to overcome Chemoresistance. Int J Mol Sci. 2018;19(12):4036.
223. Huang HC, Rizvi I, Liu J, Anbil S, Kalra A, Lee H, Baglo Y, Paz N, Hayden D, Pereira S, Pogue BW, Fitzgerald J, Hasan T. Photodynamic priming mitigates chemotherapeutic selection pressures and improves drug delivery. Cancer Res. 2018;78(2):558–71.
224. Han LF, Wang YM, Huang XX, Liu FL, Ma CY, Feng F, Zhang JG, Liu WY, Qu W, Pang H, Xue JW. Specific-oxygen-supply functionalized core-shell nanoparticles for smart mutual-promotion between photodynamic therapy and gambogic acid-induced chemotherapy. Biomaterials. 2020;257:120228.
225. Ren SZ, Wang B, Zhu XH, Zhu D, Liu M, Li SK, Yang YS, Wang ZC, Zhu HL. Oxygen self-sufficient Core-Shell metal-organic framework-based smart Nanoplatform for enhanced synergistic chemotherapy and photodynamic therapy. ACS Appl Mater Interfaces. 2020;12(22):24662–74.
226. Maeda H, Nakamura H, Fang J. The EPR effect for macromolecular drug delivery to solid tumors: improvement of tumor uptake, lowering of systemic toxicity, and distinct tumor imaging in vivo. Adv Drug Deliv Rev. 2013;65(1):71–9.
227. Park J, Fong PM, Lu J, Russell KS, Booth CJ, Saltzman WM, Fahmy TM. PEGylated PLGA nanoparticles for the improved delivery of doxorubicin. Nanomed Nanotechnol Biol Med. 2009;5(4):410–8.
228. Yin Q, Shen JN, Zhang ZW, Yu HJ, Li YP. Reversal of multidrug resistance by stimuli-responsive drug delivery systems for therapy of tumor. Adv Drug Deliv Rev. 2013;65(13–14):1699–715.

229. Zhang JSW, Su H, Mu G, Sun JH, Tan CP, Liang XJ, Ji LN, Mao ZW. Co-delivery of cisplatin prodrug and Chlorin e6 by mesoporous silica nanoparticles for chemo-photodynamic combination therapy to combat drug resistance. ACS Appl Mater Interfaces. 2016;8(21):13332–40.
230. Sun TM, Wang Y, Wang YC, Xu JB, Zhao X, Vangveravong S, Mach RH, Xia YN. Using SV119-gold Nanocage conjugates to eradicate cancer stem cells through a combination of Photothermal and chemo therapies. Adv Healthc Mater. 2014;3(8):1283–91.
231. Castillo RR, Lozano D, González B, Manzano M, Izquierdo-Barba I, Vallet-Regí M. Advances in mesoporous silica nanoparticles for targeted stimuli-responsive drug delivery: an update. Expert Opin Drug Deliv. 2019;16(4):415–39.
232. Gao P, Pan W, Li N, Tang B. Boosting cancer therapy with organelle-targeted nanomaterials. ACS Appl Mater Interfaces. 2019;11(30):26529–58.
233. Ang MJY, Yoon J, Zhou MZ, Wei HL, Goh YY, Li ZL, Feng J, Wang HF, Su QQ, Ong DST, Liu XG. Deciphering nanoparticle trafficking into glioblastomas uncovers an augmented antitumor effect of metronomic chemotherapy. Adv Mater. 2022;34(3):2106194.
234. Chen Y, Lyu R, Wang J, Cheng Q, Yu Y, Yang S, Mao C, Yang M. Metal−organic frameworks nucleated by silk fibroin and modified with tumor-targeting peptides for targeted multimodal cancer therapy. Adv Sci. 2023;10(28):2302700.
235. Hao YW, Zheng CX, Wang L, Zhang JJ, Niu XX, Song QL, Feng QH, Zhao HJ, Li L, Zhang HL, Zhang ZZ, Zhang Y. Tumor acidity-activatable manganese phosphate nanoplatform for amplification of photodynamic cancer therapy and magnetic resonance imaging. Acta Biomater. 2017;62:293–305.
236. Bennewith KL, Dedhar S. Targeting hypoxic tumour cells to overcome metastasis. BMC Cancer. 2011;11:1–6.
237. Luo C, Sun J, Liu D, Sun BJ, Miao L, Musetti S, Li J, Han XP, Du YQ, Li L, Huang L, He ZG. Self-assembled redox dual-responsive prodrug-Nanosystem formed by single Thioether-bridged paclitaxel-fatty acid conjugate for cancer chemotherapy. Nano Lett. 2016;16(9):5401–8.
238. Cheng R, Meng F, Deng C, Zhong Z. Bioresponsive polymeric nanotherapeutics for targeted cancer chemotherapy. Nano Today. 2015;10(5):656–70.
239. Du JZ, Lane LA, Nie SM. Stimuli-responsive nanoparticles for targeting the tumor microenvironment. J Control Release. 2015;219:205–14.
240. Sun BJ, Chen Y, Yu H, Wang C, Zhang XB, Zhao HQ, Chen Q, He ZG, Luo C, Sun J. Photodynamic PEG-coated ROS-sensitive prodrug nanoassemblies for core-shell synergistic chemo-photodynamic therapy. Acta Biomater. 2019;92:219–28.
241. Su T, Cheng F, Pu Y, Cao J, Lin S, Zhu G, He B. Polymeric micelles amplify tumor oxidative stresses through combining PDT and glutathione depletion for synergistic cancer chemotherapy. Chem Eng J. 2021;411:128561.
242. Wang M, He M, Zhang M, Xue S, Xu T, Zhao Y, Li D, Zhi F, Ding D. Controllable hypoxia-activated chemotherapy as a dual enhancer for synergistic cancer photodynamic immunotherapy. Biomaterials. 2023;301:122257.
243. Karimi M, Zangabad PS, BaghaeeRavari S, Ghazadeh M, Mirshekari H, Hamblin MR. Smart nanostructures for cargo delivery: uncaging and activating by light. J Am Chem Soc. 2017;139(13):4584–610.
244. Zhu T, Lin Z, Tang X, Liu J, Zhang Y, Zhong X. Programmable activation of berbamine and photosensitizers for enhanced photodynamic therapy using emission-switchable upconversion nanoparticles. Int J Pharm. 2024;659:124202.
245. Chen WT, Liu JX, Zheng CY, Bai Q, Gao Q, Zhang YN, Dong K, Lu TL. Research Progress on improving the efficiency of CDT by exacerbating tumor acidification. Int J Nanomedicine. 2022;17:2611–28.
246. Jiang F, Ding BB, Liang S, Zhao YJ, Cheng ZY, Xing BG, Ma PA, Lin J. Intelligent MoS-CuO heterostructures with multiplexed imaging and remarkably enhanced antitumor efficacy via synergetic photothermal therapy/chemodynamic therapy/immunotherapy. Biomaterials. 2021;268:120545.

247. Deng LM, Liu MZ, Sheng DL, Luo YL, Wang D, Yu X, Wang ZG, Ran HT, Li P. Low-intensity focused ultrasound-augmented Cascade chemodynamic therapy via boosting ROS generation. Biomaterials. 2021;271:120710.
248. Tan JY, Duan XH, Zhang F, Ban XH, Mao JJ, Cao MH, Han SS, Shuai XT, Shen J. Theranostic Nanomedicine for synergistic Chemodynamic therapy and chemotherapy of Orthotopic glioma. Adv Sci. 2020;7(24):2003036.
249. Zhang BC, Huang Y, Huang Y. Advances in nanodynamic therapy for cancer treatment. Nanomaterials. 2024;14(7):648.
250. Liu G, Zhu J, Guo H, Sun A, Chen P, Xi L, Huang W, Song X, Dong X. Mo_2C-derived polyoxometalate for NIR-II photoacoustic imaging-guided chemodynamic/photothermal synergistic therapy. Angew Chem Int Ed. 2019;58(51):18641–6.
251. Ma Y, Wu M, Zhang X, Xia Q, Yang J, Xu S, Pan F. Efficacy and safety of tocilizumab with inhibition of interleukin-6 in adult-onset still's disease: a meta-analysis. Mod Rheumatol. 2018;28(5):849–57.
252. Gong F, Yang N, Wang X, Zhao Q, Chen Q, Liu Z, Cheng L. Tumor microenvironment-responsive intelligent nanoplatforms for cancer theranostics. Nano Today. 2020;32:100851.
253. Wang X, Zhong X, Liu Z, Cheng L. Recent progress of chemodynamic therapy-induced combination cancer therapy. Nano Today. 2020;35:100946.
254. Yin Q, Zhang J, Zhang H, Gao J, Weng L, Liu T, Sun S, Yao Y, Chen X. Cascade nanoreactor employs mitochondrial-directed chemodynamic and δ-ALA-mediated photodynamic synergy for deep-seated oral cancer therapy. Adv Healthc Mater. 2024;13(19):e2304639.
255. Liu C, Cao Y, Cheng Y, Wang D, Xu T, Su L, Zhang X, Dong H. An open source and reduce expenditure ROS generation strategy for chemodynamic/photodynamic synergistic therapy. Nat Commun. 2020;11(1):1735.
256. Xu J, Shi R, Chen G, Dong S, Yang P, Zhang Z, Niu N, Gai S, He F, Fu Y, Lin J. All-in-one theranostic nanomedicine with ultrabright second near-infrared emission for tumor-modulated bioimaging and Chemodynamic/photodynamic therapy. ACS Nano. 2020;14(8):9613–25.
257. Liu F, Lin L, Zhang Y, Wang Y, Sheng S, Xu C, Tian H, Chen X. A tumor-microenvironment-activated nanozyme-mediated theranostic nanoreactor for imaging-guided combined tumor therapy. Adv Mater. 2019;31(40):e1902885.
258. Chen Y, Liu Y, Guo C, Yin C, Xie C, Fan Q. Self-amplified competitive coordination of MnO_2-doped CeO_2 Nanozyme for synchronously activated combination therapy. Adv Funct Mater. 2023;33(2):2209927.
259. Zhang F, Cheng K, Huang Z, Hou X, Zhang X, Zhong Z, Hu Y, Lei X, Li Y, Zhang P, Zhao Y, Xu Q. Tumor microenvironment-responsive Nanocarrier based on VO_x Nanozyme amplify oxidative stress for tumor therapy. Adv Funct Mater. 2023;33(30):2212740.
260. Gao P, Chen Y, Pan W, Li N, Liu Z, Tang B. Antitumor agents based on metal–organic frameworks. Angew Chem Int Ed. 2021;60(31):16763–76.
261. Zhang J, Chang L, Hao R, Zhang G, Liu T, Li Z, Wang T, Zeng L. Copper/gold-modified porphyrinic metal–organic frameworks nanoprobes for enhanced photodynamic/chemodynamic therapy. Chem Eng J. 2023;474:145485.
262. Shen J, Yu H, Shu Y, Ma M, Chen H. A robust ROS generation strategy for enhanced Chemodynamic/photodynamic therapy via H_2O_2/O_2 self-supply and Ca^{2+} overloading. Adv Funct Mater. 2021;31(50):2106106.
263. Valastyan S, Weinberg RA. Tumor metastasis: molecular insights and evolving paradigms. Cell. 2011;147(2):275–92.
264. Qian CN, Mei Y, Zhang J. Cancer metastasis: issues and challenges. Chin J Cancer. 2017;36(1):38.
265. Alzeibak R, Mishchenko TA, Shilyagina NY. Targeting immunogenic cancer cell death by photodynamic therapy: past, present and future. J Immunother Cancer. 2021;9(10):e001926.
266. Xie DQ, Wang QF, Wu GZ. Research progress in inducing immunogenic cell death of tumor cells. Front Immunol. 2022;13:1017400.
267. Zhang L, Montesdeoca N, Karges J, Xiao H. Immunogenic cell death inducing metal complexes for cancer therapy. Angew Chem Int Ed Engl. 2023;62(21):e202300662.

268. Sen S, Won M, Levine MS, Noh Y, Sedgwick AC, Kim JS, Sessler JL, Arambula JF. Metal-based anticancer agents as immunogenic cell death inducers: the past, present, and future. Chem Soc Rev. 2022;51(4):1212–33.
269. Wei D, Huang Y, Wang B, Ma L, Karges J, Xiao H. Photo-reduction with NIR light of nucleus-targeting Pt^{IV} nanoparticles for combined tumor-targeted chemotherapy and photodynamic immunotherapy. Angew Chem Int Ed. 2022;61(20):e202201486.
270. Li J, Ou H, Ding D. Recent Progress in boosted PDT induced immunogenic cell death for tumor immunotherapy. Chem Res Chin Univ. 2021;37(1):83–9.
271. Ji B, Wei MJ, Yang B. Recent advances in nanomedicines for photodynamic therapy (PDT)-driven cancer immunotherapy. Theranostics. 2022;12(1):434–58.
272. Yang G, Ni JS, Li YX, Zha ML, Tu Y, Li K. Acceptor engineering for optimized ROS generation facilitates reprogramming macrophages to M1 phenotype in photodynamic immunotherapy. Angew Chem Int Ed. 2021;60(10):5386–93.
273. Miao YY, Wang S, Zhang BT, Liu L. Carbon dot-based nanomaterials: a promising future nano-platform for targeting tumor-associated macrophages. Front Immunol. 2023;14:1133238.
274. Qiao CQ, Yang Z, Liu XL, Zhang RL, Xia YQ, Wang LX, Chen Z, Jia Q, Wang RH, Yang Y, Wang ZL. Post-remedial oxygen supply: a New perspective on photodynamic to suppress tumor metastasis. Nano Lett. 2022;22(20):8250–7.
275. Zhou ZG, Liu Y, Song W, Jiang X, Deng ZA, Xiong W, Shen JL. Metabolic reprogramming mediated PD-L1 depression and hypoxia reversion to reactivate tumor therapy. J Control Release. 2022;352:793–812.
276. Donohoe C, Senge MO, Arnaut LG, Gomes-da-Silva LC. Cell death in photodynamic therapy: from oxidative stress to anti-tumor immunity. Biochim Biophys Acta Rev Cancer. 2019;1872(2):188308.
277. Liu X, Lu YC, Li X, Luo LH, You J. Nanoplatform-enhanced photodynamic therapy for the induction of immunogenic cell death. J Control Release. 2024;365:1058–73.
278. Wang WM, Gao Y, Xu JJ, Zou TH, Yang B, Hu SB, Cheng X, Xia Y, Zheng QC. A NRF_2 regulated and the immunosuppressive microenvironment reversed Nanoplatform for Cholangiocarcinoma photodynamic-gas therapy. Adv Sci. 2024;11(14):2307143.
279. Warszy M, Repetowski P, Dabrowski JM. Photodynamic therapy combined with immunotherapy: recent advances and future research directions. Coord Chem Rev. 2023;495:215350.
280. Wu CS, Cui MH, Cai L, Chen C, Zhu XH, Wu YH, Liu JL, Wang HJ, Zhang Y. NIR-responsive photodynamic nanosystem combined with antitumor immune Optogenetics bacteria for precise synergetic therapy. ACS Appl Mater Interfaces. 2022;14(11):13094–106.
281. Lee HT, Lee SH, Heo YS. Molecular interactions of antibody drugs targeting PD-1, PD-L1, and CTLA-4 in Immuno-oncology. Molecules. 2019;24(6):1190.
282. Lin G, Tillman L, Luo T, Jiang X, Fan Y, Liu G, Lin W. Nanoscale metal-organic layer reprograms cellular metabolism to enhance photodynamic therapy and antitumor immunity. Angew Chem Int Ed. 2024;136(37):e202410241.
283. Kalbasi A, Ribas A. Tumour-intrinsic resistance to immune checkpoint blockade. Nat Rev Immunol. 2020;20(1):25–39.
284. Xiong MM, Zhang Y, Zhang H, Shao QQ, Hu QF, Ma JJ, Wan YQ, Guo L, Wan X, Sun HT, Yuan ZY, Wan H. A tumor environment-activated photosensitized biomimetic nanoplatform for precise photodynamic immunotherapy of colon cancer. Adv Sci. 2024;11(28):2402465.
285. Zheng R, Zhao L, Yang N, Chen Z, Kong R, Huang C, Rao X, Chen X, Cheng H, Li S. Cascade immune activation of self-delivery biomedicine for photodynamic immunotherapy against metastatic tumor. Small. 2023;19(3):2205694.

Nanomaterial-Based Cancer Photothermal Therapy

2

Yu Zhang and Wenliang Song

Abstract

Cancer photothermal therapy (PTT) has attracted considerable interest from researchers in the field of oncology due to its potential for localized, minimally invasive, and trigger-activated therapeutic effects. In this modality, nanoparticles that absorb light and convert light energy into heat have demonstrated promising outcomes in various preclinical and clinical assays. These nanoparticles can specifically ablate tumors and inhibit tumor growth through localized hyperthermia. A range of nanoparticles composed of diverse materials and nanostructures has been developed and employed in PTT. In this chapter, we discuss the fabrication and application of the most commonly utilized nanoparticles available today for cancer PTT. We introduce the unique physicochemical properties of these nanoparticles and the primary classifications of photothermal conversion mechanisms. Additionally, we describe the general synthetic procedures and morphological regulation of nanoparticles while further illustrating their significant applications in PTT. Finally, we address the future challenges and biological safety concerns associated with the utilization of these intelligent nanoparticles.

Y. Zhang (✉)
Digital and Intelligent Empowerment Biomedical Innovation Center, School of Pharmacy, Shanghai University of Medicine and Health Sciences, Shanghai, P. R. China
e-mail: zhangy_21@sumhs.edu.cn

W. Song
School of Materials and Chemistry, University of Shanghai for Science and Technology, Shanghai, P. R. China

M. Chu (ed.), *Modern Nanomaterials and Nanotechnology for Cancer Therapy*,
https://doi.org/10.1007/978-981-95-6849-9_2

Keywords

Photothermal Therapy (PTT) · Nanomaterials · Cancer Treatment · Near-Infrared (NIR) Irradiation · Photothermal Agents · Targeted Drug Delivery

Cancer photothermal therapy (PTT) has attracted considerable interest from researchers in the field of oncology due to its potential for localized, minimally invasive, and trigger-activated therapeutic effects. In this modality, nanoparticles that absorb light and convert light energy into heat have demonstrated promising outcomes in various preclinical and clinical assays. These nanoparticles can specifically ablate tumors and inhibit tumor growth through localized hyperthermia. A range of nanoparticles composed of diverse materials and nanostructures has been developed and employed in PTT. In this chapter, we discuss the fabrication and application of the most commonly utilized nanoparticles available today for cancer PTT. We introduce the unique physicochemical properties of these nanoparticles and the primary classifications of photothermal conversion mechanisms. Additionally, we describe the general synthetic procedures and morphological regulation of nanoparticles while further illustrating their significant applications in PTT. Finally, we address the future challenges and biological safety concerns associated with the utilization of these intelligent nanoparticles.

1 Introduction

Photothermal therapy (PTT), which employs a light-to-heat conversion mechanism, has been recognized as a promising cancer treatment modality due to its high selectivity, minimal invasiveness, and notable therapeutic effects [1]. Thermal therapy exposes tumors to elevated temperatures that can induce distinct tumor elimination effects. Based on the magnitude of temperature increase, thermal therapy can be categorized into irreversible injury therapy (above 48 °C), hyperthermia therapy (41–48 °C), and diathermia therapy (37–41 °C) [2–4]. PTT relies on localized hyperthermia generated from the optical energy absorbed by photothermal agents (PTAs) [5, 6]. Hyperthermia can induce various biological processes, including protein denaturation, aggregation, and cell inactivation, leading to thermal erosion of tissue. Furthermore, hyperthermia has been shown to suppress cancer metastasis by inhibiting the expression of metastasis-related growth factors [7]. The deprivation of growth factors, in conjunction with intra-/extracellular conditions such as DNA damage and signals generated from cytotoxic T cells and caspase proteases, can simultaneously activate apoptotic pathways in cells, resulting in localized cancer cell death and inhibition of early-stage metastasis [8, 9].

Compared to traditional techniques, PTT is significantly less invasive and does not cause evident destruction of healthy cells. To enhance the efficacy of PTT, nanoparticle-based PTAs with specific optical properties have been extensively employed [10–12]. Due to their biocompatibility and dispersibility, nanoparticles can be readily introduced into organisms by being loaded into functional materials.

Furthermore, the small size of nanoparticles confers intrinsic properties, including prolonged circulation times in the bloodstream and high accumulation in tumor sites. In the absence of external stimuli, nanoparticles exhibit minimal side effects on the normal biological processes of the host organism. Conversely, during PTT, nanoparticles can mediate regional thermal ablation of tumors when exposed to specific light irradiation [13]. Near-infrared (NIR) radiation, typically ranging from 700 to 1400 nm, allows for maximum penetration depth in tissue with minimal invasiveness to healthy tissues, further enhancing therapeutic effects [14, 15]. Various nanoparticles, including metals, carbons, organic molecules, and conjugated polymers, have been employed as PTAs to facilitate the PTT process [16–21]. The physicochemical properties of nanoparticles, such as morphology, size, surface charge, spectral range, conjugated molecules, and heating mechanisms, are crucial parameters that can be optimized to improve PTT performance.

In this chapter, we will focus on the construction and application of the most commonly utilized nanoparticles for cancer PTT. Firstly, we will outline the unique properties and general light-to-heat conversion mechanisms of various nanoparticles used in PTT. Following this, we will illustrate the synthetic procedures and morphological regulation of various types of nanoparticles, along with their outstanding in vivo and in vitro applications. Finally, we will discuss potential future challenges and perspectives regarding the utilization of these intelligent nanoparticles for cancer PTT. This chapter aims to provide a systematic overview of nanoparticles, from their state-of-the-art structural design to practical applications in cancer PTT.

2 Classification of Nanoparticles and Their Photothermal Conversion Mechanisms

Nanoparticles used as PTAs in PTT involve the irradiation of specific NIR light to promote the accumulation of nanoparticles within targeted tumors. The heat generated from the conversion of NIR light energy induces regional tumor ablation [22, 23]. Nanoparticles with high absorption efficacy and low luminescence quantum yields are optimal for efficient PTT performance. Metals, carbons, organic molecules, and conjugated polymers are the most commonly used nanoparticles for cancer PTT. The small sizes (typically below 200 nm) and specific morphologies (such as spheres, vesicles, tubes, and rods) of these nanoparticles facilitate their high accumulation concentrations in tumor tissues through either passive or active targeting pathways [24, 25]. Passive targeting occurs due to the enhanced permeability of blood vessels and the high proliferative rate of tumors [26]. In the tumor microenvironment (TME), the endothelial cells of blood vessels become more permeable compared to their healthy state. Rapidly proliferating cancer cells compromise vascular capillaries and lymphatic vessels, resulting in the formation of leaky vasculature that promotes the extravasation and retention of nanoparticles within the tumor stroma. In contrast to passive targeting, active targeting relies on receptor-ligand interactions [27, 28]. The surface of nanoparticles can be functionalized with specific ligands that bind to receptors overexpressed at tumor sites, enhancing the affinity of nanoparticles for cancer cells and subsequently increasing their accumulation in tumors.

The physicochemical properties of nanoparticles must be tailored to prolong their blood circulation during practical cancer PTT. It is a well-established fact that nanoparticles sized below 5 nm are rapidly cleared by the kidneys [29]. Nanoparticles that accumulate in the liver typically range below 50 nm, while those that accumulate in the spleen are generally over 200 nm. Thus, nanoparticles sized between 100 and 200 nm are optimal for intravenous administration in organisms [30]. However, it has been reported that PEGylated gold nanoparticles measuring 20–30 nm achieved higher tumor uptake compared to their larger counterparts exceeding 60 nm [31, 32]. Nanoparticles with a neutral surface charge (±10 mV) also play a crucial role in improving circulation time [33, 34]. Additionally, the dose, nanostructures, and functionalities of nanoparticles significantly impact retention effects during cancer PTT, which will be discussed in detail in Sect. 3. Furthermore, the unique properties of nanoparticles and laser parameters are related to photothermal conversion efficacy (PCE) [35, 36]. The appropriate size and morphology of nanoparticles favor prolonged circulation time in the bloodstream, and high PCE can be achieved by adjusting the time and intensity of the NIR laser, resulting in a localized hyperthermia effect.

Nanoparticle-based PTAs are among the most significant factors in light-to-heat conversion, and PCE directly determines the intensity of irradiation during the cancer PTT process. Nanoparticles induce hyperthermia by converting absorbed energy into heat and diffusing it to localized surroundings. Understanding the photothermal conversion mechanisms is crucial for the practical application of these nanoparticles in cancer PTT [37, 38]. Clarifying different mechanisms provides researchers with fundamental rules and insights for designing and adjusting nanoparticles. Due to their inherent nanostructures and properties, the photothermal conversion mechanisms of nanoparticles vary among different material types [39, 40]. Generally, the widely studied photothermal conversion mechanisms can be classified into the following categories: (1) localized surface plasmon resonance (LSPR) effect, (2) electron-hole generation and relaxation, and (3) conjugation and hyperconjugation effects.

2.1 LSPR Effect

Metallic nanoparticles absorb light energy and convert it into heat via the LSPR effect [41]. The LSPR effect is defined as the phenomenon wherein resonant photons induce a collectively oscillating behavior of conduction-band electrons on the surface of metallic nanoparticles when the photon frequency coincides with the inherent frequency of the electrons [42, 43]. In the LSPR effect, two decay pathways occur: the radiative and nonradiative decay processes, which often compete [44]. The radiative decay process enhances the electric field through re-emitted photons, while the nonradiative decay process converts hot electrons generated by intraband or interband transitions into thermal energy. The energy of hot electrons is first redistributed through electron-electron scattering and then gradually transferred from electrons to the crystal lattice and to the external surroundings via electron-phonon scattering and phonon-phonon interaction processes, respectively, resulting in thermalization of the surroundings [45]. Presently, noble metals and transition metals are frequently used as plasmonic materials, and numerous examples have been employed as PTAs in cancer PTT [46–48]. The compositions, sizes, and morphologies of these nanoparticles are closely related to the LSPR effect and the resulting PTT performance [49].

2.2 Electron-Hole Generation and Relaxation

The photothermal effect of semiconductor nanoparticles, based on transition metal oxides and transition metal chalcogenides, is primarily induced by the generation and relaxation of electron-hole pairs [50, 51]. Semiconductor nanoparticles absorb photon energy when irradiated by incident light with energy exceeding that of the bandgap. The electrons formed through photoexcitation in the valence band transfer to the conduction band, generating electron-hole pairs. These pairs subsequently relax to the band edges through nonradiative pathways in the form of phonons, releasing extra energy to the crystal lattice and thereby inducing thermal effects [52–54]. Narrow-bandgap semiconductor nanoparticles, such as CuS, Ti_2O_3, and Fe_3O_4, can achieve high PCE [55–57]. Several oxygen-deficient metal oxides, including tungsten- and molybdenum-derived oxides, also exhibit strong photothermal conversion abilities across broad bandgaps [58, 59].

2.3 Conjugation and Hyperconjugation Effects

Conjugation and hyperconjugation effects predominantly occur in conjugated polymers and carbon-based nanoparticles [60–62]. The conjugation effect refers to the delocalization of electrons induced by the overlapping of adjacent π-electrons or the interaction between π-bonds and p-orbital electrons, while hyperconjugation involves the interaction of electrons in σ-orbitals with adjacent unoccupied or partially occupied p-orbitals [63]. Both conjugation and hyperconjugation effects promote the excitation of electrons under light irradiation and accelerate their motion. The less tightly held π-electrons are readily excited from the π-orbital to the π^*-orbital and release absorbed energy as heat upon returning to the ground state [64–66]. Consequently, relaxation of excited electrons from the lowest unoccupied molecular orbital (LUMO) to the highest occupied molecular orbital (HOMO) occurs through electron-phonon coupling, resulting in a distinct photothermal conversion effect [67–69].

3 Application of Nanoparticles in Cancer Photothermal Therapy

Presently, numerous nanoparticles capable of efficient light-to-heat conversion have been employed for both in vitro and in vivo cancer PTT. During treatment, each family of nanoparticles exerts its light-to-heat conversion abilities via respective mechanisms. In subsequent sections, we will discuss the most frequently studied inorganic and organic nanoparticles for mediating tumor-centered PTT, as well as their combinations with other therapeutic techniques, highlighting essential factors that affect PTT efficiency.

3.1 Inorganic Nanoparticles

3.1.1 Metal-Based Nanoparticles

Metal-based nanoparticles, such as those composed of gold, silver, copper, tungsten, molybdenum, and palladium, are among the most extensively explored PTAs due to their tunable optical properties and high PCE. To satisfy diverse biomedical PTT requirements, the optical and physicochemical properties of these nanoparticles can be precisely controlled through variations in size, morphology, and surface modifications [70, 71]. For instance, gold nanoparticles (Au NPs) can be tailored into various shapes, including nanospheres, nanoshells, nanorods, nanostars, nanocages, nanoclusters, nanoflowers, nanobipyramids, nanocorals, and nanodumbbells (Fig. 2.1) [72–81]. Different types of Au NPs possess specific optical and dielectric properties, with some having entered preclinical or clinical trial stages [82]. The excellent biocompatibility and small size of Au NPs facilitate their penetration into tumors and safe excretion via the urinary system [83]. Additionally, the strong light absorption and high photothermal conversion ability of Au NPs enable effective heat generation and selective ablation of localized tumors, establishing Au NPs as ideal PTAs for tumor PTT [84, 85].

Au nanospheres represent the simplest nanostructure and can be synthesized using the citrate reduction method in a matter of minutes [86]. They possess unique characteristics, such as tunable sizes ranging from 10 to 150 nm and facile surface bioconjugation capabilities. The absorption peak of Au nanospheres is approximately 500–600 nm. Compared to other Au NP morphologies, nanospheres exhibit superior cellular uptake and lower cytotoxicity [87]. Furthermore, Au nanospheres exhibit inherent nonlinear optical properties, although they lack a broad tunable LSPR peak in the NIR region. Nonetheless, they can nonradiatively absorb NIR light energy via second harmonic generation and convert it to heat for PTT. Huang et al. reported that Au nanospheres conjugated with anti-EGFR antibodies can irreversibly destroy tumors under 800 nm NIR laser irradiation, requiring only 20 times lower energy than that used in PTT without Au nanospheres, illustrating their high PTT efficacy [88]. Gao et al. developed an intelligent platform for cancer PTT based on the thermally triggered assembly of Au NPs [89]. Thermally responsive elastin-like polypeptides (ELPs) were conjugated to Au NPs, yielding ELP-Au NPs.

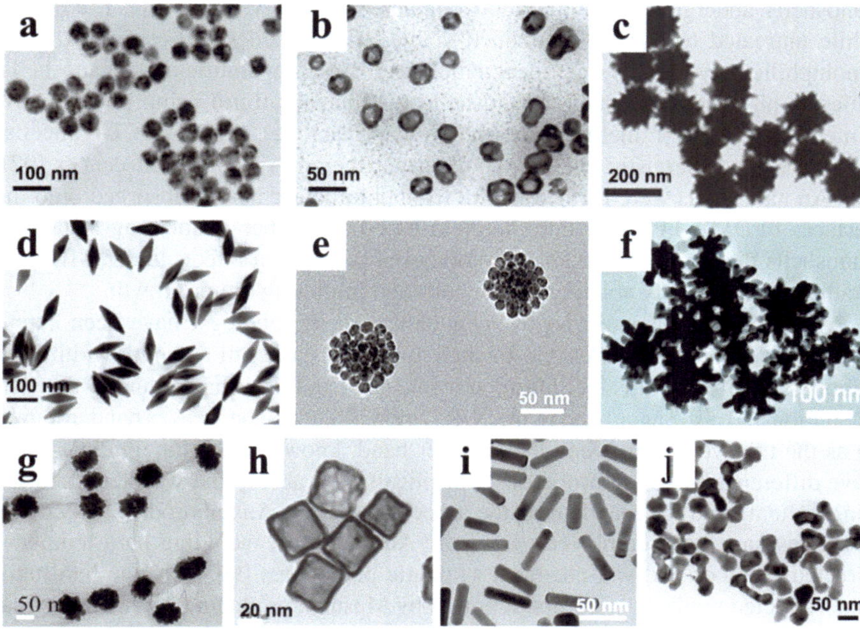

Fig. 2.1 Au NPs with different morphologies. (**a**) Nanospheres, (**b**) nanoshells, (**c**) nanostars, (**d**) nanobipyramids, (**e**) nanoclusters, (**f**) nanocorals, (**g**) nanoflowers, (**h**) nanocages, (**i**) nanorods, and (**j**) nanodumbbells. (Copyright © 2017 Springer-Verlag, 2017 and 2019 American Chemical Society, 2016 and 2019 Wiley-VCH GmbH, 2017, 2019, and 2020 Elsevier, and 2019 Royal Society of Chemistry, respectively. Reprinted with permission from Refs. [72–81])

Notably, ELP-Au NPs formed short necklace-like shapes composed of several nanospheres at elevated temperatures, facilitating strong NIR light absorption, efficient heat conversion, and high photothermal cytotoxicity.

Au nanoshells, composed of a spherical core-shell structure with a silica core and a thin Au shell, are another innovative design. Halas et al. prepared a series of Au nanoshells with tunable LSPR peaks in the NIR region. The thickness of the Au shell determines the LSPR absorption wavelength, which can be adjusted within the range of 800–1200 nm. This unique core-shell architecture and tunable NIR light absorption capability enable Au nanoshells to encapsulate and deliver therapeutic drugs to localized tumors while generating heat, making them ideal PTAs for PTT. West et al. engineered a composite composed of a thermally responsive *N*-isopropylacrylamide-*co*-acrylamide (NIPAAm-*co*-AAm) hydrogel and Au nanoshells for loading PTAs [90]. The incorporated Au nanoshells mediated temperature variations through light-to-heat conversion, causing the hydrogel to collapse once the surrounding temperature surpassed the lower critical solution temperature of NIPAAm-*co*-AAm, resulting in the burst release of guest substances within the hydrogel matrix. Notably, this research group achieved in vitro and in vivo PTT of breast cancer using Au nanoshells, with tumor cells exposed to Au nanoshells undergoing photothermally induced morbidity under NIR irradiation, while untreated tumor cells remained viable [91]. Micelles self-assembled from amphiphilic polymers serve as ideal templates for the preparation of Au nanoshells. Shieh et al. constructed an Au nanoshell-based nanoplatform using poly[2-(*N*,*N*-dimethylamino)ethyl methacrylate]-poly(ε-caprolactone) (PDMA-PCL) micellar templates for synergistic chemo-photothermal therapy of colorectal cancer [92]. The Au nanoshells were formed in situ from chloroauric anions absorbed onto the surface of PDMA-PCL micelles. Platinum-based chemotherapeutic drug-loaded Au nanoshells displayed remarkable photothermal conversion effects under NIR light irradiation, effectively ablating cancer cells and inhibiting tumor growth.

Au nanorods, characterized by their elongated morphology, have been extensively investigated as PTAs due to their tunable aspect ratio (length/width). Au nanorods exhibit two distinct plasmon resonance bands: a weaker band in the visible region and a stronger band in the NIR region [93, 94]. The weaker band, referred to as the transverse band, and the stronger band, known as the longitudinal band, have different absorption properties. Specifically, the absorption wavelength of the longitudinal band is dependent on the aspect ratio of the Au nanorods [95]. Various strategies have been employed to prepare Au nanorods, including hard template-directed methods and wet-chemical synthetic procedures [96, 97]. The hard template-directed methods were first reported by Masuda, producing Au nanorods that are relatively large (approximately 100 nm) and low in yield, which limits their potential biomedical applications. In contrast, Au nanorods synthesized via wet-chemical methods tend to be smaller, particularly when utilizing the seed growth method. Correia et al. developed multifunctional gold-core silica shell (AuMSS) nanorods for tumor-targeted PTT [98]. The surface of these Au cores was functionalized with hyaluronic acid (HA), a ligand for the CD44 receptor, and D-α-tocopherol polyethylene glycol 1000 succinate (TPGS), an inhibitor of the P-glycoprotein (P-gp). Notably, this surface functionalization did not affect the light-to-heat conversion capacity of the AuMSS while also promoting charge neutralization on the surface. Furthermore, the AuMSS demonstrated selective targeting of cancer cells and facilitated an on-demand photothermal effect, resulting in a significant increase in cancer cell mortality. Chen et al. successfully modified the conventional wetting growth mode of Au nanorod seeds to an island-growth mode by increasing the interfacial energy between the metallic growth material and the underlying seeds [99]. They developed two distinct growth modes to construct a series of Au-on-Au nanorod hybrids (Fig. 2.2) [100]. The newly formed Au domains could be precisely regulated into nanospheres and branched nanowires by controlling the interfacial energy and growth kinetics. The specific nanostructures of these hybrids enabled easy modulation of their LSPR across the entire visible-NIR spectral region. Both in vitro and in vivo experiments demonstrated that the resultant nanohybrids were uniformly distributed throughout the tumor area and exhibited high photothermal conversion efficiency, thereby prolonging the lifespan of mice without causing systemic damage to their organs.

Au nanostars possess numerous sharp protruding tips, each exhibiting a single-crystalline structure. Their light absorption properties can be precisely tuned by

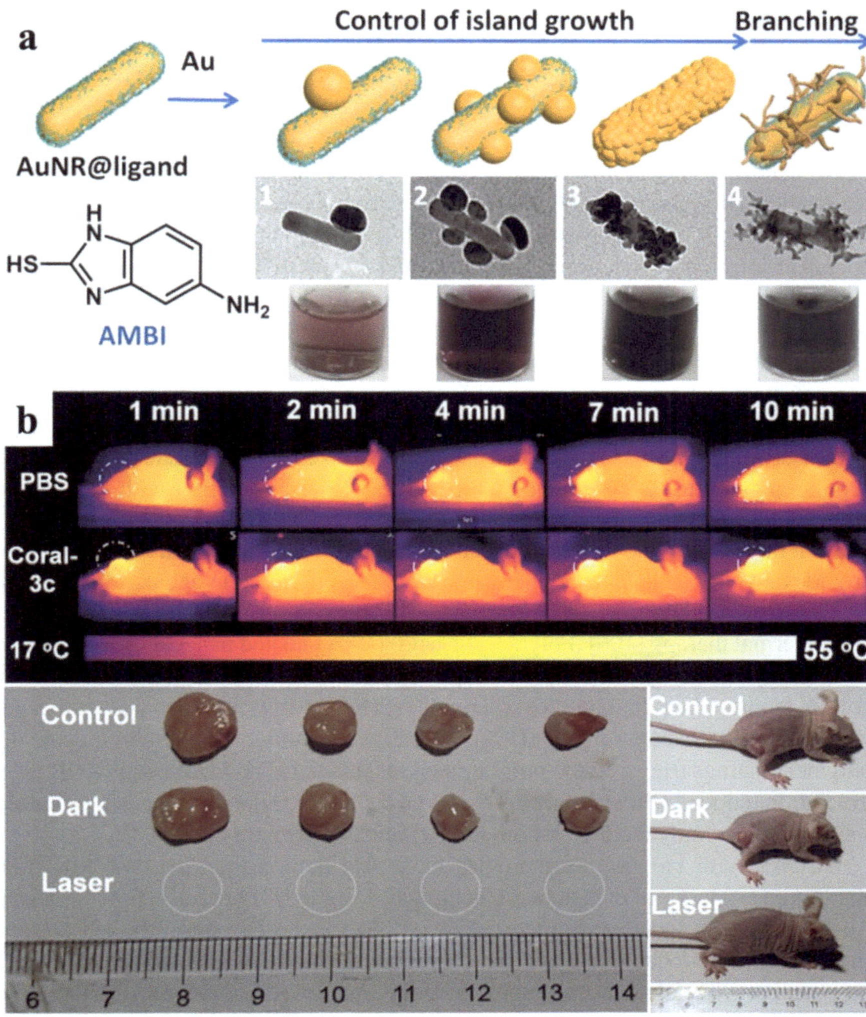

Fig. 2.2 (**a**) Representative illustrations and TEM images of a series of Au-on-Au nanorod hybrids. (**b**) In vivo PTT treatment of mice under laser irradiation and photographs of the dissected tumors after a full course of treatment. (Copyright © 2020 Wiley-VCH GmbH. Reprinted with permission from Ref. [100])

controlling parameters such as overall size, as well as the length, width, and number of tips. Khlebtsov et al. reported a surfactant-free synthetic method for the fabrication of Au nanostars with outstanding surface-enhanced Raman scattering (SERS) imaging capabilities [101]. By varying the Au seed size from 3 to 35 nm, nanostars with diameters ranging from 45 to 150 nm were obtained, while their plasmon resonance could be tuned from 630 to 900 nm. Wilhelm et al. investigated the PCE of a wide range of Au nanostars with average diameters between 25 and 150 nm and

LSPR peaks spanning from 500 to 1000 nm in different dispersed systems [74]. Their study revealed that both particle size and excitation laser wavelength exerted a significant influence on PTT efficiency in aqueous dispersions. However, this effect was markedly attenuated or even completely suppressed, once the Au nanostars were internalized by tumor cells and aggregated within endosomes.

Au nanocages are hollow nanostructures with an Au shell and an empty interior. A widely used synthetic strategy involves a galvanic replacement reaction between silver nanocubes and chloroauric acid [102]. By optimizing the wall thickness, the optical absorption of Au nanocages can be tuned from the visible to the NIR region [103]. Chen et al. demonstrated that PEGylated Au nanocages could be passively delivered to solid tumors, exhibiting prolonged circulation in the bloodstream and preferential accumulation in tumor tissue via the enhanced permeability and retention effect, thereby enhancing the PTT outcome [104]. Improving tumor-targeting capability thus represents an effective approach to increase PCE. For example, Griffin et al. engineered Au nanocages functionalized with the galectin-1-binding peptide anginex to achieve targeted PTT [105]. Upon irradiation with an 808 nm laser for 10 min, these nanocages generated sufficient heat to induce over 50% mortality in both tumor and endothelial cells. The hollow interior of Au nanocages also provides a versatile cavity for encapsulating chemotherapeutic agents. Zhang et al. developed two types of polymer-coated Au nanocages for synergistic chemo-photothermal therapy [106]. Poly(acrylic acid) (PAA)-coated nanocages were loaded with the epidermal growth factor receptor (EGFR) inhibitor erlotinib (Erl), while poly(*N*-isopropylacrylamide-*co*-acrylamide) (P(NIPAM-*co*-AM))-coated nanocages were loaded with doxorubicin (Dox). In the acidic tumor microenvironment, the polymer coatings triggered the programmed release of Erl and Dox, while NIR laser irradiation induced efficient light-to-heat conversion. This dual mechanism produced a synergistic cytotoxic effect in both MCF-7 and A431 tumor cells.

Au NPs with unique morphologies have also demonstrated strong NIR light absorption properties. For instance, Cheng et al. designed U11 peptide-functionalized Au nanoclusters, which exhibited enhanced NIR absorption and efficient conversion of light into localized heat [76]. This innovative design enabled selective PTT for pancreatic ductal adenocarcinoma, thereby improving both therapeutic specificity and efficacy. Similarly, Zhang et al. reported that Au nanoflower-decorated graphene oxide (GO) composites exhibited an outstanding PTT effect, achieving remarkable temperature elevation within a short irradiation period [78]. Au nanobipyramids, characterized by two elongated sharp tips, display superior chemical stability and tunable absorption wavelengths. Zhao et al. developed Raman reporter-bioconjugated Au nanobipyramid probes, which enabled simultaneous SERS detection and PTT for breast cancer [75]. These probes demonstrated quantitative cancer cell detection across concentrations ranging from 5 to 500 cells/mL while reducing cancer cell viability to as low as 6.44%. In another approach, Wu et al. engineered Au nanocorals composed of multibranched Au NPs to achieve selective apoptosis induction during breast cancer PTT [77]. Owing to their high surface-to-volume ratio, Au nanocorals exhibited enhanced PTT performance, enabling efficient apoptosis induction under low-power NIR irradiation for just 15 min.

Silver nanoparticles (Ag NPs) are regarded as promising candidates in the field of PTT for cancer treatment. They exhibit moderate biocompatibility, unique antiproliferative properties, and favorable physicochemical characteristics, all of which contribute to their potential for diverse biomedical applications, including cancer PTT [107]. Notably, the high optical cross section of Ag NPs underpins their strong plasmonic absorption effect [108]. Chen et al. investigated the photophysical properties of core-shell Ag@SiO_2 nanohybrids synthesized via the Stober method [109]. Upon exposure to 400 nm visible light, Ag nanoshells exhibited a remarkable photothermal effect against cancer cells. However, the plasmon resonance of conventional Ag NPs typically occurs around 410 nm, which limits their effectiveness for deep-tissue PTT applications. In contrast, anisotropic Ag NPs with tunable plasmon resonances in the NIR region offer significant advantages for cancer therapy, as NIR light penetrates more deeply into biological tissues compared to visible light. To address this, researchers have developed a variety of Ag-based PTT nanoconverters with distinctive morphologies, such as nanopentagons, nanotriangles, and nanocages. For example, Bose et al. reported a one-pot chemical strategy to synthesize folate receptor-targeted Ag nanopentagons for combined chemo-photothermal therapy of breast cancer [110]. These pentagonal Ag NPs displayed strong absorption at approximately 860 nm. The localized hyperthermia generated by the nanoparticles enhanced the efficacy of chemotherapeutic agents, resulting in selective ablation and apoptosis of cancer cells.

Ag nanotriangles have been shown to exhibit strong absorbance in the NIR region. Singh et al. evaluated their potential for the selective treatment of triple-negative breast cancer (TNBC) cells [111]. The triangular Ag nanoparticles displayed a plasmon resonance peak at approximately 933 nm, and their heat-generating capacity under NIR irradiation was systematically assessed. Furthermore, the authors optimized the dosage of Ag nanotriangles as well as the intensity of NIR and ionizing radiation (IR), aiming to maximize cytotoxicity against TNBC cells while minimizing adverse effects on surrounding healthy mammary epithelial cells. Their findings demonstrated that the combination of Ag nanotriangle-based PTT and IR enables selective ablation of TNBC cells without harming nonmalignant cells in the tumor microenvironment. To develop plasmonic Ag nanoparticles with both high photothermal conversion efficiency and robust stability under biologically harsh conditions, Wilhelm et al. fabricated triangular nanohybrids (Ag@Au NPs) consisting of Ag nanoplate cores coated with protective Au shells (Fig. 2.3) [112]. A comparative study was conducted to evaluate their heating efficiency against that of monometallic Ag nanoplates and Au nanostars across environments of increasing biological complexity. In aqueous medium, the heating efficiency followed the order: Ag nanoplates > Ag@Au NPs > Au nanostars. However, this trend was reversed in intracellular conditions. Due to intracellular biodegradation, Ag nanoplates rapidly lost their photothermal performance, whereas the Au component in Ag@Au NPs and Au nanostars conferred high structural stability and sustained heating efficiency. In vivo experiments further demonstrated that Ag@Au NPs

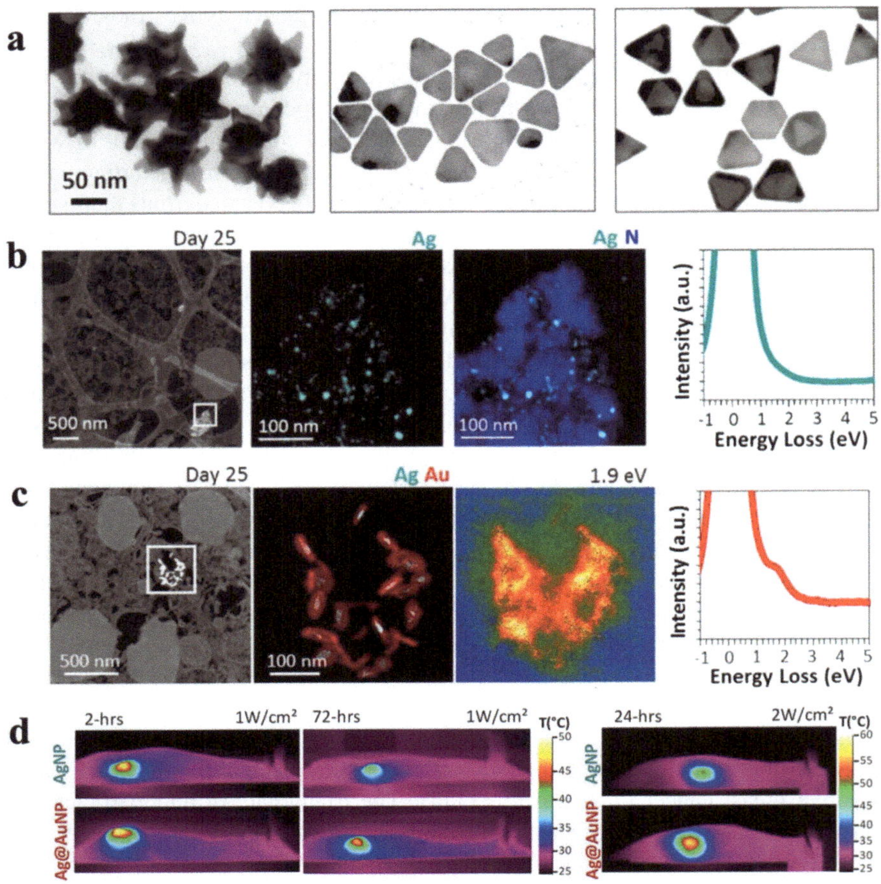

Fig. 2.3 (**a**) TEM images of Au nanostars, Ag nanoplates, and Ag@AuNPs. (**b**) Annular dark field (ADF) image, electron energy-loss spectroscopy (EELS) elemental maps, and representative low-energy-loss spectrum of Ag nanoplates in the cellular environment at day 25. (**c**) ADF image, EELS elemental maps, and plasmon resonance maps at 1.9 eV of Ag@AuNPs in the cellular environment at day 25. (**d**) Infrared thermal images of mice treated with Ag nanoplates and Ag@AuNPs, respectively. (Copyright © 2018 American Chemical Society. Reprinted with permission from Ref. [112])

induced near-complete tumor regression, while Ag nanoplates merely slowed tumor progression. Thus, the integration of an Au shell with an Ag core not only protects the structural integrity of Ag in complex biological environments but also preserves its ultrahigh photothermal efficiency.

Biomineralization is regarded as a green, versatile, and precise strategy for constructing multifunctional nanoplatforms for cancer therapy [113]. Gao et al. fabricated hierarchical mineralized Ag nanocages through octreotide-directed mineralization for PTT [114]. The physicochemical properties of these nanocages could be modulated by adjusting the composition of the ionic precursors, thereby enabling fine-tuning of their plasmonic resonance peak from the visible to the NIR region. Functionalization with octreotide conferred improved biocompatibility and photostability under 808 nm NIR laser irradiation. The resulting Ag nanocages exhibited an impressive PCE of 46.1% and achieved a tumor-killing efficacy of up to 82.7%, highlighting their potential as highly effective PTAs for cancer treatment.

The small size, low cost, and high stability of copper nanoparticles (Cu NPs), including copper sulfide (CuS), copper selenide (CuSe), and copper telluride, have attracted considerable research interest [115]. Their sophisticated crystal structures and tunable compositions endow them with superior optical and electrical properties, making Cu NPs highly promising candidates for PTAs. In particular, copper sulfide exhibits strong NIR absorption due to the d-d energy band transition of Cu^{2+} ions. Hu et al. synthesized hydrophilic cysteine-coated CuS nanoparticles (Cys-CuS NPs) for efficient cancer PTT [116]. These nanoparticles displayed strong NIR absorption and achieved a high PCE of 38.0%. Furthermore, hyperthermia generated by a low concentration (50 ppm) of Cys-CuS NPs effectively suppressed tumor growth under 980 nm laser irradiation at a safe power density of 0.72 $W{\cdot}cm^{-2}$.

To confer tumor-targeting capability to CuS, Zhu et al. developed multifunctional nanocomposites (HANPCs) by encapsulating CuS within Cy5.5-conjugated HA nanoparticles for optical and photoacoustic (PA) imaging-guided PTT [117]. The overexpression of hyaluronidase in the tumor microenvironment facilitated the time-dependent accumulation of HANPCs. Initially, the fluorescent signal of Cy5.5 was quenched by CuS; however, this quenching gradually diminished as HA was enzymatically degraded by hyaluronidase, resulting in strong fluorescence signals that delineated the tumor. The PA imaging capability of CuS in HANPCs enabled visualization of tumor blood vessel distribution and dynamic nanoparticle accumulation over time, providing valuable complementary information for therapeutic planning. In vivo experiments demonstrated strong fluorescence and PA signals at tumor sites 6 h postinjection, and the tumor inhibition rate reached 89.74% by day 5. Tang et al. engineered CuS nanoparticles functionalized with cyclic Arg-Gly-Asp (RGD) and TAT peptides for nuclear-targeted PTT and to prevent local cancer recurrence [118]. Surface decoration with RGD and TAT peptides enabled the nanocomplexes to selectively target tumor cells and penetrate the nucleus. Upon irradiation with a 980 nm laser for 5 min, the treated cancer cells exhibited 84% mortality. In vivo studies further confirmed effective solid tumor ablation after 14 days of treatment with these targeted nanocomplexes, with no tumor recurrence observed.

Selenium is essential for maintaining thyroid function and supporting the immune system. It has been employed as a therapeutic adjunct for conditions such as Hashimoto's thyroiditis and hypercholesterolemia [119]. However, selenium-based copper nanoparticles have been scarcely reported, primarily due to challenges in synthesizing high-quality, water-soluble CuSe nanoparticles (CuSe NPs) [120]. Li et al. developed a facile aqueous approach to synthesize water-soluble PEGylated $Cu_{2-x}Se$ nanoparticles for multimodal imaging-guided PTT of cancer [121]. Dimercapto PEG was conjugated to the surface of $Cu_{2-x}Se$ nanoparticles, yielding PEGylated nanoparticles ($Cu_{2-x}Se$-PEGSH) with spherical morphology and an ultrasmall size of approximately 3.6 ± 0.3 nm. The hydrophilic PEG coating conferred excellent biocompatibility, high aqueous solubility, and prolonged blood circulation, enabling efficient accumulation in targeted tumor tissues. The $Cu_{2-x}Se$-PEGSH nanoparticles exhibited high photothermal conversion efficiency, making them effective for both PA imaging and PTT. Additionally, they demonstrated superior X-ray computed tomography (CT) imaging performance due to the high X-ray attenuation coefficients of copper and selenium. The peripheral sulfhydryl groups on $Cu_{2-x}Se$-PEGSH also provide functional sites for radiolabeling with technetium-99m (^{99m}Tc), facilitating single-photon emission computed tomography imaging.

Tungsten nanoparticles, particularly tungsten oxide nanoparticles, have emerged as promising candidates for PTAs in cancer PTT. These nanoparticles exhibit notable localized LSPR effects due to their distinct outer-shell d valence electrons [122, 123]. The intensity and wavelength of NIR absorption in tungsten nanoparticles can be precisely tuned through doping, stoichiometric control, and optimization of morphology and size [124]. Inspired by the strong adhesive properties of mussels, Lee et al. developed WO_3 nanoparticles functionalized with dopamine-conjugated HA for targeted cancer PTT [125]. Under irradiation with an 808 nm laser at a safe power density of 2 $W{\cdot}cm^{-2}$, a 1 mg/mL solution of these nanoparticles reached 44 °C within 5 min, achieving a PCE of 41.96%. To further enhance aqueous biocompatibility, Yi et al. reported lentinan in situ-decorated tungsten oxide nanorods ($W_{18}O_{49}$@LTN NRs) synthesized via a one-pot solvothermal method [126]. Lentinan, with its abundant surface hydroxyl groups, significantly improved the dispersibility and biocompatibility of $W_{18}O_{49}$@LTN NRs. Under irradiation with a 980 nm laser at a low power density of 0.4 $W{\cdot}cm^{-2}$, a 125 μg/mL solution of these nanorods reached 40 °C, with a PCE of 33.86%. Doping tungsten nanoparticles with heteroatoms offers further potential to enhance PCE in the second NIR window for cancer PTT. Dong et al. developed two-dimensional ultrathin tellurium oxide/ammonium tungsten bronze nanoribbons ($TeO_2/(NH_4)_xWO_3$, TONW NRs) for multimodal imaging and second NIR window PTT [127]. PEGylated TONW NRs (PEG-TONW NRs) exhibited strong absorption in both NIR-I and NIR-II windows, attributable to the incorporation of TeO_2. In the NIR-II window, PEG-TONW NRs achieved a PCE of 43.6%, comparable to previously reported nanoparticles designed for this spectral region [128]. By integrating CT and PA imaging, PEG-TONW NRs effectively inhibited the growth and metastasis of MCF-7 tumors upon irradiation with a 1064 nm laser.

Recently, molybdenum nanoparticles, including molybdenum oxides and molybdenum disulfides, have attracted considerable attention due to their high PCE and tunable NIR absorption spectra [129]. Molybdenum oxides exhibit a range of stoichiometries, from fully stoichiometric molybdenum trioxide (MoO_3) to reduced oxides (MoO_{3-x})and semimetallic MoO_2 [130]. Liu et al. reported the synthesis of phase-tunable MoO_2 nanoparticles via a surfactant-free solvothermal approach [131]. These nanoparticles demonstrated strong NIR absorption and could serve as SERS substrates for detecting trace probe molecules. Compared to other metal-based semiconductors, the MoO_2 nanoparticles exhibited high sensitivity, achieving a minimum detection concentration of 5×10^{-8} M and a maximum enhancement factor of 1.10×10^{7}. In addition, Xu et al. developed a chiral, substoichiometric, cysteine-capped molybdenum oxide system (L/D-Cys-MoO_{3-x}) exhibiting dual responsiveness to visible and NIR light, making it highly suitable for PTT applications [132]. By adjusting the chiral ligand concentration during redox treatment, the resulting Cys-MoO_{3-x} nanoparticles displayed intense circular dichroism responses across both visible and NIR regions. Structurally and electronically, molybdenum disulfides (MoS_2) share similarities with GO and can be engineered into diverse nanostructures via surface or interlayer modifications [133]. Sundaramurthy et al. fabricated a multifunctional nanozyme by decorating flower-like MoS_2 nanoflakes with nanoceria ($NCeO_2$) for cancer PTT [134]. The incorporation of $NCeO_2$ significantly enhanced the PCE of MoS_2 nanoflakes, with increasing effect observed as the $NCeO_2$ concentration rose from 0.25 to 0.5 mg·mL^{-1}. The decorated MoS_2 nanoflakes achieved selective tumor cell eradication via reactive oxygen species (ROS) induction.

Bismuth is a low-cost, abundant, and environmentally benign metal. Bismuth nanoparticles (Bi NPs) are considered ideal PTAs owing to their excellent biosafety and high PCE [135]. The narrow direct bandgap of Bi NPs contributes to their strong NIR absorption. However, rapid surface oxidation of Bi NPs significantly diminishes their photo-absorption capabilities, limiting their practical use as PTAs [136]. To address this limitation, Li et al. reported multifunctional Bi-embedded silica nanoparticles (Bi@SiO_2) for effective PTT [137]. The incorporation of mesoporous SiO_2 enabled the entrapment of Bi nanocrystals within its pores, effectively preventing oxidation and resulting in an excellent PCE of approximately 43%. Furthermore, the high atomic number of Bi allows Bi NPs to serve as contrast agents for CT and PA imaging [138]. For example, Han et al. designed a "one-for-all" multifunctional theranostic agent comprising 1,2-dilauroyl-*sn*-glycero-3-phosphocholine (DLPC) lipid-membrane-coated Bi NPs (Bi@DLPC NPs) for dual PA/CT imaging-guided PTT [139]. Bi@DLPC NPs demonstrated a high PCE of 35% along with excellent PA and CT imaging performance.

Other metal-based nanoparticles, including platinum, palladium, and titanium oxide nanoparticles, have also been widely employed as PTAs. These nanoparticles exhibit strong absorption in the NIR region when their nanostructures and sizes are properly optimized. Cheng et al. synthesized trifolium-like platinum nanoparticles (TPNs) with small size and high photothermal conversion efficiency for PTT applications [140]. In a bone metastasis model, TPNs effectively inhibited tumor growth and prevented osteolysis, highlighting their potential as highly effective therapeutic agents. Wu et al. reported Magnéli-phase titanium oxide (Ti_8O_{15})-based nanoparticles as PTAs for PTT [141]. Compared to conventional TiO_2 nanoparticles, Ti_8O_{15} exhibited intense absorption across the UV-Vis-NIR region, enabling excellent photothermal therapeutic performance in an in vivo mouse model.

3.1.2 Carbon-Based Nanomaterials

Light-induced heating using carbon-based nanomaterials, including carbon nanotubes (CNTs), carbon nanodots (CNDs), and graphene-based nanomaterials, is a well-established phenomenon with diverse biological applications [142]. The graphitic structures of these materials confer superior light absorption across a broad

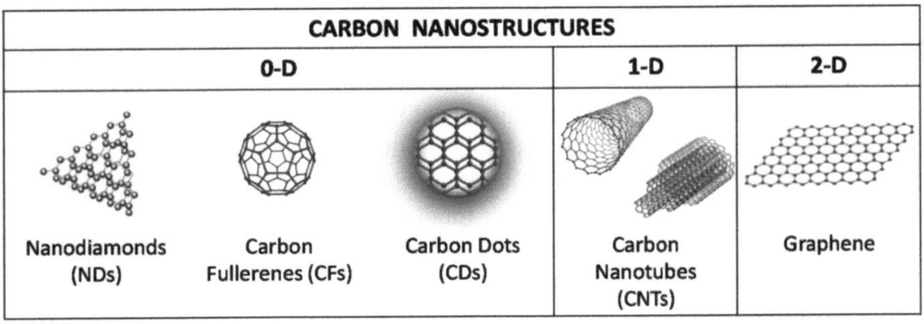

Fig. 2.4 Schematic drawing of the basic geometries of different carbon-based nanostructures explored for cancer PTT treatments. (Copyright © 2020 MDPI. Reprinted with permission from Ref. [143])

range of wavelengths. Compared to metal-based nanoparticles, carbon-based nanomaterials are relatively low cost and abundant, making them attractive candidates for various biomedical applications. As illustrated in Fig. 2.4, carbon-based nanomaterials can be engineered into diverse 0-D, 1-D, and 2-D nanostructures for applications in cancer PTT [143].

CNTs, first discovered by Iijima in 1991, are the most extensively studied carbon-based nanomaterials [144]. CNTs are typically synthesized from graphite using arc discharge or chemical vapor deposition methods. They possess cylindrical nanostructures composed of sp^2-hybridized graphite sheets, with diameters on the nanometer scale and lengths ranging from tens of nanometers to several microns. CNTs are classified as single-walled (SWCNTs) or multi-walled (MWCNTs) depending on the number of graphite layers. SWCNTs consist of a single graphite sheet rolled into a tubular structure, whereas MWCNTs comprise multiple concentric graphite cylinders. The NIR light absorption of CNTs is influenced by their diameter and the relative orientation of the graphite hexagons with respect to the tube axis [145]. To enhance therapeutic efficacy, CNTs have been functionalized with various compounds, including polymers, proteins, and enzymes, enabling the tailoring of their properties for specific biomedical applications and improved therapeutic outcomes. For example, Fan et al. developed insulin-like growth factor receptor (IGFR)-conjugated SWCNTs for targeted PTT of pancreatic cancer [146]. These IGFR-bound SWCNTs achieved active tumor targeting and exhibited high PCE, resulting in significant therapeutic effects in orthotopic pancreatic cancer-bearing mice. Studies have indicated that MWCNTs are more effective than SWCNTs in generating heat during PTT [147]. Cai et al. reported a dual-modality lymphatic theranostic agent based on manganese oxide (MnO) and PEG-coated MWCNTs (MWCNTs-MnO-PEG) for the PTT of metastatic lymph nodes [148]. MWCNTs-MnO-PEG enabled accurate mapping of regional lymph nodes and efficient eradication of metastatic nodes upon NIR laser irradiation.

CNDs, composed of a carbon core and surface functional groups, represent an emerging class of fluorescent carbon-based nanomaterials [149]. With sizes smaller than 10 nm, CNDs can be rapidly cleared from the body via renal filtration. Owing to their low cost, stable fluorescence emission, and tunable photophysical properties, CNDs have been widely employed in various biomedical applications. However, their use in PTT has been limited, as their absorption bands are primarily confined to the UV and green regions of the spectrum [150]. To overcome this limitation, numerous researchers have focused on developing novel therapeutic systems based on CNDs. Qu et al. constructed supra-CNDs with sizes ranging from 8 to 20 nm for PA imaging-guided PTT of cancer [151]. These supra-CNDs exhibited strong absorption across the visible-to-NIR region and achieved an impressive NIR photothermal conversion efficiency exceeding 50%. PA imaging demonstrated that supra-CNDs accumulated effectively at tumor sites under 655 nm laser irradiation at 1 $W{\cdot}cm^{-2}$, with the local tumor temperature reaching 65 °C within 300 s, sufficient for irreversible tumor ablation. Huang et al. reported highly crystalline CNDs (HCCNDs) with tunable fluorescence emissions and excellent PA/photothermal performance for dual-modal imaging-guided PTT of glioma [152]. Due to their small size (6–8 nm) and high hydrophilicity, HCCNDs efficiently crossed the glioma blood–brain barrier and selectively accumulated in tumor tissue, achieving potent antitumor effects upon 808 nm NIR laser irradiation at 2 $W{\cdot}cm^{-2}$.

Graphene-based nanomaterials, including GO and reduced GO (rGO), are 2D materials primarily composed of sp^2-hybridized carbon atoms [153]. These graphene derivatives can be readily functionalized with various oxygen-containing groups, such as epoxide, hydroxyl, and carboxyl groups. Compared to GO, rGO exhibits stronger NIR absorption and higher PCE [154]. For instance, Robinson et al. reported that amphiphilic PEGylated rGO demonstrated a sixfold increase in NIR absorption capacity relative to GO [155]. Correia et al. developed HA-conjugated rGO (HA-rGO) with optimized size and enhanced NIR absorption for targeted cancer PTT [156]. Upon irradiation with an 808 nm NIR laser (1.7 $W{\cdot}cm^{-2}$) for 5 min, 75 $\mu g{\cdot}mL^{-1}$ of HA-rGO induced a temperature rise of approximately 33 °C. Furthermore, the large surface area of graphene-based nanomaterials allows efficient loading of various molecules through electrostatic, hydrophobic, and π-π interactions, enabling combinatorial PTT with other therapeutic modalities [157]. For example, Roy et al. prepared poly(allylamine hydrochloride)-modified rGO for combinational chemo-PTT of breast cancer. The modified rGO exhibited strong absorption under 808 nm NIR laser irradiation, resulting in a temperature increase of up to 45 °C (6 $W{\cdot}cm^{-2}$ for 6 min).

3.2 Organic Nanoparticles

3.2.1 Small Organic Molecule-Based Nanoparticles

NIR fluorescent dye-based small organic molecules have attracted significant attention due to their high sensitivity, low cytotoxicity, and excellent biocompatibility. These NIR fluorescent dyes, with emission wavelengths ranging from 740 to 1700 nm, can penetrate deeply into pathological tissues while minimizing damage to normal tissues [158]. Various classes of fluorophores are used in NIR dyes, including cyanines, phthalocyanines, methylene blue, and porphyrins. Among them, indocyanine green (ICG) has been approved by the U.S. Food and Drug Administration for biomedical applications owing to its well-established clinical safety and favorable pharmacokinetic properties [159]. In addition, ICG's strong NIR imaging capability and high PCE make it an attractive PTA for cancer PTT. However, ICG is prone to photodegradation in aqueous solution, leading to a concurrent loss of both its absorption and fluorescence properties. Moreover, ICG undergoes photothermal degradation within the effective temperature range used in PTT. These limitations significantly restrict its clinical utility.

To address these limitations, physical encapsulation and chemical modification strategies have been employed to enhance the photostability and thermal stability of ICG. For example, Tan et al. reported ultrasmall PEGylated chitosan nanocarriers loaded with ICG for tumor PTT [160]. These nanocarriers, with an ultrasmall size of approximately 5 nm, were prepared via a water-in-oil reverse microemulsion method, and ICG was incorporated through electrostatic interactions during the self-assembly process. The loading efficiency of ICG was determined to be 1.93 wt%. The resulting nanocarriers exhibited good photostability and thermal stability, along with low cytotoxicity. Upon 808 nm laser irradiation, cells incubated with the nanoparticles showed markedly reduced viability (~15%). Unlike exogenous compounds, serum albumin can actively target tumor cells via interaction with the gp60 receptor and facilitate ICG delivery in tumor PTT. Studies have demonstrated that the rational integration of ICG into albumin-based nanoparticles can enhance the photoluminescence quantum yield (PLQY). For instance, Guo et al. constructed a phototheranostic nanocomplex (NC) based on bovine serum albumin (BSA) and ICG for enhanced tumor imaging-guided PTT [161]. Compared to the PLQY of free ICG (2.7%), the NC, with an ICG loading ratio of 0.5%, achieved a higher PLQY of 16.8%, which is advantageous for tumor PTT. In vitro studies indicated that the NC exhibited enhanced tumor accumulation and efficient NIR fluorescence, resulting in significant suppression of tumor growth.

Compared to structurally similar ICG, heptamethine cyanine (Cy7) dyes exhibit higher photostability, thermal stability, and PCE. Xie et al. reported a theranostic nanomedicine containing a high content of Cy7-based polymer (CyP) for PTT of cervical carcinoma [162]. CyP was co-assembled with PEG2k-*block*-poly(D,L-lactide)2k (PEG-PLA) to form spherical nanoparticles (CyP@PEG-PLA NPs) with an average diameter of 50 nm. The CyP loading content reached 44.2 wt% when CyP and PEG-PLA were co-assembled at a 1:1 (w/w) ratio. Upon 808 nm light irradiation for 30 min, the tumor site in CyP@PEG-PLA NP-treated HeLa tumor-bearing mice reached 44 °C, inducing irreversible tumor cell damage. In addition, Song et al. developed an anti-photobleaching nanocomplex incorporating simultaneous photoinduced electron transfer (PET) and aggregation-caused quenching effects for enhanced tumor PTT [163]. The Cy7 derivative DTZCy, featuring an intramolecular PET effect, was encapsulated into silica nanoparticles via a reverse microemulsion method, yielding DTZCy@SiO_2 NPs (Fig. 2.5). The DTZCy@SiO_2 nanocomplex exhibited a high PCE of 34.6% and excellent photothermal stability. In 4T1 tumor-bearing BALB/c mice, the tumor temperature rapidly increased to

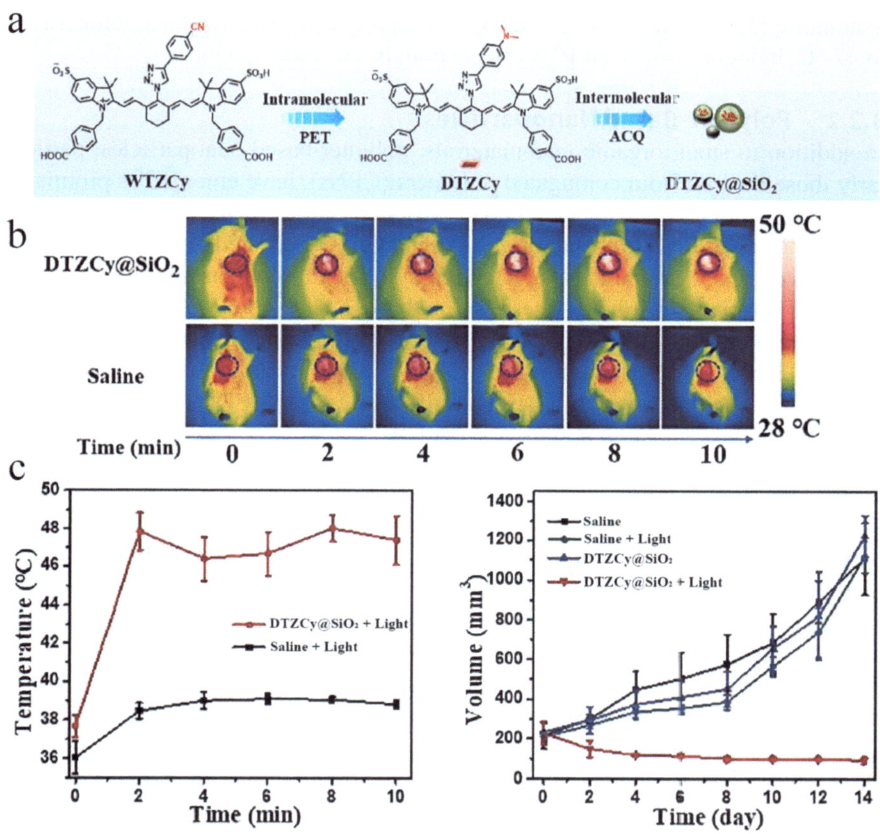

Fig. 2.5 (**a**) Schematic illustration of the preparation of DTZCy@SiO_2 NPs. (**b**) In vivo fluorescence imaging of 4T1 tumor-bearing mice that were intratumorally injected with saline and DTZCy@SiO_2 NPs, respectively. (**c**) Temperature change curves and tumor growth curves of 4T1 tumor-bearing mice with various formulations. (Copyright © 2022 Elsevier. Reprinted with permission from Ref. [163])

approximately 47 °C under 808 nm laser irradiation (0.3 W·cm^{-2}) and was maintained for more than 30 min, providing sufficient heat for tumor ablation. Compared to control groups, tumor volumes in the DTZCy@SiO_2 NP-treated group were significantly reduced after 14 days of treatment, demonstrating effective inhibition of tumor growth.

Self-assembly of small organic molecules without the use of polymeric surfactants represents a straightforward approach for the preparation of nanotherapeutics with high drug-loading capacity. Xie et al. reported the development of cyanine-curcumin co-assembled nanoparticles (CCNPs) for NIR imaging-guided tumor PTT [164]. The IR-780-C4 (Cyc4) dye was co-assembled with hydrophobic curcumin via a single-step reprecipitation method, yielding CCNPs with a high Cyc4 loading content of approximately 70%. The resulting CCNPs exhibited excellent stability under physiological conditions and a high PCE of 52%. Upon 808 nm laser irradiation (1.5 W·cm^{-2}) for 5 min, CCNPs rapidly increased the local temperature to 42 °C, inducing enhanced PTT effects both in vitro and in vivo.

3.2.2 Polymer-Based Nanoparticles

In addition to small organic nanomaterials, polymer-based nanoparticles, particularly those derived from conjugated polymers (CPNs), have emerged as promising PTAs due to their excellent biocompatibility and biodegradability, superior light-harvesting capability, photostability, and high PCE. Conjugated polymers are representative organic compounds characterized by delocalized electronic structures and extended π-conjugated backbones, which have been widely employed as NIR absorbing agents in cancer PTT [165]. Notable examples of CPNs include polyaniline (PANI), polydopamine (PDA), polypyrrole (PPy), poly(3,4-ethylenedioxythiophene)/poly(4-styrenesulfonate) (PEDOT:PSS), and donor-acceptor conjugated polymers (D-A CPs).

PANI was the first conducting polymer to be employed as a PTA for PTT. PANI is considered one of the most promising PTAs due to its facile synthesis, low cost, good biocompatibility, and high electrical conductivity. Owing to its abundant π-electrons, PANI exhibits high chemical stability and strong photothermal properties. During the doping process with dopants such as strong acids or Lewis acids, the bandgap between the valence and conduction bands of PANI is reduced, facilitating electron mobility and lowering the excitation energy level [166]. The transition of PANI from the emeraldine base (EB) to emeraldine salt (ES) form induces a pronounced redshift in the optical absorbance peak from the visible to the NIR region. Compared to EB PANI, ES-PANI exhibits stronger NIR absorption, resulting in greater heat generation. Biological dopants, including protons and alkali ions present in cancer cells, can convert EB PANI to ES PANI, enabling selective photothermal ablation of cancer cells under NIR irradiation. Under 808 nm laser irradiation for 5 min (2.45 W·cm^{-2}), ES-PANI raised the temperature at lesion sites to 60 °C. To further enhance the hydrophilicity and photothermal performance of PANI, Zhou et al. synthesized F127-capped PANI nanoparticles (F-PANIPs) with a small size of 48.5 ± 1.5 nm [167]. Compared to the photothermal conversion efficiency (PCE) of typical inorganic nanoparticles (22%), F-PANIPs achieved a higher PCE of 48.5% under 808 nm irradiation.

More recently, Yang et al. reported tumor pH-responsive PANI-based nanoparticles for enhanced PA imaging-guided PTT of tumors [168]. Biocompatible and hydrophilic BSA and Fe^{3+} ions were employed to polymerize aniline under acidic conditions (Fig. 2.6). In situ-generated hydrophobic PANI readily associated with BSA via hydrophobic interactions. This co-assembly of hydrophobic PANI with BSA resulted in the formation of stable BSA-PANI nanoparticles. Notably, a self-doping effect was observed, arising from intermolecular acid-base interactions between the carboxyl groups of BSA and the imine groups of PANI. This effect reduced the degree of deprotonation of PANI and facilitated the conversion of BSA-PANI from the EB to the ES state at relatively high pH values. In vitro studies demonstrated that BSA-PANI nanoparticles exhibited stronger PA signals and higher PCE at pH 5.4 and 6.5 compared to pH 7.4. Upon 808 nm laser irradiation

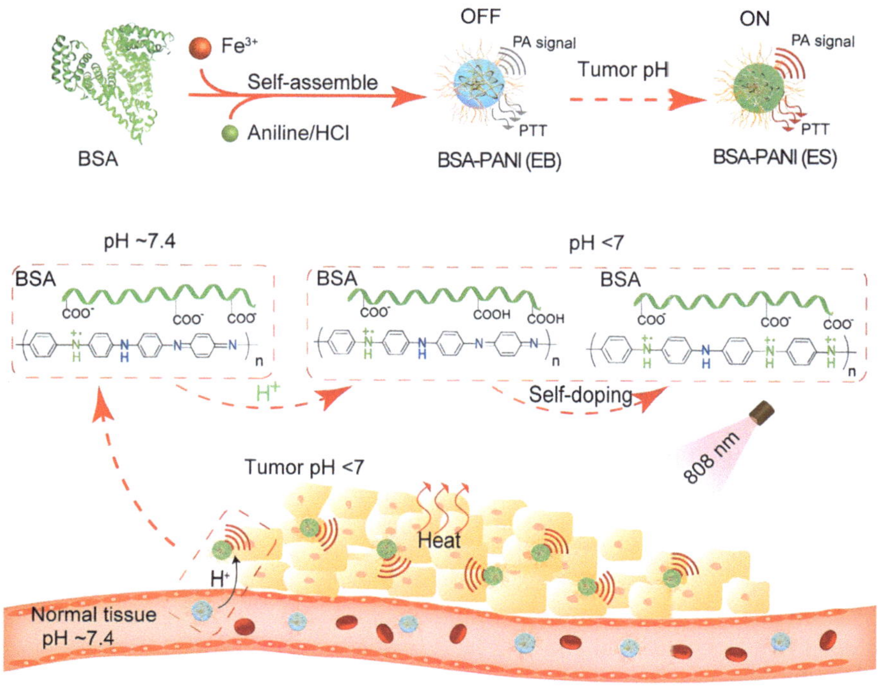

Fig. 2.6 Schematic illustration of the preparation of BSA-PANI nanoparticles and self-doping effect induced transformation from EB to ES state for PA-guided augmented PTT of tumors. (Copyright © 2019 Wiley-VCH GmbH. Reprinted with permission from Ref. [168])

(1 W·cm^{-2}) for 5 min in tumor-bearing mice, the temperature at the tumor site in the BSA-PANI nanoparticle–treated group reached 56 °C, sufficient to induce tumor tissue ablation.

Melanin, a natural biomacromolecular pigment, performs diverse functions in living organisms, including UV protection and antioxidant activity. PDA-based nanoparticles, as synthetic mimics of natural melanin, have demonstrated great potential for both in vitro and in vivo applications in PTT. PDA-based nanoparticles exhibit strong NIR absorption, rendering them highly suitable for tumor-targeted PTT. Cheng et al. developed dual-peptide-decorated PDA nanoparticles for tumor-targeted and autophagy-promoted PTT of breast cancer [169]. Functionalization of PDA nanoparticles with a Beclin 1-derived peptide induced upregulation of autophagy in breast cancer cells. Additionally, RGD peptides were conjugated to the surface of PEGylated PDA nanoparticles, resulting in multifunctional nanoparticles (PPBR) with enhanced cellular uptake and high PCE. It was demonstrated that RGD significantly facilitated the internalization of PPBR by cancer cells overexpressing integrin αvβ3. Upon 808 nm NIR laser irradiation (1.0 W·cm^{-2}, 7 min), the temperature at the tumor site rapidly increased to approximately 43 °C, achieving effective inhibition of tumor growth.

PPy, a representative conductive polymer, has been extensively explored in both organic electronics and biomedical applications. Yang et al. were the first to demonstrate the photothermal potential of PPy for cancer PTT. The pseudo-polycondensation of pyrrole is the most common method for synthesizing PPy with tunable sizes. PPy-based nanoparticles with diameters of 50–60 nm can be readily prepared using Fe^{3+} as a catalyst and polyvinyl alcohol (PVA) or polyvinylpyrrolidone (PVP) as stabilizing agents. The resulting nanoparticles exhibit strong NIR absorption in the 700–1200 nm range. PPy nanoparticles with ultrasmall sizes have been reported as promising photothermal agents and contrast agents for PA imaging-guided cancer PTT in both NIR I and NIR II windows [170]. Wu et al. developed PEGylated ultrasmall PPy-based nanoparticles (PPy-PEG NPs) with renal-clearable properties for fluorescence/PA/NIR II multimodal imaging-guided tumor PTT [171]. PPy-PEG NPs with particle sizes ranging from 2 to 120 nm were synthesized by adjusting the feed ratio of the steric stabilizer (Fig. 2.7). The average hydrodynamic diameter of PPy-PEG NPs could be reduced to 4.8 nm, satisfying the requirements for renal filtration. At a concentration of 150 $\mu g \cdot mL^{-1}$, PPy-PEG NPs increased the temperature of tumor sites to 50 and 52 °C under 808 and 1064 nm laser irradiation for 10 min, respectively. The PCE of PPy-PEG NPs was calculated to be 33.35% and 41.97% at 808 nm and 1064 nm, respectively. In vitro cytotoxicity assays revealed that 200 $\mu g \cdot mL^{-1}$ of PPy-PEG NPs reduced tumor cell viability to 17.09% and 15.74% upon 808 and 1064 nm laser irradiation, respectively. In vivo fluorescence/PA/NIR multimodal imaging studies demonstrated that ultrasmall PPy-PEG NPs achieved efficient tumor accumulation, remarkable tumor growth inhibition, and effective renal clearance.

PEDOT:PSS is another type of conductive polymer that has been widely investigated in organic electronics. Liu et al. employed a layer-by-layer strategy to prepare stealth-like PEGylated PEDOT:PSS nanoparticles (PEDOT:PSS-PEG) with an average diameter of 80–90 nm for NIR-mediated PTT of cancer [172]. The prolonged blood circulation half-life of PEDOT:PSS-PEG (21.4 ± 3.1 h) facilitated substantial accumulation within tumors. Under 808 nm laser irradiation for 5 min (0.5 $W \cdot cm^{-2}$), PEDOT:PSS-PEG increased the tumor surface temperature to 51 °C, efficiently eliminating 4T1 tumors with minimal adverse effects. To achieve synergistic therapeutic effects, PEDOT:PSS has also been integrated with other modalities within a single platform. For example, Kang et al. reported PEDOT:PSS-conjugated magnetic nanoparticles (MNP@PEDOT:PSS) with a uniform core-shell nanostructure for magnetic resonance (MR) and PA imaging-guided PTT of cancer [173]. Building on the excellent photothermal-magnetic therapeutic performance of MNP@PEDOT:PSS, the same team further developed an "all-in-one" nanoparticle comprising MNPs, PEDOT:PSS, Cy7, and 2-deoxyglucose (2-DG)-PEG for trimodality imaging-guided PTT [174]. Guided by trimodal imaging, these nanoparticles achieved substantial cytoplasmic accumulation in tumor cells and effectively ablated tumors, resulting in enhanced therapeutic efficacy.

D-A CPs are composed of electron-rich donor monomers and electron-deficient acceptor monomers. The overlap of π-orbitals between donors and acceptors

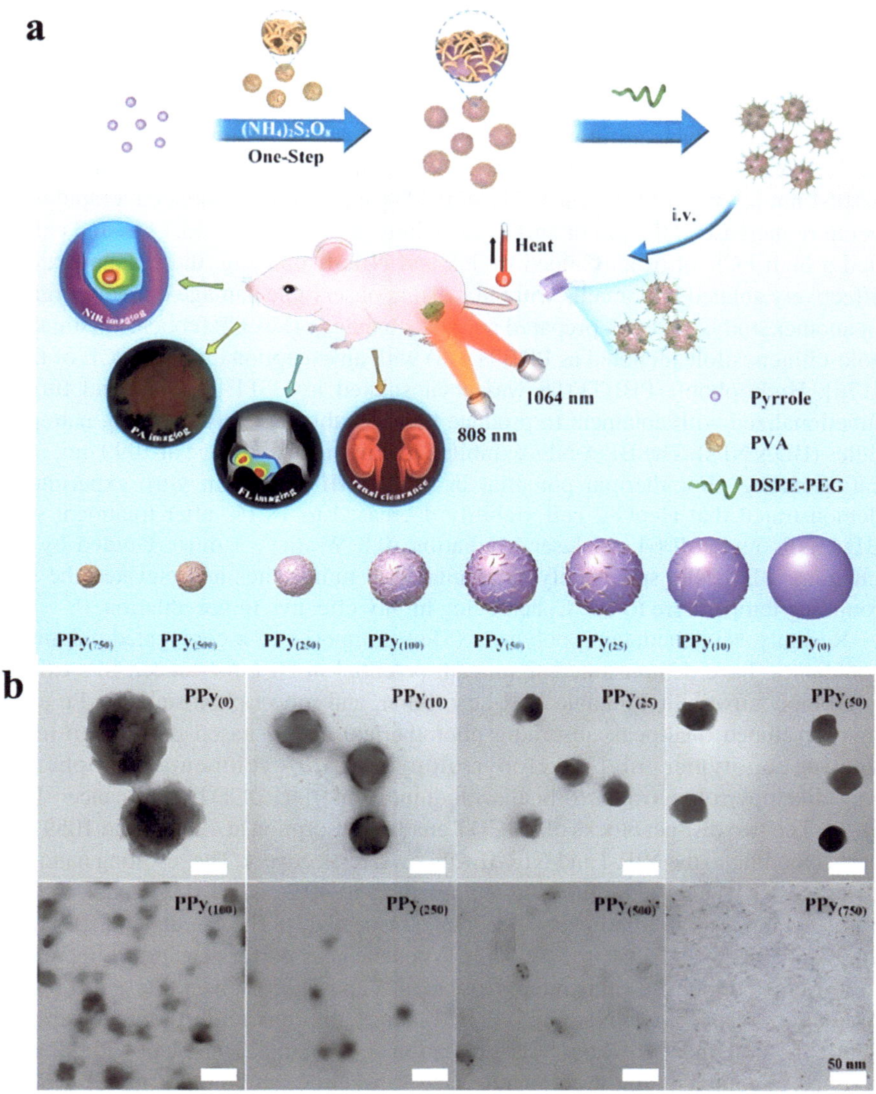

Fig. 2.7 (**a**) Schematic illustration of the fabrication of ultrasmall PPy-PEG NPs and their application in fluorescence/PA/NIR II multimodal imaging-guided PTT of tumor. (**b**) TEM images of PPy NPs with various particulate sizes from 2 to 120 nm. (Copyright © 2021 Wiley-VCH GmbH. Reprinted with permission from Ref. [171])

generates new electronic states between HOMO and LUMO, thereby modulating the bandgap of the entire D-A CP system. The difference in electronic density between donors and acceptors determines the bandgap of D-A CPs. Polymers with lower bandgaps exhibit absorption at longer wavelengths in the NIR region, enabling deeper tissue penetration and reduced adverse effects, which makes them highly promising for PTT of cancer. Consequently, considerable research has focused on tuning the bandgap of D-A CPs by varying the combinations of donor and acceptor monomers, donor-to-acceptor ratios, and molecular weights. Xie et al. synthesized narrow-bandgap D-A CPs (TBDOPV-DT) using 2,2-bithiophene as the donor and thiophene-fused benzodifurandione-based oligo(phenylenevinylene) as the acceptor for PA imaging-guided PTT in the NIR-II window [175]. Upon irradiation with a 1064 nm laser at 1.96 W·cm^{-2}, TBDOPV-DT nanoparticles induced a rapid temperature increase at the tumor surface, reaching approximately 53.7 °C, and exhibited a high PCE of 50%. Guided by PA and NIR-II imaging, these nanoparticles effectively ablated tumor cells without causing observable damage to major organs. In another study, Liu et al. prepared a narrow-bandgap D-A CP (cobenzobisthiadiazole-dithienosilole, denoted as PBBTDTS) with an exceptionally high PCE of 65% [176]. Hydrophobic PBBTDTS was encapsulated in DSPE-PEG-N_3 and further functionalized with aptamers to produce highly stable, actively targeting nanoparticles (BDA NPs). The BDA NPs exhibited maximum absorbance at 1090 nm, indicating strong photothermal potential in the NIR-II region. In vitro experiments demonstrated that HepG-2 cell viability decreased to 18.1% after treatment with BDA NPs under 1064 nm laser irradiation (0.8 W·cm^{-2}, 5 min). Guided by PA imaging, BDA NPs specifically accumulated at tumor sites and elevated the surrounding temperature to 63 °C, achieving highly effective tumor ablation.

Recently, semiconducting polymers (SPs), composed of π-conjugated polymeric backbones, have demonstrated significant potential in PTT for cancer. SPs exhibit good biocompatibility, tunable NIR absorption, and robust photostability. Pu et al. first developed dual-peak absorbing photothermal agents based on the semiconducting copolymer poly[(diketopyrrolopyrrole-*alt*-cyclopentadithiophene)-*ran*-(diketopyrrolopyrrole-*alt*-thiadiazoloquinoxaline)] (PDCDT) for cancer PTT [177]. The two distinct blocks of PDCDT enabled absorption at 728 nm and 1059 nm, corresponding to the NIR-I and NIR-II windows, respectively. The resulting nanoparticles (SPNI-II) elevated tumor site temperatures to 40.2 and 41.0 °C under 808 nm and 1064 nm laser irradiation at 0.3 W·cm^{-2}, respectively. The calculated PCEs were 44.9% at 808 nm and 43.4% at 1064 nm. When the laser power density was increased to 1 W·cm^{-2}, tumor site temperatures rose to 68.1 and 71.7 °C under 808 and 1064 nm irradiation, respectively. Notably, SPNI-II achieved a maximum tissue penetration depth of 5 mm under 1064 nm laser irradiation at 1 W·cm^{-2}. Treatment with SPNI-II resulted in complete tumor elimination and substantial inhibition of tumor growth, demonstrating markedly enhanced therapeutic efficacy.

3.3 Hybrid Nanoparticles

Hybrid nanoparticles, composed of both inorganic and organic components, have emerged as an intriguing class of nanoplatforms for cancer PTT. These hybrid nanoparticles integrate multiple organic and inorganic constituents into a single system in a modular manner. When optimally constructed, hybrid nanoparticles can synergistically combine the advantageous properties of both purely inorganic and organic components, resulting in diversified and tunable physical and chemical characteristics suitable for combination cancer therapies [178]. The specific therapeutic functions of hybrid nanoparticles are determined by their well-defined inorganic and organic compositions. Furthermore, their light absorption properties are critically dependent on the geometry and intrinsic characteristics of the constituent inorganic and organic materials. For instance, Fan et al. synthesized highly stable hybrid nanomaterials by conjugating cyanine 5.5 (Cy5.5) with SWCNTs for PTT of breast cancer [179]. Guided by the NIR imaging capability of Cy5.5, the SWCNTs accurately targeted tumor sites and effectively ablated cancer cells through photothermal energy conversion, achieving significant tumor growth suppression in mice. Similarly, Yu et al. developed multifunctional core-shell nanohybrids based on PEGylated polypyrrole-coated bismuth nanoparticles (Bi@PPy-PEG NHs) for PA/CT dual-imaging-guided PTT [180]. Bi@PPy-PEG NHs exhibited enhanced photothermal conversion, with a PCE of approximately 46.3%, surpassing that of single Bi nanoparticles (~30.4%).

Ferrotherapy, which induces ferroptosis, has emerged as a promising strategy for cancer treatment. To achieve optimal therapeutic efficacy, the iron dosage typically needs to exceed 75 $mg{\cdot}kg^{-1}$ body weight via intravenous administration. Pu et al. first designed hybrid ferrotherapeutic nanoagents based on iron-chelated semiconducting polycomplex nanoparticles (SPFeN) for PA imaging-guided photothermal ferrotherapy of cancer [181]. SPFeN were self-assembled from an amphiphilic semiconducting polycomplex (SPC) that functions both as a photothermal nanotransducer and as an iron ion chelator. The SPC readily chelates iron ions to form SPFe, and its amphiphilic properties facilitate the self-assembly of SPFe into homogeneous SPFeN. Guided by PA imaging, SPFeN effectively accumulated at tumor sites. Elevated levels of reductive biomolecules, such as glutathione and ascorbic acid, in the acidic TME induced in situ ferroptosis and generated hydroxyl radicals via Fenton reactions. Upon 808 nm laser irradiation, the localized heat produced by SPFeN not only accelerated the Fenton reaction but also enhanced the overall therapeutic effect. Compared to normal physiological conditions, SPFeN facilitated a fivefold increase in hydroxyl radical generation. The PCE of SPFeN was calculated to be 42%. By integrating iron ions with semiconducting polymers, SPFeN effectively ablated tumor cells and inhibited tumor growth, achieving a synergistic photothermal-ferrotherapy effect with a reduced iron dosage.

Au NPs possess tremendous potential in PTT for cancer. However, unsupported Au NPs typically exhibit limited light-to-heat conversion due to their low stability. Encapsulation of organic components on the surface of Au NPs can prevent aggregation of the Au cores and enhance their accumulation at tumor sites. Gao et al. synthesized yolk-shell nanozymes (Au@HCNs), in which porous hollow carbon nanospheres coated the Au NP cores [182]. Upon 808 nm NIR laser irradiation, Au@HCNs exhibited intrinsic enzyme-like activities and enhanced ROS generation, which are favorable for cancer PTT. The PCE of Au@HCNs was calculated to be 26.8%, surpassing that of single-component Au-based PTAs, such as Au nanoshells (~13%) and Au nanorods (~22%) [183, 184]. Lu et al. designed Au nanostar@PANI core-shell nanohybrids (AuNSs@PANI) with high PCE for PA imaging-guided PTT at low dosages [185]. The photothermal conversion properties of AuNSs@PANI could be optimized by precisely controlling the relative composition. When the thickness of the PANI shell was approximately 20 nm, AuNSs@PANI achieved an exceptional PCE of 78.6%. Consequently, a single injection of these nanohybrids effectively ablated tumors with an extremely low Au dosage of 0.095 mg/kg. In another study, Wang et al. prepared pH-responsive PANI-gelatin/Cu (PANI-Gel/Cu) nanohybrids for precise cancer PTT [186]. Owing to the self-doping effect between the carboxyl groups in gelatin and the imine groups of the PANI backbone, the photothermal performance and PA signal of PANI-Gel/Cu were significantly enhanced in the acidic TME.

4 Conclusions and Future Perspectives

In summary, this chapter provides a comprehensive overview of nanoparticles, including their classification, photothermal conversion mechanisms, and diverse applications in PTT. A broad spectrum of nanoparticles, such as metallic nanoparticles, carbon-based nanomaterials, small organic molecules, polymer-based nanoparticles, and hybrid systems, have been investigated as PTAs. The primary objective of these nanoheaters is to achieve high PCE and effective tumor ablation while minimizing adverse effects on normal tissues. Different photothermal conversion mechanisms, including LSPR, electron-hole pair generation and relaxation, and conjugation and hyperconjugation effects, have been systematically reviewed. For each category of nanoparticle, we have summarized the key design strategies and photothermal properties that contribute to their therapeutic performance. Particular emphasis has been placed on optimizing optical properties by modulating nanoparticle size, morphology, and doping with additional components. Furthermore, this chapter details the present state of the art in the application of these PTAs, highlighting their efficacy in both in vitro and in vivo cancer treatments.

The small sizes and specific morphologies of nanoparticles enable their penetration through narrow capillaries and facilitate accumulation within target tumor tissues. NIR laser irradiation further enhances therapeutic efficacy by allowing deeper tissue penetration with minimal invasiveness to surrounding healthy tissues, thereby promoting effective tumor ablation. Despite these promising prospects, the clinical translation of nanoparticles remains a formidable challenge due to unresolved biosafety concerns. To achieve optimal photothermal performance, nanoparticles are often engineered with specific components; however, their potential long-term cytotoxicity and persistence in biological systems may hinder clinical applicability. Thus, striking an appropriate balance between maximizing photothermal efficacy and ensuring biosafety is of critical importance. Comprehensive studies are required to rigorously evaluate the safety, biocompatibility, and degradation pathways of photothermal nanomaterials across diverse tumor models. Moreover, considerations of biodegradability and scalable synthesis are paramount to enable their successful transition from laboratory research to clinical practice.

References

1. Zou Y, Li M, Xiong T, Zhao XZ, Du JJ, Fan JL, Peng XJ. A single molecule drug targeting photosensitizer for enhanced breast cancer photothermal therapy. Small. 2020;16:1907677.
2. Chakravarty P, Marches R, Zimmerman NS, Swafford ADE, Bajaj P, Musselman IH, Pantano P, Draper RK, Vitetta ES. Thermal ablation of tumor cells with antibody-functionalized single-walled carbon nanotubes. Proc Natl Acad Sci USA. 2008;105:8697–702.
3. Johannsen M, Gneveckow U, Eckelt L, Feussner A, Waldofner N, Scholz R, Deger S, Wust P, Loening SA, Jordan A. Clinical hyperthermia of prostate cancer using magnetic nanoparticles: presentation of a new interstitial technique. Int J Hyperth. 2005;21:637–47.
4. Rijt SH, Sadler PJ. Current applications and future potential for bioinorganic chemistry in the development of anticancer drugs. Drug Discov Today. 2009;14:1089–97.
5. Mendes R, Pedrosa P, Lima JC, Fernandes AR, Baptista PV. Photothermal enhancement of chemotherapy in breast cancer by visible irradiation of gold nanoparticles. Sci Rep. 2017;7(1):10872.
6. Han HS, Choi KY. Advances in nanomaterial-mediated photothermal cancer therapies: toward clinical applications. Biomedicine. 2021;9:305.
7. Zou LL, Wang K, He B, Zeng LJ, Tan T, Cao HQ, He XY, Zhang ZW, Guo SR, Li YP. Current approaches of photothermal therapy in treating cancer metastasis with nanotherapeutics. Theranostics. 2016;6:762–72.
8. Pfeffer CM, Singh ATK. Apoptosis: a target for anticancer therapy. Int J Mol Sci. 2018;19:448.
9. Eskiizmir G, Baskin Y, Yapici K. Graphene-based nanomaterials in cancer treatment and diagnosis. In: Fullerens, graphenes and nanotubes; 2018. p. 331–74. https://doi.org/10.1016/B978-0-12-813691-1.00009-9.
10. Melo-Diogo DD, Pais-Silva C, Dias DR, Moreira AF, Correia IJ. Strategies to improve cancer photothermal therapy mediated by nanomaterials. Adv Healthc Mater. 2017;6:1700073.
11. Johannsen M, Thiesen B, Wust P, Jordan A. Magnetic nanoparticle hyperthermia for prostate cancer. Int J Hyperth. 2010;26:790–5.
12. Kong GH, Braun RD, Dewhirst MW. Hyperthermia enables tumor-specific nanoparticle delivery: effect of particle size. Cancer Res. 2000;60:4440–5.
13. Jaque D, Maestro LM, Rosal B, Haro-Gonzalez P, Benayas A, Plaza JL, Rodríguez EM, Solé JG. Nanoparticles for photothermal therapies. Nanoscale. 2014;6:9494–530.
14. Smith AM, Mancini MC, Nie S. Second window for in vivo imaging. Nat Nanotechnol. 2009;4:710–1.
15. Weissleder R. A clearer vision for in vivo imaging. Nat Biotechnol. 2001;19:316–7.
16. Chen YJ, Khan AR, Yu DX, Zhai YJ, Ji JB, Shi YK, Zhai GX. Pluronic F127-functionalized molybdenum oxide nanosheets with pH-dependent degradability for chemo-photothermal cancer therapy. J Colloid Interface Sci. 2019;553:567–80.
17. Huang XH, El-Sayed I, Qian W, El-Sayed MA. Cancer cell imaging and photothermal therapy in the near-infrared region by using gold nanorods. J Am Chem Soc. 2006;128:2115–20.
18. Rocha U, Kumar KU, Jacinto C, Villa I, Sanz-Rodríguez F, Cruz DI, Juarranz A, Carrasco E, van Veggel FCJM, Bovero E, Solé JG, Jaque D. Neodymium-doped LaF_3 nanoparticles for fluorescence bioimaging in the second biological window. Small. 2014;10:1141–54.
19. Dong HF, Cao Y. Carbon nanomaterials for optical bioimaging and phototherapy. In: Carbon nanomaterials for bioimaging, bioanalysis, and therapy; 2019. p. 43–62. https://doi.org/10.1002/9781119373476.ch3.
20. Estelrich J, Busquets MA. Iron oxide nanoparticles in photothermal therapy. Molecules. 2018;23:1567.
21. Moreira AF, Rodrigues CF, Reis CA, Costa EC, Correia IJ. Gold-core silica shell nanoparticles application in imaging and therapy: a review. Micropor Mesopor Mat. 2018;270:168–79.
22. Greish K. Enhanced permeability and retention (EPR) effect for anticancer nanomedicine drug targeting. In: Cancer nanotechnology, vol. 624. Springer; 2010. p. 25–37. Totowa, NJ: Humana Press. https://doi.org/10.1007/978-1-60761-609-2_3

23. Chen JQ, Ning CY, Zhou ZN, Yu P, Zhu Y, Tan GX, Mao CB. Nanomaterials as photothermal therapeutic agents. Prog Mater Sci. 2019;99:1–26.
24. Moghimi SM, Hunter AC, Murray JC. Long-circulating and target-specific nanoparticles: theory to practice. Pharmacol Rev. 2001;53:283–318.
25. Wang B, He X, Zhang ZY, Zhao YL, Feng WY. Metabolism of nanomaterials in vivo: blood circulation and organ clearance. Acc Chem Res. 2013;46:761–9.
26. Matsumoto Y, Nichols JW, Toh K, Momoto T, Cabral H, Miura Y, Christie RJ, Yamada N, Ogura T, Kano MR, Matsumura Y, Nishiyama N, Yamasoba T, Bae YH, Kataoka K. Vascular bursts enhance permeability of tumour blood vessels and improve nanoparticle delivery. Nat Nanotechnol. 2016;11:533–8.
27. Heo DN, Yang DH, Moon H-J, Lee JB, Bae MS, Lee SC, Lee WJ, Sun I-C, Kwon IK. Gold nanoparticles surface-functionalized with paclitaxel drug and biotin receptor as theranostic agents for cancer therapy. Biomaterials. 2012;33:856–66.
28. Chitgupi U, Qin YR, Lovell JF. Targeted nanomaterials for phototherapy. Nanotheranostics. 2017;1:38–58.
29. Jain RK, Stylianopoulos T. Delivering nanomedicine to solid tumors. Nat Rev Clin Oncol. 2010;7:653–64.
30. Moreira AF, Dias DR, Correia IJ. Stimuli-responsive mesoporous silica nanoparticles for cancer therapy: a review. Micropor Mesopor Mat. 2016;236:141–57.
31. Liu Y, Ashton JR, Moding EJ, Yuan HK, Register JK, Fales AM, Choi JY, Whitley MJ, Zhao XG, Qi Y, Ma Y, Vaidyanathan G, Zalutsky MR, Kirsch DG, Badea CT, Vo-Dinh T. A plasmonic gold nanostar theranostic probe for in vivo tumor imaging and photothermal therapy. Theranostics. 2015;5:946–60.
32. Perrault SD, Walkey C, Jennings T, Fischer HC, Chan WCW. Mediating tumor targeting efficiency of nanoparticles through design. Nano Lett. 2009;9:1909–15.
33. Ernsting MJ, Murakami M, Roy A, Li S-D. Factors controlling the pharmacokinetics, biodistribution and intratumoral penetration of nanoparticles. J Control Release. 2013;172:782–94.
34. Li S-D, Huang L. Pharmacokinetics and biodistribution of nanoparticles. Mol Pharm. 2008;5:496–504.
35. Kumari S, Sharma N, Sahi SV. Advances in cancer therapeutics: conventional thermal therapy to nanotechnology-based photothermal therapy. Pharmaceutics. 2021;13:1174.
36. Li J-L, Gu M. Gold-nanoparticle-enhanced cancer photothermal therapy. IEEE J Sel Top Quantum Electron. 2009;16:989–96.
37. Zhang Y, Lu YM, Xu YX, Zhou ZK, Li YC, Ling W, Song WL. Bioinspired drug delivery systems: from synthetic polypeptide vesicles to outer membrane vesicles. Pharmaceutics. 2023;15:368.
38. Xu DX, Li ZD, Li LS, Wang J. Insights into the photothermal conversion of 2D MXene nanomaterials: synthesis, mechanism, and applications. Adv Funct Mater. 2020;30:2000712.
39. Zhu LL, Gao MM, Peh CKN, Ho GW. Solar-driven photothermal nanostructured materials designs and prerequisites for evaporation and catalysis applications. Mater Horiz. 2018;5:323–43.
40. Zhu LL, Gao MM, Peh CKN, Ho GW. Recent progress in solar-driven interfacial water evaporation: advanced designs and applications. Nano Energy. 2019;57:507–18.
41. Zijlstra P, Paulo PMR, Orrit M. Optical detection of single non-absorbing molecules using the surface plasmon resonance of a gold nanorod. Nat Nanotechnol. 2012;7:379–82.
42. Chen HJ, Shao L, Ming T, Sun ZH, Zhao CM, Yang BC, Wang JF. Understanding the photothermal conversion efficiency of gold nanocrystals. Small. 2010;6:2272–80.
43. Yu HK, Peng YS, Yang Y, Li Z-Y. Plasmon-enhanced light-matter interactions and applications. NPJ Comput Mater. 2019;5:45.
44. Webb JA, Bardhan R. Emerging advances in nanomedicine with engineered gold nanostructures. Nanoscale. 2014;6:2502–30.
45. Bisoyi HK, Urbas AM, Li Q. Soft materials driven by photothermal effect and their applications. Adv Opt Mater. 2018;6:1800458.

46. Song JB, Yang XY, Yang Z, Lin LS, Liu YJ, Zhou ZY, Shen ZY, Yu GC, Dai YL, Jacobson O, Munasinghe J, Yung B, Teng G-J, Chen XY. Rational design of branched nanoporous gold nanoshells with enhanced physico-optical properties for optical imaging and cancer therapy. ACS Nano. 2017;11:6102–13.
47. Hu M, Chen JY, Li Z-Y, Au L, Hartland GV, Li XD, Marquez M, Xia YN. Gold nanostructures: engineering their plasmonic properties for biomedical applications. Chem Soc Rev. 2006;35:1084–94.
48. Shi YG, Liu MY, Deng FJ, Zeng GJ, Wan Q, Zhang XY, Wei Y. Recent progress and development on polymeric nanomaterials for photothermal therapy: a brief overview. J Mater Chem B. 2017;5:194–206.
49. Tian Y, Zhang JP, Tang SW, Zhou L, Yang WL. Polypyrrole composite nanoparticles with morphology-dependent photothermal effect and immunological responses. Small. 2016;12:721–6.
50. Gao MM, Zhu LL, Peh CKN, Ho GW. Solar absorber material and system designs for photothermal water vaporization towards clean water and energy production. Energy Environ Sci. 2019;12:841–64.
51. Beard MC, Midgett AG, Hanna MC, Luther JM, Hughes BK, Nozik AJ. Comparing multiple exciton generation in quantum dots to impact ionization in bulk semiconductors: implications for enhancement of solar energy conversion. Nano Lett. 2010;10:3019–27.
52. Barati F, Grossnickle M, Su SS, Lake RK, Aji V, Gabor NM. Hot carrier-enhanced interlayer electron-hole pair multiplication in 2D semiconductor heterostructure photocells. Nat Nanotechnol. 2017;12:1134–9.
53. Tisdale WA, Williams KJ, Timp BA, Norris DJ, Aydil ES, Zhu XY. Hot-electron transfer from semiconductor nanocrystals. Science. 2010;328:1543–7.
54. Heeger AJ. Semiconducting polymers: the third generation. Chem Soc Rev. 2010;39:2354–71.
55. Tian QW, Tang MH, Sun YG, Zou RJ, Chen ZG, Zhu MF, Yang SP, Wang JL, Wang JH, Hu JQ. Hydrophilic flower-like CuS superstructures as an efficient 980 nm laser-driven photothermal agent for ablation of cancer cells. Adv Mater. 2011;23:3542–7.
56. Wang J, Li YY, Deng L, Wei NN, Weng YK, Dog S, Qi DP, Qiu J, Chen XD, Wu T. High-performance photothermal conversion of narrow-bandgap Ti_2O_3 nanoparticles. Adv Mater. 2017;29:1603730.
57. Chen R, Wu ZJ, Zhang TQ, Yu TC, Ye MM. Magnetically recyclable self-assembled thin films for highly efficient water evaporation by interfacial solar heating. RSC Adv. 2017;7:19849–55.
58. Huang QQ, Hu S, Zhuang J, Wang X. MoO_{3-X}-based hybrids with tunable localized surface plasmon resonances: chemical oxidation driving transformation from ultrathin nanosheets to nanotubes. Chem Eur J. 2012;18:15283–7.
59. Wu C-M, Naseem S, Chou M-H, Wang J-H, Jian Y-Q. Recent advances in tungsten-oxide-based materials and their applications. Front Mater. 2019;6:49.
60. Qiu S-Y, Xu H, Li L, Xu H-T, Meng L-K, Pang H-S, Tang C, Pang Z-Q, Xiao J, Wang X, Ye S-H, Fan Q-L, Huang W. Influence of π-π hyperconjugation effect on thermal, morphological, and photoelectronic properties of non-conjugated pyrene derivatives. J Phys Chem C. 2017;121:9230–41.
61. Feixas F, Matito E, Poater J, Sola M. Understanding conjugation and hyperconjugation from electronic delocalization measures. J Phys Chem A. 2011;115:13104–13.
62. Hyun JY, Kimura Y, Yamane H. Competitive effects of stereocomplexation and hyper-conjugation of triethoxysilyl-terminated poly (D-lactide) in poly (L-lactide) matrices. Macromol Mater Eng. 2015;300:1123–32.
63. Larrañaga O, Romero-Nieto C, de Cózar A. Dismantling the hyperconjugation of π-conjugated phosphorus heterocycles. Chem Eur J. 2019;25:9035–44.
64. Haque MH, Sohn H. Comparison of aggregation-induced emission enhancement effect between π-σ^* and σ-σ^* conjugation on silole and application in explosives detection. Dyes Pigments. 2022;201:110175.

65. Zou QL, Abbas M, Zhao LY, Li SK, Shen GZ, Yan XH. Biological photothermal nanodots based on self-assembly of peptide-porphyrin conjugates for antitumor therapy. J Am Chem Soc. 2017;139:1921–7.
66. He YL, Cao YY, Wang YP. Progress on photothermal conversion in the second NIR window based on conjugated polymers. Asian J Org Chem. 2018;7:2201–12.
67. Bader RFW, Slee TS, Cremer D, Kraka E. Description of conjugation and hyperconjugation in terms of electron distributions. J Am Chem Soc. 1983;105:5061–8.
68. Fernández I, Frenking G. Direct estimate of the strength of conjugation and hyperconjugation by the energy decomposition analysis method. Chem Eur J. 2006;12:3617–29.
69. Song QY, Xu MM, Zhang BL, He MX, Guo X, Nie J, Xing Y, Liang XB, Chang YC. Near-infrared-I to III absorption and emission via core engineering of open-shelled organic mixed-valence systems. Adv Healthc Mater. 2023;12:2300484.
70. Jebelli A, Oroojalian F, Fathi F, Mokhtarzadeh A, Guardia M. Recent advances in surface plasmon resonance biosensors for microRNAs detection. Biosens Bioelectron. 2020;169:112599.
71. Ahmad R, Fu J, He NY, Li S. Advanced gold nanomaterials for photothermal therapy of cancer. J Nanosci Nanotechnol. 2016;16:67–80.
72. Huang XH, Jain PK, El-Sayed IH, El-Sayed MA. Plasmonic photothermal therapy (PPTT) using gold nanoparticles. Lasers Med Sci. 2008;23:217–28.
73. Zhang L, Chen P, Loiseau A, Brouri D, Casale S, Salmain M, Boujday S, Liedberg B. Spatially controlled reduction and growth of silver in hollow gold nanoshell particles. J Phys Chem C. 2019;123:10614–21.
74. Hou GH, Qian JM, Xu WJ, Sun TT, Wang JL, Wang YP, Suo AL. Multifunctional PEG-b-polypeptide-decorated gold nanorod for targeted combined chemo-photothermal therapy of breast cancer. Colloids Surf B Biointerfaces. 2019;181:602–11.
75. Espinosa A, Silva AKA, Sánchez-Iglesias A, Grzelczak M, Péchoux C, Desboeufs K, Liz-Marzán LM, Wilhelm C. Cancer cell internalization of gold nanostars impacts their photothermal efficiency in vitro and in vivo: toward a plasmonic thermal fingerprint in tumoral environment. Adv Healthc Mater. 2016;5:1040–8.
76. Huang SN, Duan SF, Wang J, Bao SJ, Qiu XJ, Li CM, Liu Y, Yan LJ, Zhang ZZ, Hu YR. Folic-acid-mediated functionalized gold nanocages for targeted delivery of Anti-miR-181b in combination of gene therapy and photothermal therapy against hepatocellular carcinoma. Adv Healthc Mater. 2016;26:2532–44.
77. Li H, Wang P, Deng YX, Zeng MY, Tang Y, Zhu W-H, Cheng YS. Combination of active targeting, enzyme-triggered release and fluorescent dye into gold nanoclusters for endomicroscopy-guided photothermal/photodynamic therapy to pancreatic ductal adenocarcinoma. Biomaterials. 2017;139:30–8.
78. Liu J, Zhai FF, Zhou H, Yang WR, Zhang SS. Nanogold flower-inspired nanoarchitectonics enables enhanced light-to-heat conversion ability for rapid and targeted chemo-photothermal therapy of a tumor. Adv Healthc Mater. 2019;8:1801300.
79. Feng J, Chen LM, Xia YZ, Xing J, Li ZH, Qian QP, Wang Y, Wu AG, Zeng LY, Zhou YL. Bioconjugation of gold nanobipyramids for SERS detection and targeted photothermal therapy in breast cancer. ACS Biomater Sci Eng. 2017;3:608–18.
80. Xing YF, Kang TL, Luo XJ, Zhu JT, Wu P, Cai CX. Coral-shaped Au nanostructures for selective apoptosis induction during photothermal therapy. J Mater Chem B. 2019;7:6224–31.
81. Zhang YF, Song TJ, Feng T, Wan YL, Blum NT, Liu CB, Zheng CQ, Zhao ZY, Jiang T, Wang JW, Li Q, Lin J, Tang LH, Huang P. Plasmonic modulation of gold nanotheranostics for targeted NIR-II photothermal-augmented immunotherapy. Nano Today. 2020;35:100987.
82. Riley RS, Day ES. Gold nanoparticle-mediated photothermal therapy: applications and opportunities for multimodal cancer treatment. Wiley Interdiscip Rev Nanomed Nanobiotechnol. 2017;9:e1449.
83. Vines JB, Yoon J-H, Ryu N-E, Lim D-J, Park H. Gold nanoparticles for photothermal cancer therapy. Front Chem. 2019;7:167.

84. Liu Y, Crawford BM, Vo-Dinh T. Gold nanoparticles-mediated photothermal therapy and immunotherapy. Immunotherapy. 2018;10:1175–88.
85. Granja A, Pinheiro M, Sousa CT, Reis S. Gold nanostructures as mediators of hyperthermia therapies in breast cancer. Biochem Pharmacol. 2021;190:114639.
86. Turkevich J, Stevenson PC, Hillier J. A study of the nucleation and growth processes in the synthesis of colloidal gold. Discuss Faraday Soc. 1951;11:55–75.
87. Cobley CM, Chen JY, Cho EC, Wang LHV, Xua YN. Gold nanostructures: a class of multifunctional materials for biomedical applications. Chem Soc Rev. 2011;40:44–56.
88. Huang XH, Qian W, El-Sayed IH, El-Sayed MA. The potential use of the enhanced nonlinear properties of gold nanospheres in photothermal cancer therapy. Laser Surg Med. 2007;39:747–53.
89. Sun MM, Peng D, Hao HJ, Hu J, Wang DL, Wang K, Liu J, Guo XM, Wei Y, Gao WP. Thermally triggered in situ assembly of gold nanoparticles for cancer multimodal imaging and photothermal therapy. ACS Appl Mater Interfaces. 2017;9:10453–60.
90. Sershen SR, Westcott SL, Halas NJ, West JL. Temperature-sensitive polymer-nanoshell composites for photothermally modulated drug delivery. J Biomed Mater Res. 2000;51:293–8.
91. Hirsch LR, Stafford RJ, Bankson JA, Sershen SR, Rivera B, Price RE, Hazle JD, Halas NJ, West JL. Nanoshell-mediated near-infrared thermal therapy of tumors under magnetic resonance guidance. Proc Natl Acad Sci. 2003;100:13549–54.
92. Lee S-Y, Shieh M-J. Platinum(II) drug-loaded gold nanoshells for chemo-photothermal therapy in colorectal cancer. ACS Appl Mater Interfaces. 2020;12:4254–64.
93. Lee K-S, El-Sayed MA. Dependence of the enhanced optical scattering efficiency relative to that of absorption for gold metal nanorods on aspect ratio, size, end-cap shape, and medium refractive index. J Phys Chem B. 2005;109:20331–8.
94. Luo X, Wang X, Zhang L, Song L, Sun Z, Zhao Y, Su F, Huang Y. Engineering miniature gold nanorods with tailorable plasmonic wavelength in NIR region via ternary surfactants mediated growth. Nano Res. 2023;16:5087–97.
95. Huang XH, El-Sayed MA. Gold nanoparticles: optical properties and implementations in cancer diagnosis and photothermal therapy. J Adv Res. 2010;1:13–28.
96. Payne EK, Shuford KL, Park SH, Schatz GC, Mirkin CA. Multipole plasmon resonances in gold nanorods. J Phys Chem B. 2006;110:2150–4.
97. Keul HA, Moller M, Bockstaller MR. Structural evolution of gold nanorods during controlled secondary growth. Langmuir. 2007;23:10307–15.
98. Jacinto TA, Rodrigues CF, Moreira AF, Miguel SP, Costa EC, Ferreira P, Correia IJ. Hyaluronic acid and vitamin E polyethylene glycol succinate functionalized gold-core silica shell nanorods for cancer targeted photothermal therapy. Colloids Surf B Biointerfaces. 2020;188:110778.
99. Feng YH, Wang YW, Song XH, Xing SX, Chen HY. Depletion sphere: explaining the number of Ag islands on Au nanoparticles. Chem Sci. 2017;8:430–6.
100. Jia J, Liu GY, Xu WJ, Tian XL, Li SB, Han F, Feng YH, Dong XC, Chen HY. Fine-tuning the homometallic interface of Au-on-Au nanorods and their photothermal therapy in the NIR-II window. Angew Chem Int Edit. 2020;59:14443–8.
101. Khlebtsov B, Panfilova E, Khanadeev V, Khlebtsov N. Improved size-tunable synthesis and SERS properties of Au nanostars. J Nanopart Res. 2014;16:2623.
102. Sun YG, Xia YN. Mechanistic study on the replacement reaction between silver nanostructures and chloroauric acid in aqueous medium. J Am Chem Soc. 2004;126:3892–901.
103. Chen JY, McLellan JM, Siekkinen A, Xiong YJ, Li ZY, Xia YN. Facile synthesis of gold-silver nanocages with controllable pores on the surface. J Am Chem Soc. 2006;128:14776–7.
104. Chen JY, Glaus C, Laforest R, Zhang Q, Yang MX, Gidding M, Welch MJ, Xia YN. Gold nanocages as photothermal transducers for cancer treatment. Small. 2010;6:811–7.
105. Jenkins SV, Nedosekin DA, Miller EK, Zharov VP, Dings RPM, Chen JY, Griffin RJ. Galectin-1-based tumour-targeting for gold nanostructure-mediated photothermal therapy. Int J Hyperth. 2018;34:19–29.

106. Feng YL, Cheng Y, Chang Y, Jian H, Zheng RX, Wu XQ, Xu KQ, Wang L, Ma XM, Li X, Zhang HY. Time-staggered delivery of erlotinib and doxorubicin by gold nanocages with two smart polymers for reprogrammable release and synergistic with photothermal therapy. Biomaterials. 2019;217:119327.
107. Boca SC, Potara M, Gabudean A-M, Juhem A, Baldeck PL, Astilean S. Chitosan-coated triangular silver nanoparticles as a novel class of biocompatible, highly effective photothermal transducers for in vitro cancer cell therapy. Cancer Lett. 2011;311:131–40.
108. Díeza I, Ras RHA. Fluorescent silver nanoclusters. Nanoscale. 2011;3:1963–70.
109. Manivannan K, Cheng C-C, Anbazhagan R, Tsai H-C, Chen J-K. Fabrication of silver seeds and nanoparticle on core-shell Ag@SiO_2 nanohybrids for combined photothermal therapy and bioimaging. J Colloid Interface Sci. 2019;537:604–14.
110. Bose P, Priyam A, Kar R, Pattanayak SP. Quercetin loaded folate targeted plasmonic silver nanoparticles for light activated chemo-photothermal therapy of DMBA induced breast cancer in Sprague Dawley rats. RSC Adv. 2020;10:31961–78.
111. Sears J, Swanner J, Fahrenholtz CD, Snyder C, Rohde M, Levi-Polyachenko N, Singh R. Combined photothermal and ionizing radiation sensitization of triple-negative breast cancer using triangular silver nanoparticles. Int J Nanomedicine. 2021;16:851–65.
112. Espinosa A, Curcio A, Cabana S, Radtke G, Bugnet M, Kolosnijaj-Tabi J, Péchoux C, Alvarez-Lorenzo C, Botton GA, Silva AKA, Abou-Hassan A, Wilhelm C. Intracellular biodegradation of Ag nanoparticles, storage in ferritin, and protection by a Au shell for enhanced photothermal therapy. ACS Nano. 2018;12:6523–35.
113. Wang WC, Liu XF, Zheng XJ, Jin HJ, Li XM. Biomineralization: an opportunity and challenge of nanoparticle drug delivery systems for cancer therapy. Adv Healthc Mater. 2020;9:2001117.
114. Bian K, Zhang XW, Liu K, Yin T, Liu H, Niu K, Cao WW, Gao DW. Peptide-directed hierarchical mineralized silver nanocages for anti-tumor photothermal therapy. ACS Sustain Chem Eng. 2018;6:7574–88.
115. Gawande MB, Goswami A, Felpin F-X, Asefa T, Huang XX, Silva R, Zou XX, Zboril R, Varma RS. Cu and Cu-based nanoparticles: synthesis and applications in catalysis. Chem Rev. 2016;116:3722–811.
116. Liu XJ, Li B, Fu FF, Xu KB, Zou RJ, Wang Q, Zhang BJ, Chen ZJ, Hu JQ. Facile synthesis of biocompatible cysteine-coated CuS nanoparticles with high photothermal conversion efficiency for cancer therapy. Dalton Trans. 2014;43:11709–15.
117. Zhang LW, Gao S, Zhang F, Yang K, Ma QJ, Zhu L. Activatable hyaluronic acid nanoparticle as a theranostic agent for optical/photoacoustic image-guided photothermal therapy. ACS Nano. 2014;8:12250–8.
118. Li N, Sun QQ, Yu ZZ, Gao XN, Pan W, Wan XY, Tang B. Nuclear-targeted photothermal therapy prevents cancer recurrence with near infrared triggered copper sulfide nanoparticles. ACS Nano. 2018;12:5197–206.
119. Rayman MP. Selenium in cancer prevention: a review of the evidence and mechanism of action. Proc Nutr Soc. 2005;64:527–42.
120. Liu XJ, Wang Q, Li C, Zou RJ, Li B, Song GS, Xu KB, Zheng Y, Hu JQ. $Cu_{2-X}Se$@$mSiO_2$-PEG core-shell nanoparticles: a low-toxic and efficient difunctional nanoplatform for chemo-photothermal therapy under near infrared light radiation with a safe power density. Nanoscale. 2014;8:4361–70.
121. Zhang S, Sun C, Zeng J, Sun Q, Wang G, Wang Y, Wu Y, Dou S, Gao M, Li Z. Ambient aqueous synthesis of ultrasmall PEGylated $Cu_{2-X}Se$ nanoparticles as a multifunctional theranostic agent for multimodal imaging guided photothermal therapy of cancer. Adv Mater. 2016;28:8927–36.
122. Zou HQ, Tao JX, Shi H, He J, Zhou ZY, Zhang C. Platelet-mimicking nanoparticles co-loaded with W18O49 and metformin alleviate tumor hypoxia for enhanced photodynamic therapy and photothermal therapy. Acta Biomater. 2018;80:296–307.

123. Liu P, Wang YR, An L, Tian QW, Lin JM, Yang SP. Ultrasmall WO_{3-X}@γ-poly-l-glutamic acid nanoparticles as a photoacoustic imaging and effective photothermal-enhanced chemodynamic therapy agent for cancer. ACS Appl Mater Interfaces. 2018;10:38833–44.
124. Li B, Zhang YX, Zou RJ, Wang Q, Zhang BJ, An L, Yin F, Hua YQ, Hu JQ. Self-assembled WO_{3-X} hierarchical nanostructures for photothermal therapy with a 915 nm laser rather than the common 980 nm laser. Dalton Trans. 2014;43:6244–50.
125. Sharker SM, Kim SM, Lee JE, Choi KH, Shin G, Lee S, Lee KD, Jeong JH, Lee H, Park SY. Functionalized biocompatible WO_3 nanoparticles for triggered and targeted in vitro and in vivo photothermal therapy. J Control Release. 2015;217:211–20.
126. Tian YL, Yi WH, Bai L, Zhang P, Si JH, Hou X, Deng YT, Hou J. Lentinan in-situ coated tungsten oxide nanorods as a nanotherapeutic agent for low power density photothermal cancer therapy. Int J Biol Macromol. 2019;137:904–11.
127. Cheng YR, Yang F, Xiang GL, Zhang K, Cao Y, Wang DD, Dong HF, Zhang XJ. Ultrathin tellurium oxide/ammonium tungsten bronze nanoribbon for multimodality imaging and second near-infrared region photothermal therapy. Nano Lett. 2019;19:1179–89.
128. Ding XG, Liow CH, Zhang MX, Huang RJ, Li CY, Shen H, Liu MY, Zou Y, Gao N, Zhang ZJ, Li YG, Wang QB, Li SZ, Jiang J. Surface plasmon resonance enhanced light absorption and photothermal therapy in the second near-infrared window. J Am Chem Soc. 2014;136:15684–93.
129. Odda AH, Xu YC, Lin J, Wang G, Ullah N, Zeb A, Liang K, Wen LP, Xu A-W. Plasmonic MoO_{3-X} nanoparticles incorporated in Prussian blue frameworks exhibit highly efficient dual photothermal/photodynamic therapy. J Mater Chem B. 2019;7:2032–42.
130. Castro IA, Datta RS, Ou JZ, Castellanos-Gomez A, Sriram S, Daeneke T, Kalantar-zadeh K. Molybdenum oxides-from fundamentals to functionality. Adv Mater. 2017;29:1701619.
131. Zhan Y, Liu YL, Zu HR, Guo YX, Wu SS, Yang HY, Liu ZM, Lei BF, Zhuang JL, Zhang XJ, Huang D, Hu CF. Phase-controlled synthesis of molybdenum oxide nanoparticles for surface enhanced Raman scattering and photothermal therapy. Nanoscale. 2018;10:5997–6004.
132. Li YW, Miao ZW, Shang ZW, Cai YC, Cheng JJ, Xu XQ. A visible- and NIR-light responsive photothermal therapy agent by chirality-dependent MoO_{3-X} nanoparticles. Adv Funct Mater. 2020;30:1906311.
133. Song GS, Hao JL, Liang C, Liu T, Gao M, Cheng L, Hu JQ, Liu Z. Degradable molybdenum oxide nanosheets with rapid clearance and efficient tumor homing capabilities as a therapeutic nanoplatform. Angew Chem Int Edit. 2016;55:2122–6.
134. Murugan C, Murugan N, Sundramoorthy AK, Sundaramyrthy A. Nanoceria decorated flower-like molybdenum sulphide nanoflakes: an efficient nanozyme for tumour selective ROS generation and photo thermal therapy. Chem Commun. 2019;55:8017–20.
135. Liu J, Zheng XP, Yan L, Zhou LJ, Tian G, Yin WY, Wang LM, Liu Y, Hu ZB, Gu ZJ, Chen CY, Zhao YL. Bismuth sulfide nanorods as a precision nanomedicine for in vivo multimodal imaging-guided photothermal therapy of tumor. ACS Nano. 2015;9:696–707.
136. Yu N, Wang ZJ, Zhang JL, Liu ZX, Zhu B, Yu J, Zhu MF, Peng C, Chen ZG. Thiol-capped Bi nanoparticles as stable and all-in-one type theranostic nanoagents for tumor imaging and thermoradiotherapy. Biomaterials. 2018;161:279–91.
137. Chen T, Cen D, Ren ZH, Wang YF, Cai XJ, Huang J, Silvio LD, Li X, Han GR. Bismuth embedded silica nanoparticles loaded with autophagy suppressant to promote photothermal therapy. Biomaterials. 2019;211:119419.
138. Zhang X-D, Chen J, Min YH, Park GB, Shen X, Song S-S, Sun Y-M, Wang H, Long W, Xie JP, Gao K, Zhang LF, Fan SJ, Fan FY, Jeonget U. Metabolizable Bi_2Se_3 nanoplates: biodistribution, toxicity, and uses for cancer radiation therapy and imaging. Adv Funct Mater. 2014;24:1718–29.
139. Yang CY, Guo CS, Guo W, Zhao XL, Liu SP, Han XJ. Multifunctional bismuth nanoparticles as theranostic agent for PA/CT imaging and NIR laser-driven photothermal therapy. ACS Appl Nano Mater. 2018;1:820–30.

140. Wang CP, Cai XP, Zhang JS, Wang XY, Wang Y, Ge HYF, Yan WJ, Huang Q, Xiao JR, Zhang Q, Cheng YY. Trifolium-like platinum nanoparticle-mediated photothermal therapy inhibits tumor growth and osteolysis in a bone metastasis model. Small. 2015;11:2080–6.
141. Ou G, Li ZW, Li DK, Cheng LC, Liu ZL, Wu H. Photothermal therapy by using titanium oxide nanoparticles. Nano Res. 2016;9:1236–43.
142. Dongare PD, Alabastri A, Pedersen S, Zodrow KR, Hogan NJ, Neumann O, Wu JJ, Wang TX, Deshmukh A, Elimelech M, Li QL, Nordlander P, Halas NJ. Nanophotonics-enabled solar membrane distillation for off-grid water purification. Proc Natl Acad Sci USA. 2017;114:6936–41.
143. Fedel M. Hemocompatibility of carbon nanostructures. C J Carbon Res. 2020;6:12.
144. Pai C-L, Chen Y-C, Hsu C-Y, Su H-L, Lai P-S. Carbon nanotube-mediated photothermal disruption of endosomes/lysosomes reverses doxorubicin resistance in MCF-7/ADR cells. J Biomed Nanotechnol. 2016;12:619–29.
145. Ahmadi H, Ramezani M, Yazdian-Robati R, Behnam B, Azarkhiavi KR, Nia AH, Mokhtarzadeh A, Riahi MM, Razavi BM, Abnous K. Acute toxicity of functionalized single wall carbon nanotubes: a biochemical, histopathologic and proteomics approach. Chem Biol Interact. 2017;275:196–209.
146. Lu G-H, Shang W-T, Deng H, Han Z-Y, Hu M, Liang X-Y, Fang C-H, Zhu X-H, Fan Y-F, Tian J. Targeting carbon nanotubes based on IGF-1R for photothermal therapy of orthotopic pancreatic cancer guided by optical imaging. Biomaterials. 2019;195:13–22.
147. Singh R, Torti SV. Carbon nanotubes in hyperthermia therapy. Adv Drug Deliv Rev. 2013;65:2045–60.
148. Wang S, Zhang Q, Yang P, Yu XR, Huang L-Y, Shen S, Cai SJ. Manganese oxide-coated carbon nanotubes as dual-modality lymph mapping agents for photothermal therapy of tumor metastasis. ACS Appl Mater Interfaces. 2016;8:3736–43.
149. Wen YM, Jia QY, Nan FC, Zheng XL, Liu WM, Wu JS, Ren HH, Ge JC, Wang PF. Pheophytin derived near-infrared-light responsive carbon dot assembly as a new phototheranotic agent for bioimaging and photodynamic therapy. Chemistry. 2019;14:2162–8.
150. Shen YZ, Shuhendler AJ, Ye DJ, Xu J-J, Chen H-Y. Two-photon excitation nanoparticles for photodynamic therapy. Chem Soc Rev. 2016;45:6725–41.
151. Xu GY, Bao X, Chen JQ, Zhang BH, Li D, Zhou D, Wang XY, Liu CB, Wang YL, Qu SN. In vivo tumor photoacoustic imaging and photothermal therapy based on supra-(carbon nanodots). Adv Healthc Mater. 2019;8:1800995.
152. Qian M, Du YL, Wang SS, Li CY, Jiang HL, Shi W, Chen J, Wang Y, Wagner E, Huang RQ. Highly crystalline multicolor carbon nanodots for dual-modal imaging-guided photothermal therapy of glioma. ACS Appl Mater Interfaces. 2018;10:4031–40.
153. Hasanzadeh M, Shadjou N, Mokhtarzadeh A, Ramezani M. Two dimension (2-D) graphene-based nanomaterials as signal amplification elements in electrochemical microfluidic immune-devices: recent advances. Mater Sci Eng C. 2016;68:482–93.
154. Zhang JL, Yang HJ, Shen GX, Cheng P, Zhang JY, Guo SW. Reduction of graphene oxide vial-ascorbic acid. Chem Commun. 2010;46:1112–4.
155. Robinson JT, Tabakman SM, Liang YY, Wang HL, Casalongue HS, Vinh D, Dai HJ. Ultrasmall reduced graphene oxide with high near-infrared absorbance for photothermal therapy. J Am Chem Soc. 2011;133:6825–31.
156. Lima-Sousa R, Melo-Diogo D, Alves CG, Costa EC, Ferreira P, Louro RO, Correia IJ. Hyaluronic acid functionalized green reduced graphene oxide for targeted cancer photothermal therapy. Carbohydr Polym. 2018;200:93–9.
157. Roy S, Sarkar A, Jaiswal A. Poly (allylamine hydrochloride)-functionalized reduced graphene oxide for synergistic chemo-photothermal therapy. Nanomedicine. 2019;14:255–74.
158. Song WL, Zhang Y, Yu D-G, Tran CH, Wang M, Varyambath A, Kim J, Kim I. Efficient synthesis of folate-conjugated hollow polymeric capsules for accurate drug delivery to cancer cells. Biomacromolecules. 2020;22:732–42.

159. Yuan ZW, Gui LJ, Zheng JR, Chen YS, Qu SS, Shen YZ, Wang FW, Er M, Gu YQ, Chen HY. GSH-activated light-up near-infrared fluorescent probe with high affinity to αvβ3 integrin for precise early tumor identification. ACS Appl Mater Interfaces. 2018;10:30994–1007.
160. Song XJ, Wu H, Li S, Wang YF, Ma XJ, Tan MQ. Ultrasmall chitosan-genipin nanocarriers fabricated from reverse microemulsion process for tumor photothermal therapy in mice. Biomacromolecules. 2015;16:2080–90.
161. An FF, Yang Z, Zheng MC, Mei T, Deng GW, Guo P, Li YN, Sheng RL. Rationally assembled albumin/indocyanine green nanocomplex for enhanced tumor imaging to guide photothermal therapy. J Nanobiotechnol. 2020;18:49.
162. Lin WH, Li Y, Zhang W, Liu S, Xie ZG, Jing XB. Near-infrared polymeric nanoparticles with high content of cyanine for bimodal imaging and photothermal therapy. ACS Appl Mater Interfaces. 2016;8:24426–32.
163. Feng E, Jiao L, Tang SL, Chen MM, Lv SB, Liu DP, Song JT, Zheng DY, Peng XJ, Song FL. Anti-photobleaching cyanine-based nanoparticles with simultaneous PET and ACQ effects for improved tumor photothermal therapy. Chem Eng J. 2022;432:134355.
164. Zhang JX, Liu S, Hu XL, Xie ZG, Jing XB. Cyanine-curcumin assembling nanoparticles for near-infrared imaging and photothermal therapy. ACS Biomater Sci Eng. 2016;2:1942–50.
165. Wang YX, Feng LH, Wang S. Conjugated polymer nanoparticles for imaging, cell activity regulation, and therapy. Adv Funct Mater. 2019;29:1806818.
166. Yang J, Choi J, Bang D, Kim E, Lim E-K, Park H, Suh J-S, Lee K, Yoo K-H, Kim E-K, Huh Y-M, Haam S. Convertible organic nanoparticles for near-infrared photothermal ablation of cancer cells. Angew Chem Int Edit. 2011;50:441–4.
167. Zhou J, Lu ZG, Zhu XJ, Wang XJ, Liao Y, Ma ZF, Li FY. NIR photothermal therapy using polyaniline nanoparticles. Biomaterials. 2013;34:9584–92.
168. Tian QW, Li YP, Jiang SS, An L, Lin JM, Wu HX, Huang P, Yang SP. Tumor pH-responsive albumin/polyaniline assemblies for amplified photoacoustic imaging and augmented photothermal therapy. Small. 2019;15:1902926.
169. Zhou ZJ, Yan Y, Wang L, Zhang Q, Cheng YY. Melanin-like nanoparticles decorated with an autophagy-inducing peptide for efficient targeted photothermal therapy. Biomaterials. 2019;203:63–72.
170. Feng W, Han XG, Wang RY, Gao X, Hu P, Yue WW, Chen Y, Shi JL. Nanocatalysts-augmented and photothermal-enhanced tumor-specific sequential nanocatalytic therapy in both NIR-I and NIR-II biowindows. Adv Mater. 2019;31:1805919.
171. Zeng WW, Wu XX, Chen T, Sun SJ, Shi ZF, Liu J, Ji XY, Zeng XW, Guan J, Mei L, Wu MY. Renal-clearable ultrasmall polypyrrole nanoparticles with size-regulated property for second near-infrared light-mediated photothermal therapy. Adv Funct Mater. 2021;31:2008362.
172. Cheng L, Yang K, Chen Q, Liu Z. Organic stealth nanoparticles for highly effective in vivo near-infrared photothermal therapy of cancer. ACS Nano. 2012;6:5605–13.
173. Yan H, Zhao LY, Shang WT, Liu ZQ, Xie WS, Qiang C, Xiong ZY, Zhang RR, Li BH, Sun XD, Kang FY. General synthesis of high-performing magneto-conjugated polymer core-shell nanoparticles for multifunctional theranostics. Nano Res. 2017;10:704–17.
174. Yan H, Shang WT, Sun XD, Zhao LY, Wang JY, Xiong ZY, Yuan J, Zhang RR, Huang QL, Wang K, Li BH, Tian J, Kang FY, Feng S-S. "All-in-one" nanoparticles for trimodality imaging-guided intracellular photo-magnetic hyperthermia therapy under intravenous administration. Adv Funct Mater. 2018;28:1705710.
175. Sun TT, Dou J-H, Liu S, Wang X, Zheng XH, Wang YP, Pei J, Xie ZG. Second near-infrared conjugated polymer nanoparticles for photoacoustic imaging and photothermal therapy. ACS Appl Mater Interfaces. 2018;10:7919–26.
176. Wei ZW, Xin FL, Zhang J, Wu M, Qiu T, Lan YT, Qiao SY, Liu XL, Liu JF. Donor-acceptor conjugated polymer-based nanoparticles for highly effective photoacoustic imaging and photothermal therapy in the NIR-II window. Chem Commun. 2020;56:1093–6.
177. Jiang YY, Li JC, Zhen X, Xie C, Pu KY. Dual-peak absorbing semiconducting copolymer nanoparticles for first and second near-infrared window photothermal therapy: a comparative study. Adv Mater. 2018;30:1705980.

178. He CB, Lu JQ, Lin WB. Hybrid nanoparticles for combination therapy of cancer. J Control Release. 2015;219:224–36.
179. Liang XY, Shang WT, Chi CW, Zeng CT, Wang K, Fang CH, Chen QS, Liu HY, Fan YF, Tian J. Dye-conjugated single-walled carbon nanotubes induce photothermal therapy under the guidance of near-infrared imaging. Cancer Lett. 2016;383:243–9.
180. Yang SS, Li ZL, Wang YL, Fan XL, Miao ZH, Hu Y, Li Z, Sun Y, Besenbacher F, Yu M. Multifunctional Bi@PPy-PEG core-shell nanohybrids for dual modal imaging and photothermal therapy. ACS Appl Mater Interfaces. 2018;10:1605–15.
181. He SS, Jiang YY, Li JC, Pu KY. Semiconducting polycomplex nanoparticles for photothermal ferrotherapy of cancer. Angew Chem Int Edit. 2020;59:10633–8.
182. Fan L, Xu XD, Zhu CH, Han J, Gao LZ, Xi JQ, Guo R. Tumor catalytic-photothermal therapy with yolk-shell gold@carbon nanozymes. ACS Appl Mater Interfaces. 2018;10:4502–11.
183. Zeng J, Goldfeld D, Xia YN. Plasmon-assisted optofluidic system for measuring the photothermal conversion efficiencies of gold nanostructures and controlling an electrical switch. Angew Chem Int Edit. 2013;52:4169–73.
184. Hessel CM, Pattani VP, Rasch M, Panthani MG, Koo B, Tunnell JW, Korgel BA. Copper selenide nanocrystals for photothermal therapy. Nano Lett. 2011;11:2560–6.
185. Wang YZ, Yang YJ, Yang L, Lin YH, Tian Y, Ni QQ, Wang SJ, Ju HX, Guo JX, Lu GM. Gold nanostar@polyaniline theranostic agent with high photothermal conversion efficiency for photoacoustic imaging guided anticancer phototherapy at a low dosage. ACS Appl Mater Interfaces. 2022;14:28570–80.
186. Li J-W, Zhou Y, Xu J, Gao F, Si Q-K, Wang J-Y, Zhang F, Wang L-P. Water-soluble and degradable gelatin/polyaniline assemblies with a high photothermal conversion efficiency for pH-switchable precise photothermal therapy. ACS Appl Mater Interfaces. 2022;14:52670–83.

Nanomaterial-Mediated Cancer Electrodynamic Therapy

3

Zhengya Yue and Tiedong Sun

Abstract

Electrodynamic therapy (EDT) is an innovative approach in cancer treatment that leverages the unique properties of nanomaterials to induce the generation of reactive oxygen species (ROS) within the tumor microenvironment (TME). This method, which operates under the influence of an external electric field, has shown promise in overcoming the limitations associated with traditional cancer therapies. The use of platinum nanoparticles (Pt NPs) as catalysts is particularly noteworthy, as they facilitate the production of cytotoxic hydroxyl radicals when subjected to an alternating current (AC) electric field. This mechanism, distinct from that of electrochemotherapy (EChT), enables a more homogeneous and less invasive treatment effect across the tumor mass. The integration of EDT with other therapeutic modalities, such as chemotherapy and immunotherapy, further enhances its potential for synergistic antitumor effects. Despite the advancements, challenges remain in optimizing nanomaterial design and understanding the complex interactions within the tumor microenvironment.

Keywords

Electrodynamic Therapy (EDT) · Nanomaterials · Reactive Oxygen Species (ROS) · Tumor Microenvironment (TME) · Cancer Therapy

---

Z. Yue · T. Sun (✉)
College of Chemistry, Chemical Engineering and Resource Utilization, Northeast Forestry University, Harbin, P. R. China
e-mail: tiedongsun@nefu.edu.cn

M. Chu (ed.), *Modern Nanomaterials and Nanotechnology for Cancer Therapy*,
https://doi.org/10.1007/978-981-95-6849-9_3

1 Introduction

The utilization of an external electric field in the treatment of tumors dates back to the late nineteenth century [1], but it was when Humphrey and Seal reported positive therapeutic effects in their experiments treating a mouse sarcoma model with direct current in 1959 that the field attracted the interest of many researchers [2]. This pioneering work also led to the creation of electrochemical therapy (EChT), the precursor to electrodynamic therapy (EDT). Following that, EChT has been extensively reported in the application of animal tumor models and preliminary clinical studies. The main therapeutic approach involves placing electrodes around the tumor lesion to apply continuous low-voltage direct current (DC), which induces changes in the local microenvironment, leading to the death of tumor cells. Due to its high efficacy and minimally invasive nature, EChT was once considered the best method for treating localized tumors. However, the development and application of EChT were at one point stagnant due to a lack of systematic research on the underlying mechanisms of treatment and the unpredictability of therapeutic outcomes.

With the in-depth research of reactive oxygen species (ROS) in the field of tumor treatment and the development of nanotechnology, strategies aimed at utilizing ROS for cancer therapy have garnered widespread attention. These include photodynamic therapy (PDT), which uses light to activate a photosensitizer; sonodynamic therapy (SDT), which employs ultrasound; and the more recently proposed chemodynamic therapy (CDT) and radiodynamic therapy (RDT), which utilize chemical and radiation energy, respectively [3–5]. The commonality among these methods is their therapeutic mechanism, which induces the generation of ROS and subsequently triggers apoptosis or necrosis in cancer cells by damaging cellular components such as lipids, proteins, and DNA.

The focus on ROS is particularly significant because of their ability to disrupt the cellular redox balance, which is a critical factor in the survival and proliferation of cancer cells. Thus, to enhance these "dynamic therapies (DTs)," researchers often employ nanoparticles with small size, large surface area, and abundant surface defects, as they provide catalytic activity conducive to ROS generation. However, it is important to note that these "DTs" heavily rely on the medium to activate the dynamic reaction and the subsequent production of free radicals. Indeed, the selection of the type and parameters of the external physical fields is vital for the generation of ROS, which ultimately determines the effectiveness of DTs.

However, the integration of DTs with electronic current and nanotechnology for cancer treatment represents an unexplored frontier. Chinese researcher Zhuang Liu and coworkers in 2019 have proposed an innovative cancer treatment concept, EDT, which utilizes a catalytic reaction driven by a square-wave alternating current (AC) electric field applied to platinum nanoparticles (Pt NPs) to generate ROS to induce tumor cell death [6]. It is distinct from both previously reported ECHT (direct placement of electrodes) and other DT strategies such as PDT and RDT. As shown in Fig. 3.1, this chapter aims to summarize the current progress and challenges in the field of EDT, including the basic principle and electric sensitizers.

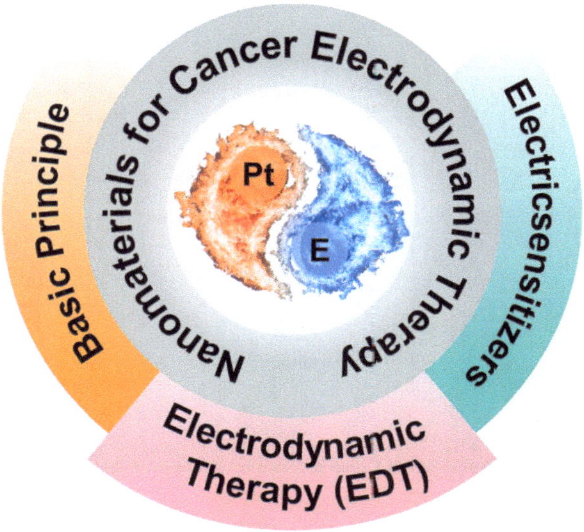

Fig. 3.1 Schematic illustration of the electrodynamic therapy (EDT), including the basic principle and electric sensitizers (ESs)

2 The Basic Principle of EDT

As mentioned above, EDT represents a burgeoning domain in the realm of nanomaterial-based cancer research, stemming from the innovative principles of free radical-driven DT and ECHT. The fundamental mechanism of EDT hinges on the activation of chemical reactions within the TME by nanomaterials, facilitated by an applied electric field, resulting in the demise of cancer cells. Central to this approach is the generation of cytotoxic hydroxyl radicals (•OH) in situ by Pt NPs under the influence of an electric field, thereby achieving efficacious cytotoxicity against tumor cells.

Unlike EChT, which relies on direct electrode insertion and is limited by pH changes near the electrode, EDT applies an alternating current (AC) electric field to activate Pt NPs at the tumor site. This activation induces a Faraday cage effect, promoting water molecule dissociation and the generation of hydroxyl radicals, which are pivotal for the direct and indirect cytotoxic effects on tumor cells (Fig. 3.2a). Conversely, EDT eschews direct electrode insertion, opting instead for the application of an AC electric field over the tumor region to activate the Pt NPs, inciting a sequence of chemical reactions that yield highly reactive hydroxyl radicals. These radicals engage in reactions with intracellular biomolecules such as lipids, proteins, and DNA, culminating in cellular structural and functional devastation and, ultimately, tumor cell demise. Hence, EDT offers superior treatment efficacy and reduced side effects in comparison to EChT.

Furthermore, compared to other traditional DTs, EDT also has significant advantages. Although the mechanism of ROS production in EDT is still under investigation, the traditional view is that the electric field directly activates the dissociation of water molecules on the surface of Pt NPs, generating hydroxyl radicals (Fig. 3.2b).

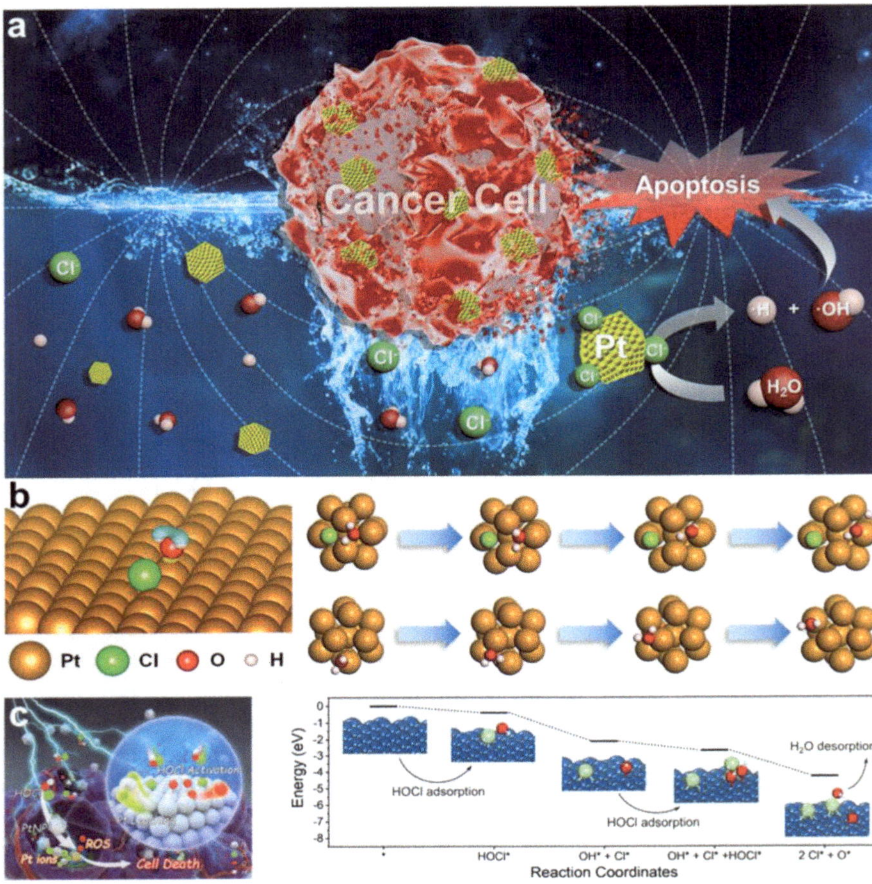

Fig. 3.2 (**a**) The schematic diagram of platinum-mediated electrodynamic therapy. (Copyright © 2019 WILEY-VCH Verlag GmbH & Co. KGaA, Weinheim [6]). (**b**) The charge density diagram of H_2O molecule absorbed on Pt-cluster surface process through molecular dynamics simulation with or without chloride atom. (**c**) Schematic diagram of the heterogeneous catalysis-mediated electrodynamic therapy hypothesis and corresponding reaction path and the relative free energy diagram. (Copyright © 2023 Wiley-VCH GmbH Reprinted with permission from Ref. [7])

However, this theoretical assumption has obvious flaws because, in a stable external electric field, the free charges on the surface of Pt NPs will quickly redistribute and reach electrostatic equilibrium, making it difficult for electrochemical reactions to occur on these equipotential surfaces.

As shown in Fig. 3.2c, recent studies suggest that the reaction mechanism of EDT is more likely a traditional heterogeneous catalytic process. Pt NPs exhibit excellent catalytic activity, capable of activating molecules such as oxygen or peroxides, initiating oxidation processes. Free reactive chlorine (especially hypochlorous acid) produced during electrolysis plays a decisive role in EDT. Experiments and density functional theory (DFT) calculations have confirmed the basic reaction mechanism of Pt NPs catalyzing HOCl to produce oxy radicals [7].

Although the therapeutic mechanisms of EDT are still under development, it is clear that EDT has achieved enhanced efficacy compared to conventional DTs. EDT has made considerable progress in overcoming the intrinsic limitations of the tumor microenvironment (TME), such as hypoxia, acidic pH, and low hydrogen peroxide substrate concentrations. It is important to acknowledge that, according to the current research landscape, interactions between EDT and the TME may still exist and warrant further investigation. Gaining a deeper understanding of these factors is essential for propelling the development of advanced nanomaterials for EDT [6].

3 Electric Sensitizers in Electrodynamic Therapy

All DTs are performed with the involvement of sensitizers. Just like the PS in photodynamic therapy, which, after absorbing photons and transferring the energy to the reactants, forms various ROS that are further used to eliminate tumors or bacteria, EDT is no exception. Therefore, this section provides a brief introduction to electric sensitizers (ESs) in EDT.

Since conventional small molecules as sensitizers for various dynamic therapies usually have the disadvantages of poor hydrophobicity, low bioavailability, rapid excretion, and short duration of action, along with the rapid development of nanomedicine and biomaterials in the treatment of diseases, multifunctional nanomaterials, which have been carefully designed and constructed, are more suitable to be used as sensitizers and mediators for the realization of DTs or enhancement of the corresponding therapeutic effects. How nanomaterials are designed and constructed as ES often depends on the specific physicochemical properties or biological effects of the nanomaterials themselves. Some nanomaterials have intrinsic sensitizing properties in which the electric field can be achieved to induce or enhance the production of ROS.

The design and construction of nanomaterials as ES depend on their specific physicochemical properties or biological effects. Some nanomaterials possess intrinsic sensitizing characteristics that can induce or enhance the production of ROS under electric fields. For instance, Pt NPs, as one of the earliest and most extensively studied ESs, demonstrate the Faraday effect under electric fields and hole doping conditions. Moreover, the dissociation of water molecules facilitated by chloride ions leads to the generation of cytotoxic hydroxyl radicals (•OH).

According to the d-band center theory, the reactivity of Pt NPs is related to their unfilled d-orbitals, which are conducive to forming coordination bonds and thus participate in catalytic reactions that convert water into ROS. Other noble metals such as palladium (Pd) and iridium (Ir) have also been reported to possess similar activity. Traditional DTs are heavily dependent on substrates within the TME, such as hydrogen peroxide and oxygen, which are often severely lacking, greatly limiting their clinical application. Therefore, the unique activity of certain ES has profound significance for the development of DT. Under the influence of an external electric field, these ESs can directly and efficiently produce ROS on demand from water, overcoming and improving the TME and holding the potential for highly effective treatment.

Pt NPs and other ESs are presently at the forefront of research. It is noteworthy that although Pt, Pd, and Ir have all been reported to exhibit similar activities, Pt demonstrates superior activity under the same conditions. As a result, most current research on ES revolves around Pt-based nanomaterials. The goal is to create nanomaterials that can effectively penetrate the TME, overcome its unique challenges, and generate ROS in a controlled and efficient manner to achieve optimal therapeutic outcomes (Fig. 3.3).

Although the excellent ROS generation capacity and antitumor activity and prospects for antitumor applications of Pt NP-based EDT are generally recognized, there are still some limitations. First, even though Pt NPs are considered to be the best performing ES as well as homogeneously dispersed, it is unknown whether variations in its structure will have an impact on its performance and whether there are other factors affecting therapeutic efficacy during the process. Second, the efficacy of single therapy is often very limited due to the complex TME. Tumors are not homogeneous, and different cells may have varying sensitivities to ROS. Strategies to address this heterogeneity, such as combination therapies, may be necessary to improve outcomes. Therefore, similar to other DT, a plethora of work related to EDT and numerous novel ES are continually being reported in order to further improve the production efficiency or effective utilization of ROS. In the following sections, these ES will be introduced in detail.

Considering that ROS may be able to make cancer cells more susceptible to the effects of chemotherapy, combining EDT with chemotherapy is also one of the

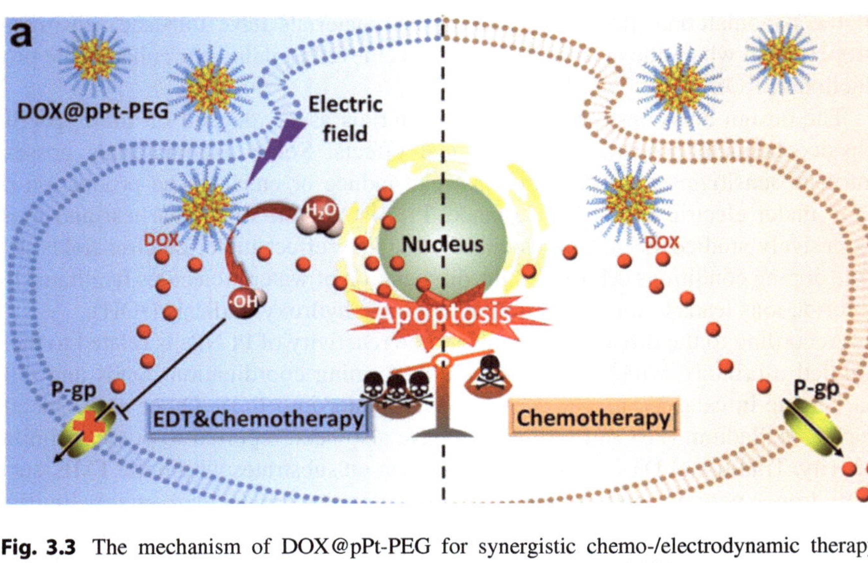

Fig. 3.3 The mechanism of DOX@pPt-PEG for synergistic chemo-/electrodynamic therapy. (Copyright © 2020 Elsevier Ltd. Reprinted with permission from Ref. [8])

current hot topics. Therefore, Zhuang Liu and his colleagues improved Pt NPs and prepared porous Pt NPs, which could be used in loading with the classical chemotherapy drug doxorubicin (DOx) to achieve chemo-/electrodynamic synergistic treatment [8]. The large amount of intracellular ROS during EDT can induce the suppression of P-glycoprotein (P-gp) expression and promote the accumulation of drugs in the cell, thus enhancing the effect of DOx. Furthermore, combining Pt with other nanomaterials may yield serendipitous benefits (Fig. 3.4). Zhuang Liu and his colleagues have also developed mesoporous silica nanoparticles (MSNs) decorated with Pt NPs and doxorubicin (DOx), termed as SDCP NPs, for drug delivery and synergistic chemo-electrodynamic therapy of cancer [10]. Due to the high specific surface area, tunable pore structure, large pore volume, and easy functionalization of MSNs, they can perfectly achieve the loading of EDT and DOx. Additionally,

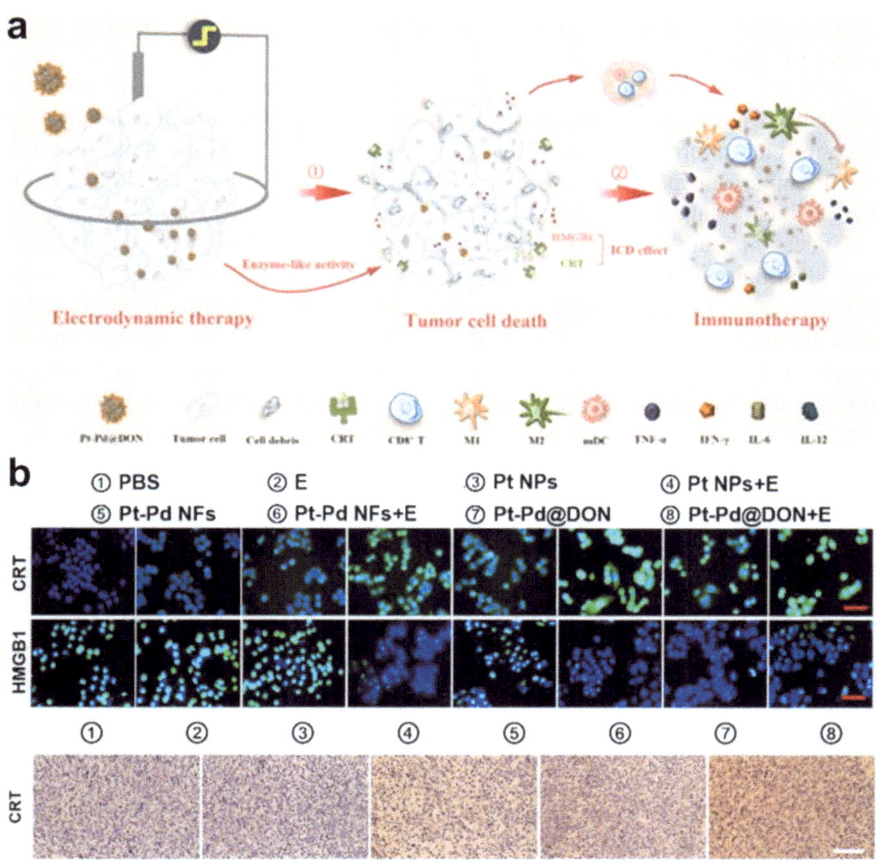

Fig. 3.4 (**a**) The diagrammatic representation of Pt-Pd@DON-mediated electrodynamic immunotherapy. (**b**) The immunohistochemical analysis of CRT exposure and HMGB1 release. (Copyright © 2022, American Chemical Society. Reprinted with permission from Ref. [9])

this pore structure facilitates the accelerated release of DOx in the acidic tumor microenvironment, thereby enhancing biocompatibility and reducing the risk of adverse biological responses.

Furthermore, dynamic therapies that involve external field intervention can rapidly and effectively kill tumors during treatment, but they often tend to recur after treatment is discontinued. Combining EDT with immunotherapy holds the promise of achieving a more sustained tumor suppression effect. The main reason is that EDT can enhance tumor immunogenicity: EDT-induced immunogenic cell death (ICD) effect promotes the release of tumor antigens and activates the immune system, thus enhancing tumor immunogenicity. Meng Yu's group presented an innovative nano-therapeutic platform that combines EDT with a glutamine antagonist (DON), resulting in the development of a novel nanocarrier Pt–Pd@DON for cancer treatment [9]. The platform is not only capable of effectively inducing immunogenic cell death (ICD) and inhibiting tumor growth in both in vitro and in vivo models but also activating a systemic antitumor immune response, thereby providing a new strategy for clinical treatment.

Additionally, to overcome the challenges posed by the low oxidative level and the immunosuppressive microenvironment that often lead to suboptimal outcomes in various antitumor strategies, researchers have designed a phospholipid-modified Pt-Pd nanocrystal platform, co-delivering the immunomodulator IPI549, to form the LPPI nanoplatform [11] (Fig. 3.5).

Under exposure to an electric field, LPPI operates independently of oxygen, instead utilizing water molecules to generate abundant ROS for EDT of tumors. Moreover, the ROS produced further induce the disintegration of the outer phospholipid membrane of LPPI, leading to the release of the immunomodulator and the suppression of myeloid-derived suppressor cells (MDSCs), thereby triggering an immune cascade reaction. The innovative approach of LPPI lies in its ability to generate reactive oxygen species (ROS) independent of oxygen, which is a

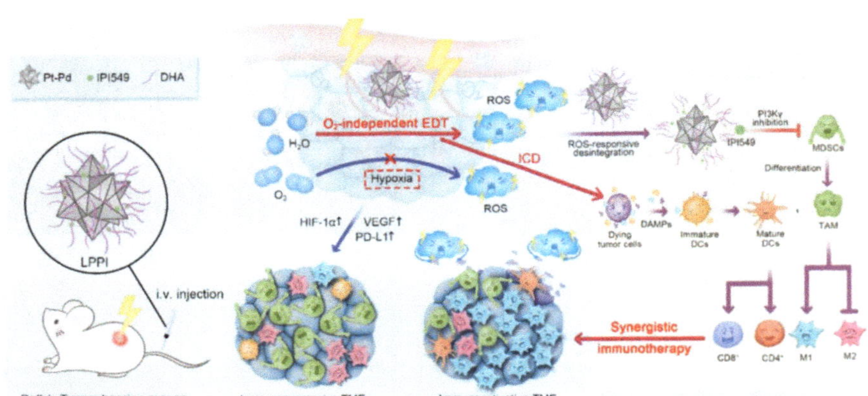

Fig. 3.5 The diagrammatic representation of the LPPI process in the initiation of non-oxygen-dependent electro-immune therapy. (Copyright © 2024 Elsevier B.V. Reprinted with permission from Ref. [11])

significant advantage over traditional therapies. Furthermore, by reshaping the immune microenvironment and inhibiting the function of myeloid-derived suppressor cells (MDSCs), LPPI not only targets the tumor directly but also stimulates a systemic immune response, potentially improving patient compliance due to its single-dose, long-term effectiveness. This research is also driven by the urgent clinical need for new cancer therapies that are both efficient and have low toxicity, and it contributes to the advancement of scientific knowledge in the fields of nanomedicine and immunotherapy (Fig. 3.6).

In addition to chemotherapy and immunotherapy, starvation therapy, which utilizes the distinctive catalytic characteristics of glucose oxidase against β-D-glucose to deprive tumor cells of essential nutrients, has gained prominence due to its inherent biocompatibility and biodegradability. Notably, Pt NPs serve not only as enzyme substitutes but also possess peroxidase-like activity, effectively catalyzing the decomposition of H_2O_2 to produce oxygen. The combination of porous Pt nanospheres (pPt) and glucose oxidase molecules has the potential to improve the hypoxic microenvironment and achieve synergistic enzyme EDT and starvation therapy. Accordingly, GaoRong Han's group synthesized uniformly porous platinum nanospheres (pt) by a one-step method and successfully loaded glucose oxidase molecules onto pt to form pt@GOx (PtG) nanocomposites for oxygen-induced starvation/EDT synergistic therapy [12] (Fig. 3.7).

The PtG NPs employ the enzymatic action of GOx to oxidize glucose, resulting in the generation of hydrogen peroxide (H_2O_2). Concurrently, the pPt catalyzes the decomposition of H_2O_2 into oxygen (O_2), thereby replenishing the oxygen levels

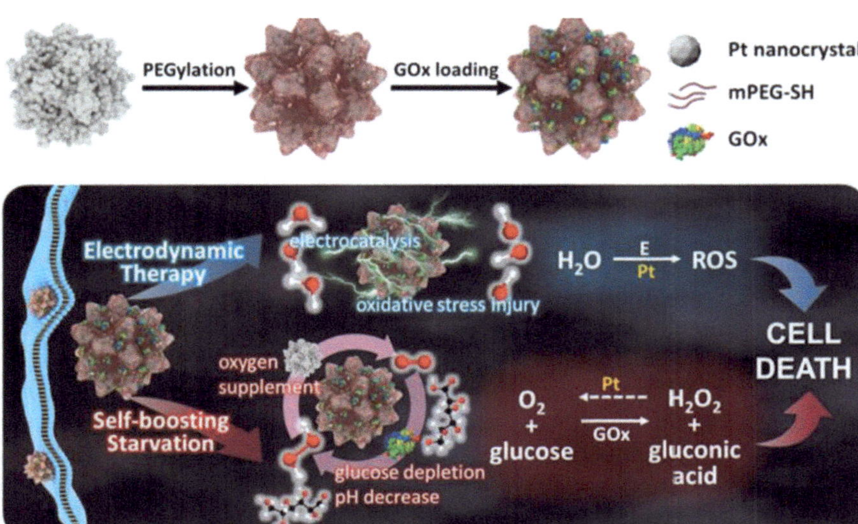

Fig. 3.6 The diagrammatic representation of pPt@GOx-mediated starvation/electrodynamic tumor therapy. (Copyright © 2020 The Authors. Published by WILEY-VCH Verlag GmbH & Co. KGaA, Weinheim. Reprinted with permission from Ref. [12])

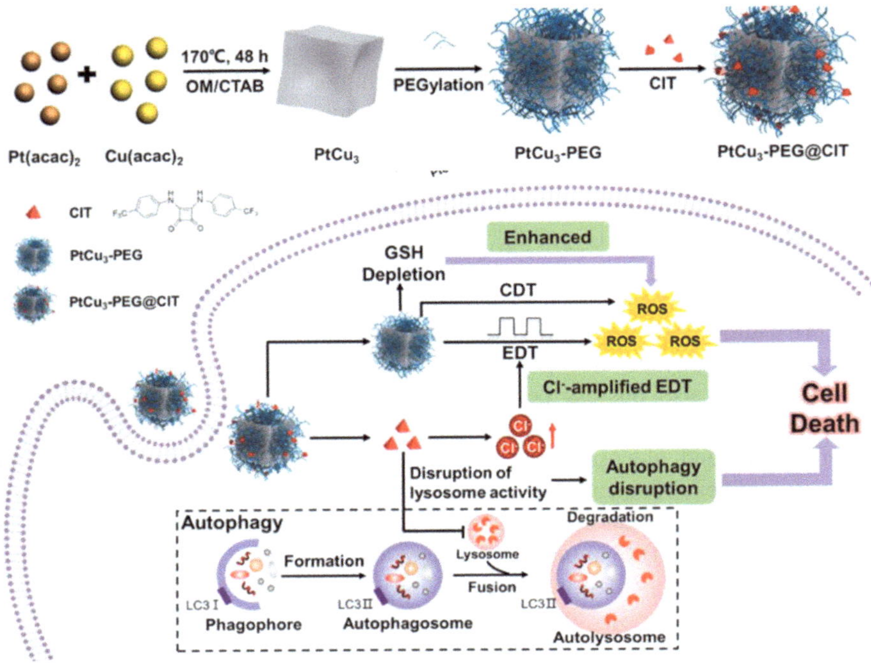

Fig. 3.7 The diagrammatic representation of CIT amplified EDT. (© 2021 The Authors. Publishing services by Elsevier B.V. on behalf of KeAi Communications Co. Ltd. Reprinted with permission from Ref. [13])

within the tumor microenvironment and enhancing the catalytic efficiency of GOx, which in turn induces tumor cell starvation. Meanwhile, under the influence of alternating electric field, pPt induces a large amount of intracellular reactive oxygen species (ROS), leading to oxidative stress and apoptosis in tumor cells.

In addition to developing combination therapies, improving or enhancing the efficiency of ES to produce ROS or increase their utilization is also a major direction of current research. As mentioned earlier, the concentration of Cl^- plays a crucial role in the process of water dissociation by Pt NPs, thereby significantly affecting ROS production. This implies that, for electrodynamic therapy EDT, increasing the intracellular concentration of Cl^- is expected to promote the generation of ROS in EDT and subsequently enhance the tumor suppression rate.

Based on this concept, a potential strategy is to utilize a chloride ion transporter (CIT) to facilitate the transport of extracellular Cl^- into the cell, increasing the intracellular Cl^- concentration to improve the efficiency of EDT. Consequently, the research group led by Xiang Li's group $PtCu_3$-PEG@CIT nanoparticles to achieve CIT-amplified EDT [13]. However, considering the limited availability of Cl^- in biological tissues (approximately 120 mM), in subsequent work, the Xiang Li research group combined the classic ES Pt NPs with potassium chloride nanocrystals (KCl) and amorphous calcium carbonate ($CaCO_3$) to construct a new type of enzyme substitute, KCCP NPs [14]. In this system, KCCP NPs dissolve after entering tumor cells, allowing a large amount of exogenous Cl^- and Ca^{2+} to enter the tumor cells through the salt particles, causing a sharp increase in osmotic pressure and affecting cellular functions as well as the dysfunction of many biomembranes. More importantly, the rise in intracellular Cl^- concentration is very conducive to promoting EDT.

Beyond the development of combination therapies, improving the intrinsic ROS generation or the consumption of reducing agents such as glutathione (GSH) by enhancing the environment is also a primary direction in the development of new ES. As previously mentioned, the concentration of Cl^- is crucial in the process of water dissociation by Pt NPs, playing a significant role in ROS production. This implies that, for EDT, increasing the intracellular Cl^- concentration could promote ROS generation in EDT and subsequently enhance the tumor suppression rate. Based on this concept, a potential strategy is to use a chloride ion transporter (CIT) to facilitate the transport of extracellular Cl^- into the cell, thereby increasing the intracellular Cl^- concentration to improve the efficiency of EDT. Therefore, the research group led by Xiang Li's group constructed $PtCu_3$-PEG@CIT NPs to achieve CIT-amplified EDT. However, considering the limited availability of Cl^- in biological tissues (approximately 120 mM), Xiang Li's group subsequently combined the typical ES, Pt NPs, with potassium chloride nanocrystals (KCl) and amorphous calcium carbonate (Ca_2CO_3) to construct a new type of ES, KCCP NPs. In this system, KCCP NPs dissolve after entering tumor cells, allowing a large amount of exogenous Cl^- and Ca^{2+} to enter the tumor cells through the salt particles, causing a sharp increase in osmotic pressure and affecting cellular functions as well as the dysfunction of many biomembranes. More importantly, the rise in intracellular Cl^- concentration also further promotes EDT (Fig. 3.8).

In excess of transporting Cl^-, enhancing the ability of ES to produce ROS or consume reducing agents by altering their structure or active sites is also viable. Danyang Li and colleagues synthesized monodispersed, raspberry-like PtMnIr nanozymes as ES through a simple one-pot hydrothermal method [15]. The PtMnIr nanozyme not only possesses EDT activity but also exhibits multiple enzyme-mimicking activities, including catalase (CAT), oxidase (OXD), superoxide dismutase (SOD), peroxidase (POD), and glutathione peroxidase (GPx). This multi-enzyme activity allows for the continuous production of reactive oxygen species (ROS), which is crucial for disrupting the oxidative balance of tumor cells and promoting cell death. Furthermore, through GPx-like activity, the PtMnIr nanoenzyme can reduce the GSH levels within tumor cells, thereby enhancing the cytotoxic effects of ROS. The substantial accumulation of reactive oxygen species (ROS) induces apoptosis and non-ferroptosis-dependent iron death (ferroptosis), thereby inhibiting tumor growth. Furthermore, PtMnIr can effectively inhibit tumor metastasis. It can activate the immune system through immunogenic cell death (ICD) and Mn^{2+}-induced cGAS-STING pathway, thereby suppressing tumor metastasis. By intravenous injection of B16-F10 cells to establish a lung metastasis model, after treatment with PtMnIr and electric field stimulation, the lung tissue was

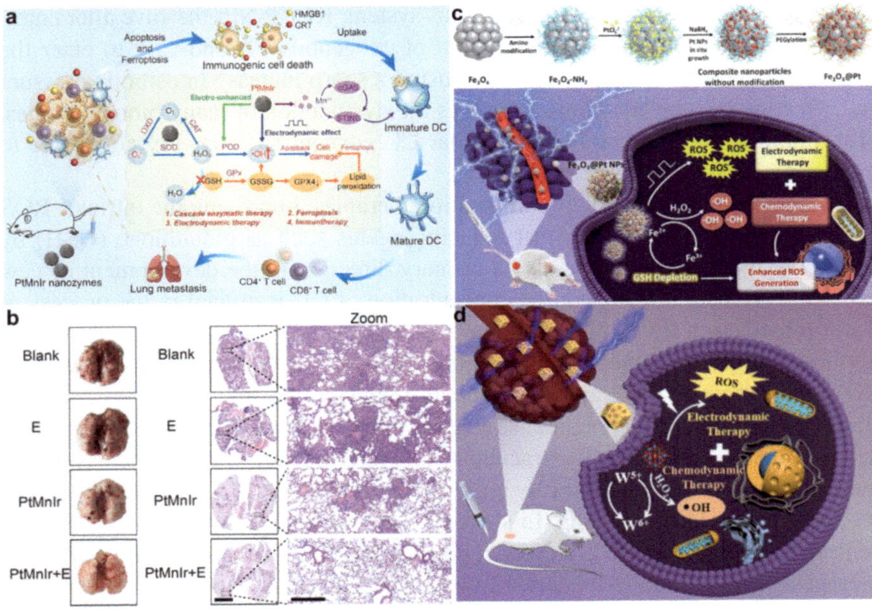

Fig. 3.8 (**a**) The mechanism of PtMnIr nanozymes for electrodynamic therapy. (**b**) Digital photographs of lung metastasis nodules in PtMnIr + E vs. other experimental group sets. (Copyright © 2023 Wiley-VCH GmbH. Reprinted with permission from Ref. [15]). (**c**) The mechanism of Fe_3O_4@Pt NPs for synergistic chemo-/electrodynamic therapy. (Copyright © 2025 BioMed Central Ltd. Reprinted with permission from Ref. [16]). (**d**) POM@ZIF-8 NPs for synergistic chemo-/electrodynamic therapy. (Copyright © 2022, American Chemical Society. Reprinted with permission from Ref. [17])

dissected, and a visible reduction in the number of lung metastatic nodules was observed.

Chen et al. also developed Fe_3O_4@Pt NPs as ES for EDT. The Fe_3O_4@Pt NPs were constructed by decorating iron oxide (Fe_3O_4) nanoparticles with platinum nanocrystals, and under the acidic conditions of tumor cells, the Fe^{3+} ions released by Fe_3O_4@Pt nanoparticles enhance the antitumor effects of ROS by consuming GSH [16]. In contrast, Gui Chen and others designed a Pt-Pd nanocarrier (Pt-Pd@DON) loaded with the glutamine antagonist 6-diazo-5-oxo-L-norleucine (DON), which can reduce the production of GSH by inhibiting glutamine-dependent enzymes, thereby further promoting the antitumor effects of ROS [9].

Despite the vigorous development of electrodynamic therapy (EDT) over the past 5 years, the exploration of non-Pt-based ES has been stagnant. It was not until 2022 that Tiedong Sun et al. ventured into the realm of non-platinum-based ES, marking the first reported use of such ES for tumor treatment research [17]. They utilized polyoxometalate (POM) NPs to modify zeolitic imidazolate framework-8 (ZIF-8), encapsulating spherical POM NPs with a diameter of approximately 8 nm in situ within ZIF-8 through a one-step method, resulting in the creation of POM@ZIF-8 nanoparticles. POM@ZIF-8 nanoparticles are capable of harnessing POM

NPs to produce reactive oxygen species (ROS) through two pathways: the electric field in EDT and the Fenton reaction in chemodynamic therapy (CDT). Furthermore, in their subsequent research, they discovered that POM could also effectively deplete glutathione (GSH), thereby increasing the utilization rate of ROS. This approach demonstrates the potential of non-platinum-based ES to enhance the effectiveness of ROS in tumor treatment strategies.

4 Conclusion

The field of electrodynamic therapy (EDT) has made significant strides in recent years, offering a novel strategy for cancer treatment. EDT's ability to generate ROS through the catalytic activity of nanomaterials, such as PtNPs, under an electric field has been extensively studied. This approach has demonstrated efficacy in inducing tumor cell death while minimizing damage to healthy tissues. The potential for combining EDT with other therapies, such as chemodynamic therapy (CDT), has also been explored, showing enhanced antitumor effects. However, further research is needed to optimize the design of nanomaterials and to better understand their interactions with the tumor microenvironment. The clinical translation of EDT is hindered by challenges such as the need for standardized medical devices and a deeper understanding of the therapeutic mechanisms. Despite these hurdles, the future of EDT holds promise, with the potential to revolutionize cancer treatment through its targeted and minimally invasive approach.

References

1. Nilsson E, von Euler H, Berendson J, Thörne A, Wersäll P, Näslund I, Lagerstedt AS, Narfström K, Olsson JM. Electrochemical treatment of tumours. Bioelectrochemistry. 2000;51(1):1–11.
2. Humphrey CE, Seal EH. Biophysical approach toward tumor regression in mice. Science. 1959;130(3372):388–90.
3. Hu H, Feng W, Qian X, Yu L, Chen Y, Li Y. Emerging nanomedicine-enabled/enhanced nano-dynamic therapies beyond traditional photodynamics. Adv Mater. 2021;33(12):e2005062.
4. Zhang Q, Luo Q, Liu Z, Sun M, Dong X. Nano-ROS-generating approaches to cancer dynamic therapy: lessons from nanoparticles. Chem Eng J. 2023;457:141225.
5. Zhang B, Huang Y, Huang Y. Advances in nanodynamic therapy for cancer treatment. Nanomaterials. 2024;14(7):648.
6. Gu T, Wang Y, Lu Y, Cheng L, Feng L, Zhang H, Li X, Han G, Liu Z. Platinum nanoparticles to enable electrodynamic therapy for effective cancer treatment. Adv Mater. 2019;31(14):e1806803.
7. Lu Z, Zhou L, Yao Y, Li F, Hong Z, Wu Y, Li X. Reaction mechanisms in platinum-mediated electrodynamic therapy. Adv Funct Mater. 2023;33(24):2214749.
8. Chen T, Gu T, Cheng L, Li X, Han G, Liu Z. Porous Pt nanoparticles loaded with doxorubicin to enable synergistic Chemo-/Electrodynamic Therapy. Biomaterials. 2020;255:120202.
9. Chen G, Xu Q, Feng Z, Xu Q, Zhang X, Yang Y, Zhang Y, Liang X-J, Yu Z, Yu M. Glutamine antagonist synergizes with electrodynamic therapy to induce tumor regression and systemic antitumor immunity. ACS Nano. 2022;16(1):951–62.

10. Gu T, Chen T, Cheng L, Li X, Han G, Liu Z. Mesoporous silica decorated with platinum nanoparticles for drug delivery and synergistic electrodynamic-chemotherapy. Nano Res. 2020;13(8):2209–15.
11. Wang M, Chen G, Hu B, Zhang F, Xu Q, Li L, Xi Q, Liu J, Yu Z, Cao P, Wang Y, Yu M. Electrically activated polymetallic nanocrystals for long-term tumor suppression via oxygen-independent ROS generation and electro-immunotherapy. J Control Release. 2024;370:677–90.
12. Lu Z, Gao J, Fang C, Zhou Y, Li X, Han G. Porous Pt nanospheres incorporated with GOx to enable synergistic oxygen-inductive starvation/electrodynamic tumor therapy. Adv Sci. 2020;7(17):2001223.
13. Chen T, Han G, Li X. Platinum-copper alloy nanoparticles armored with chloride ion transporter to promote electro-driven tumor inhibition. Bioact Mater. 2022;12:143–52.
14. Chen T, Fu Y, Zhang R, Han G, Li X. KCl-$CaCO_3$ nanoclusters armoured with Pt nanocrystals for enhanced electro-driven tumor inhibition. Biomater Sci. 2022;10(2):376–80.
15. Li D, Ha E, Zhou Z, Zhang J, Zhu Y, Ai F, Yan L, He S, Li L, Hu J. "Spark" PtMnIr nanozymes for electrodynamic-boosted multienzymatic tumor immunotherapy. Adv Mater. 2024;36(13):e2308747.
16. Chen T, Chu Q, Li M, Han G, Li X. Fe_3O_4@Pt nanoparticles to enable combinational electrodynamic/chemodynamic therapy. J Nanobiotechnology. 2021;19(1):206.
17. Song Y, Sun Y, Tang M, Yue Z, Ni J, Zhao J, Wang W, Sun T, Shi L, Wang L. Polyoxometalate modified by zeolite imidazole framework for the pH-responsive electrodynamic/chemodynamic therapy. ACS Appl Mater Interfaces. 2022;14(4):4914–20.

Nanomaterial-Mediated Cancer Sonodynamic Therapy

4

Manyu Wang, Jie Liu, Junjie Ma, Ziyue Song, and Maoquan Chu

Abstract

Nanomaterial-based sonodynamic therapy (SDT) has gained increasing attention in recent years. These ultrasound-sensitive nanomaterials encompass nanoparticles loaded with organic sonosensitizer molecules, inorganic nano-sonosensitizers, and organic-inorganic hybrid nano-sonosensitizers. Upon ultrasound exposure, sonosensitizers are activated by the sonoluminescence emitted during cavitation microbubble collapse, leading to the production of reactive oxygen species (ROS) and free radicals, analogous to the photodynamic effect. The mechanical forces and high temperatures generated by ultrasound-induced cavitation may be also important contributors to SDT. Moreover, ultrasound exhibits superior tissue penetration compared to external excitation light. This chapter introduces the concept and mechanisms of SDT, highlights different categories of nano-sonosensitizers for tumor SDT, and discusses the advantages, limitation, and present challenges associated with SDT.

Keywords

Nano-sonosensitizers · Ultrasound cavitation · Sonodynamic effect · Cancer therapy

---

M. Wang · J. Liu · J. Ma · Z. Song · M. Chu (✉)
School of Life Sciences and Technology, Tongji University, Shanghai, China
e-mail: mqchu98@tongji.edu.cn

M. Chu (ed.), *Modern Nanomaterials and Nanotechnology for Cancer Therapy*,
https://doi.org/10.1007/978-981-95-6849-9_4

1 The Concept of SDT

Sonodynamic therapy (SDT) was first introduced by Yumita et al. in 1989 [1]. Their experiments demonstrated that ultrasound alone could cause damage to tumor tissue, whereas the combination of ultrasound with injected sonosensitizers markedly enhanced cancer cell-killing efficiency. SDT is therefore recognized as a special tumor treatment modality that is based on the interaction between ultrasound and sonosensitizers.

SDT can be defined as a process in which ultrasound activates sonosensitizers and nearby oxygen molecules to generate reactive oxygen species (ROS), mechanical force, and localized thermal effects, ultimately leading to cancer cell damage. Given that mechanical wave can penetrate deep into tissue (up to~40 cm), SDT overcomes the limitation of poor tissue penetration in photodynamic therapy (PDT), making it suitable for treating deep-seated lesions or tumors. Furthermore, by adjusting the frequency, duration, and intensity of ultrasound, SDT can be precisely targeted to tumor tissues. Compared with PDT, SDT therefore holds greater promise for the treatment of cancers and inflammatory diseases.

2 The Mechanisms of Tumor SDT

The fundamental principle of tumor SDT is to use low-intensity ultrasound to activate the sonosensitizers that have been pre-injected into the tumor region. Upon ultrasound exposure, these sonosensitizers are activated that interact with intracellular oxygen molecules to produce cytotoxic ROS, such as singlet oxygen ($^1O_2$), superoxide anion (O_2^-), and hydroxyl radicals (OH). These ROS attack organelles, DNA, cell membranes, and proteins, thereby inducing apoptosis or necrosis in tumor cells. The mechanisms underlying nanomaterial-based tumor SDT can be categorized into four primary types:

1. ROS-mediated oxidative damage mechanism: SDT activates sonosensitizers to generate ROS, which oxidize specific intracellular chemical clusters. This process initiates the activation of inflammatory factors and induces DNA mutations, ultimately leading to tumor cell apoptosis. The incorporation of nanoparticles (NPs) can enhance ROS generation efficiency, thereby improving the antitumor efficacy of SDT.
2. The cavitation effect of ultrasound: The acoustic energy from ultrasound rapidly increases mechanical pressure, producing microbubbles and causing sonoporation. When ultrasound exceeding a certain threshold acts on liquid microbubbles, they undergo rapid compression and collapse in alternating high-pressure and low-pressure environments. The implosion of these ultrasonic cavitation bubbles produces shear stress, shock waves, and localized high temperatures (e.g., up to 1000 K), which can induce sonochemical reactions (sonochemical effects) and thermal decomposition. Thermal decomposition causes sonosensitizers and water to thermally decompose, producing free radicals and releasing energy that

kills tumor cells [2]. Additionally, the sonoporation effect easily disrupts cell membranes, increases microvascular permeability, and enhances drug transport [3], facilitating the entry of high-molecular-weight substances such as sonosensitizers into cells. Moreover, the thermal effect induced by ultrasound cavitation can directly damage cancer cells and inhibit tumor growth once the local temperature reaches 42.5–53 °C.

3. The inhibition of tumor angiogenesis: SDT can destroy the vascular barrier and suppress the expression of vascular endothelial growth factor (VEGF), thereby reducing the blood supply to tumor tissues. The combination with NPs can significantly inhibit VEGF expression, thereby effectively preventing the formation of new tumor blood vessels and inhibiting tumor growth.
4. Activation of antitumor immunity: The aforementioned ROS, mechanical forces, and thermal effects can collectively activate the body's antitumor immune response. This includes altering the tumor immune microenvironment, inducing immunogenic cell death, and promoting the activation of immune cells.

3 Nanoparticle-Based Sonosensitizers for Tumor SDT

Nanomaterial-based sonosensitizers can be broadly classified into three categories: organic molecular sonosensitizer-loaded NPs, inorganic nano-sonosensitizers, and organic-inorganic hybrid nano-sonosensitizers.

3.1 NPs Loaded with Organic Sonosensitizers for Tumor SDT

1. *Porphyrin Derivative-Loaded NPs*

Porphyrin and its derivatives—including hematoporphyrin (Hp), Photofrin, hematoporphyrin monomethyl ether (HMME), protoporphyrin IX (PpIX), ATX-70, DCPH-P-Na(I), and related analogs [4–8], are among the earliest studied and most widely used organic sonosensitizers. The conjugated macrocycle in their structure confers unique optoelectronic properties. However, porphyrin-based sonosensitizers face challenges such as poor water solubility, low chemical stability, and strong skin photosensitivity, which limit their clinical applications [9]. Researchers have effectively improved their pharmacokinetic properties through encapsulation within hydrophilic nanocarrier (e.g., liposomes, dendrimers) or via structural modifications (e.g., introducing hydrophilic groups) [10, 11].

A single nanoparticle can encapsulate two or more kinds of drugs in addition to organic sonosensitizers. Therefore, the sonosensitizer-loaded NPs usually exhibit multifunctionality in bioimaging and therapy. For example (Fig. 4.1), Luo research group developed a multifunctional nanoplatform named "Deep-Penetrating Nano-Bomb" (DPNB)—PFP@tLyP-1-LIP-H(Gd), synthesized through a one-step film hydration method [12]. First, lipids such as dipalmitoylphosphatidylcholine (DPPC) and distearoylphosphatidylethanolamine-polyethylene glycol-maleimide (DSPE-PEG3400-mal)

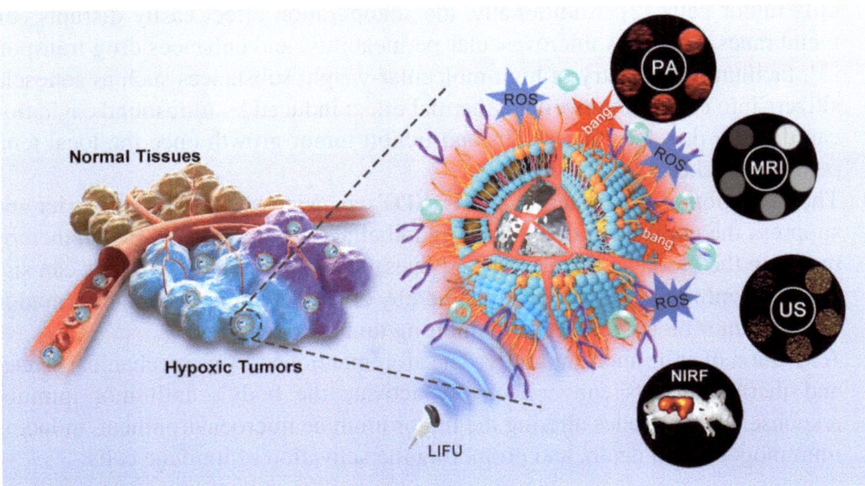

Fig. 4.1 Schematic illustration of drug-loaded organic NPs for tumor SDT and multimodal imaging [12]. (Copyright @ 2022 Luo et al. Reprinted with permission from Ref. [12])

were mixed with the sonosensitizer hematoporphyrin monomethyl ether gadolinium (H(Gd)) to form a lipid film via rotary evaporation. After hydration with phosphate-buffered saline (PBS), perfluoropentane (PFP) was added, and a phase transition was induced by ice-bath pulsed ultrasound, followed by centrifugal purification to obtain NPs. Additionally, the targeting peptide (tLyP-1) was conjugated to the liposome surface via maleimide-thiol coupling to enhance its active targeting and deep-penetrating capability toward tumors with high neuropilin-1 receptor (NRP-1) expression. Under low-intensity focused ultrasound (LIFU) exposure, PFP undergoes a "liquid-gas" phase transition (acoustic droplet vaporization effect), generating cavitation effects and hydroxyl radicals, which synergize with the sonodynamic effect of H(Gd) to efficiently kill tumor cells in hypoxic microenvironments. Meanwhile, this nanoplatform integrates four-modal imaging functions: gadolinium ion-based magnetic resonance imaging (MRI), H(Gd)-based photoacoustic imaging (PA), PFP phase transition-based ultrasound imaging, and near-infrared fluorescence (NIRF) imaging, enabling precise guidance of the treatment process. In vitro and in vivo experiments confirmed that tLyP-1 modification significantly enhances the penetration and accumulation of NPs in deep tumor tissues, achieving >80% tumor suppression under hypoxic conditions without notable systemic toxicity [12].

2. *Phthalocyanine Compound-Loaded NPs*

Phthalocyanine compounds, such as zinc phthalocyanine (ZnPc), copper phthalocyanine, nickel phthalocyanine, iron phthalocyanine, and cobalt phthalocyanine, have emerged as representative second-generation sonosensitizers following porphyrins, owing to their high triplet quantum yield and long excited-state lifetime

[13]. These compounds demonstrate high tumor accumulation efficiency, making them suitable for deep-seated tumor SDT. Their efficiency is often improved by incorporation into hydrophilic nanocarriers.

In 2022, Zhao et al. synthesized three phthalocyanine-artesunate conjugates (ZnPcT1A, ZnPcT2A, and ZnPcT4A) by covalently linking artesunate (ARS, an artemisinin derivative) with mono-, di-, and tetra-triethylene glycol-modified zinc phthalocyanine (ZnPcT1, ZnPcT2, ZnPcT4) via esterification [13]. The synthesis was catalyzed by 1-ethyl-3-(3-dimethylaminopropyl)carbodiimide hydrochloride (EDCl) and 4-dimethylaminopyridine (DMAP) in DMF or dichloromethane, with the products verified by ^1H nuclear magnetic resonance (^1H NMR) and high-resolution mass spectrometry (HRMS).

The study found that these conjugates spontaneously formed nanoscale aggregates in aqueous solution (e.g., ZnPcT2A with an average particle size of ~135 nm). For the first time, the phenomenon of "aggregation-enhanced sonodynamic activity (AESA)" was proposed: aggregated conjugates significantly enhanced ROS generation by promoting ultrasound cavitation effects. For instance, ZnPcT2A in the aggregated state produced 60 times more ROS than in the dispersed state, with its sonodynamic activity far exceeding that of the known sonosensitizer protoporphyrin IX (PpIX).

In vitro and in vivo experiments demonstrated that aggregated ZnPcT2A exhibited potent anticancer activity (98% tumor suppression rate) and excellent biocompatibility (hemolysis rate <5%). When combined with sono-photodynamic therapy (SPDT), it effectively penetrated deep tissues to inhibit tumor growth. Moreover, the AESA effect was also observed in other organic sonosensitizers such as chlorin e6 (Ce6) and PpIX, indicating its universality. This discovery challenges the conventional strategy of avoiding molecular aggregation in sonosensitizer design, providing a new direction for developing high-efficiency sonodynamic therapeutic materials [13].

3. *Xanthene Dyes*

Xanthene-based sonosensitizers, exemplified by typical compounds such as erythrosin B, rose bengal (RB), eosin, and rhodamine, are a class of bioactive molecules derived from the xanthene core—a tricyclic aromatic system comprising two benzene rings connected by an oxygen bridge and a methylene group. Their unique conjugated electronic structure confers remarkable photophysical properties (e.g., strong absorption spectra and fluorescence emission). Furthermore, properties such as solubility and targeting capability can be tuned through chemical modifications, such as halogenation or amination [14].

A typical drug, rose bengal (RB), has been widely employed in experimental tumor and infection models due to its high singlet oxygen quantum yield and low physiological toxicity. It can be activated by ultrasound to induce bactericidal [14] and anticancer [15]. The design of nanomaterials plays a critical role in improving therapeutic efficacy. By combining xanthene compounds with nanomaterials, efficient delivery and activation of sonosensitizers can be achieved. For example,

Nikolitsa Nomikou and colleagues [15] developed a sonodynamic therapeutic material called lipid microbubble-sonosensitizer conjugate (MB-RB). Using carbodiimide coupling chemistry, the researchers covalently attach carboxyl-functionalized RB sonosensitizer onto the surface of amine-modified lipid microbubbles (MBs). Specifically, the MBs were fabricated by incorporating amine-containing DSPE-PEG lipids, while the sonosensitizer RB was modified with carboxyl groups (Fig. 4.2a) via a substitution reaction between bromooctanoic acid and RB sodium salt, resulting in the formation of a stable MB-RB complex through covalent bonding (Fig. 4.2b).

Experimental results demonstrated that compared to free sonosensitizers, this complex exhibited significantly enhanced singlet oxygen generation under ultrasound stimulation (1 MHz, 1.5 W/cm^2), along with stronger cytotoxicity (72% reduction in RIF-1 cell viability) and improved antitumor effects in vivo (18% reduction in murine prostate tumor volume) (Fig. 4.3). The authors noted that covalent conjugation not only enhanced the proximity effect between MBs and sonosensitizers but also promoted efficient ROS generation via ultrasonic cavitation mechanisms—thereby addressing the penetration depth limitations of conventional PDT. This study provides a promising strategy for noninvasive treatment of deep-seated tumors and validates the critical role of covalent coupling in enhancing sonodynamic effects [15].

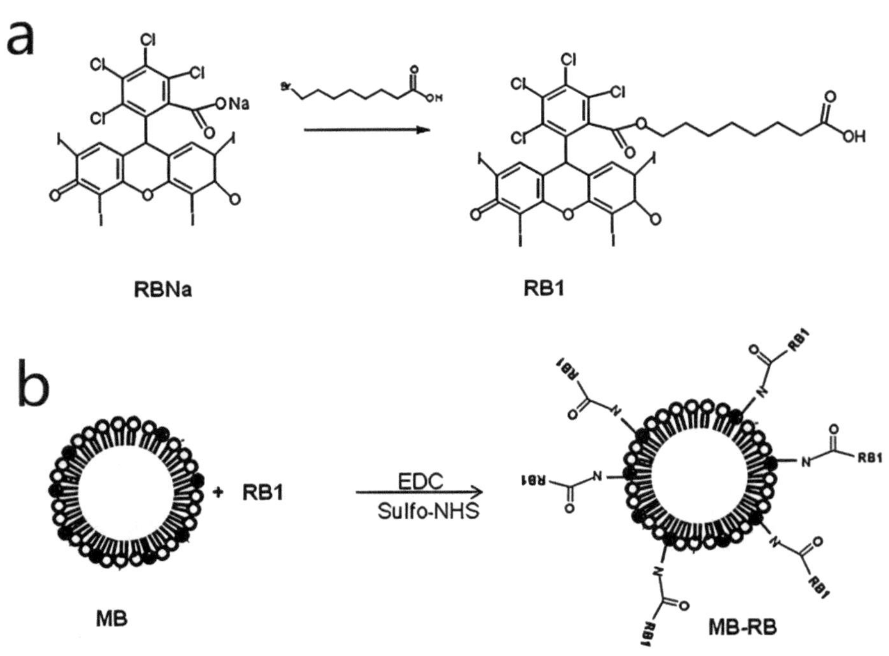

Fig. 4.2 Schematic illustrations of (**a**) the RB modified with carboxyl groups and (**b**) formation of MB-RB nanoconjugate [15]. (Copyright @ The Royal Society of Chemistry 2012. Reprinted with permission from Ref. [15])

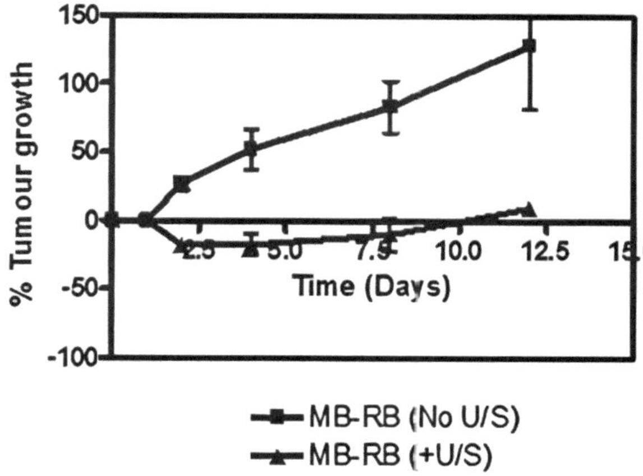

Fig. 4.3 Ultrasound-triggered MB-RB for inhibiting tumor growth and a control [15]. (Copyright @ The Royal Society of Chemistry 2012. Reprinted with permission from Ref. [15])

4. *Nano-sonosensitizers from Natural Products*

Natural products have become an important source for the development of sonosensitizers due to their low toxicity and multi-target effects. Fluorescent natural products containing benzene rings or heterocyclic structures, such as curcumin (CUR), hypericin, chlorophyll, caryophyllin, catechins, can also serve as sonosensitizers.

However, these organic molecules often face challenges such as poor water solubility and susceptibility to photooxidation and photobleaching, which may reduce their therapeutic efficacy. Additionally, the rapid systemic clearance of free forms further restricts their therapeutic window [10]. To address these issues, researchers have designed nanoparticle-based drug delivery systems, including liposomal and polymeric nanoparticle encapsulations, to enhance the stability and bioavailability of natural sonosensitizers [11].

For example, curcumin (CUR), a natural polyphenolic compound extracted from turmeric, exhibits significant antitumor activity with low toxicity toward normal cells. CUR can be directly acted as a sonosensitizer to induce tumor cell apoptosis. However, CUR is a water-insoluble drug. To improve its cytotoxic efficacy, CUR is usually incorporated into NPs. For example, Zhu et al. prepared CUR-loaded poly(lactic-co-glycolic acid) microbubbles (CUR-PLGA-MB) and demonstrated that these nanocomposites exhibited great potential as both sono−/photosensitizers and showed much higher cell-killing efficiency compared to single treatment [17].

As another example, chlorophyll is commonly incorporated into hydrophilic NPs to improve its therapeutic efficiency, as it is a water-soluble drug. In 2020, Bosca et al. prepared three types of chlorophyll-loaded NPs—liposomes (Chlor-Lipo), solid lipid NPs (Chlor-SLN), and PLGA nanospheres (Chlor-PLGA NSs) by using thin-film dispersion-extrusion, microemulsion cold dilution, and nanoprecipitation methods, respectively. Dynamic light scattering characterization revealed average particle sizes of 165 nm, 251 nm, and 140 nm, respectively, with encapsulation efficiencies all exceeding 97%. In vitro release experiments demonstrated that

PLGA NSs exhibited sustained-release properties (18% release over 24 h). Following ultrasound treatment (1.5 MHz, 1.5 W/cm^2 for 5 min), PLGA NSs achieved a 34.8% inhibition rate against PC-3 cells at 48 h, significantly higher than that of free chlorophyll (11.2%) and other carriers (48.4% for liposomes). A 3D tumor spheroid model further verified that the PLGA NSs, in combination with ultrasound, enhanced chlorophyll uptake (Fig. 4.4) and reduced the volume of DU-145 tumor spheroid by 60% compared to untreated controls (Fig. 4.5). This study confirmed that natural chlorophyll delivered via PLGA nanocarriers can significantly enhance sonodynamic therapeutic effects [20].

Hypocrellin, a natural product derived from the traditional Chinese medicinal fungus *Hypocrella bambusae*, is characterized by its high singlet oxygen quantum yield, excellent biocompatibility, and rapid metabolic properties. However, the clinical application of hypocrellin B (HB) is hindered by its poor water solubility and insufficient near-infrared (NIR) absorption, limiting its clinical applications. Zheng et al. reported the application of multifunctional NPs (APHB NPs) based on hypocrellin B derivatives for NIR fluorescence (NIR FL) imaging-guided SDT (Fig. 4.6)

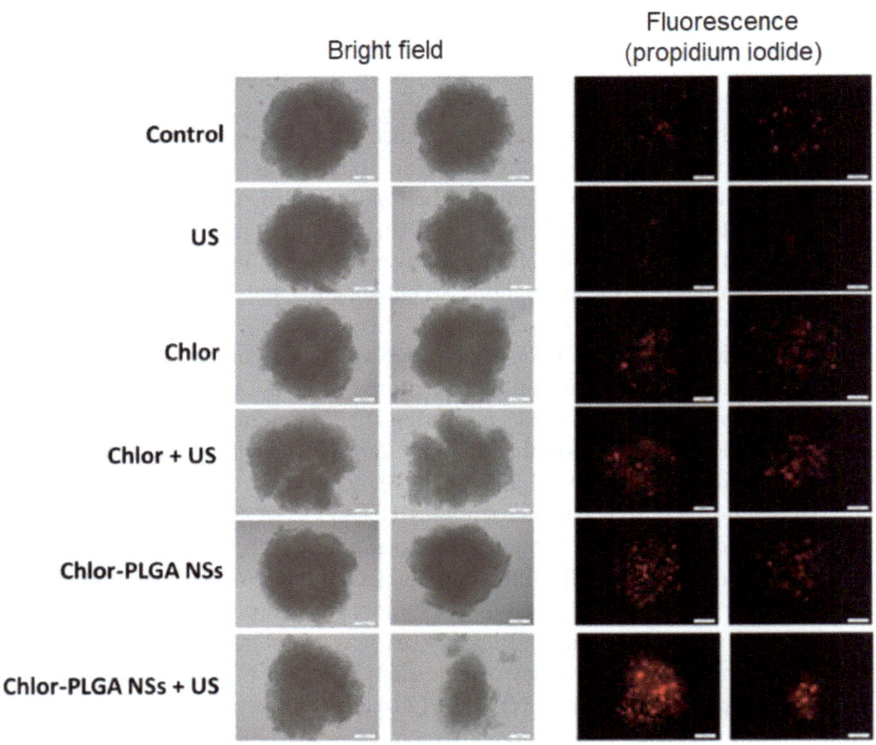

Fig. 4.4 Bright field and fluorescence images of tumor spheroids at 24 h after each treatment (Chlor means chlorophyll) [20]. (Copyright @ 2020 by the authors. Licensee MDPI, Basel, Switzerland. Reprinted with permission from Ref. [20])

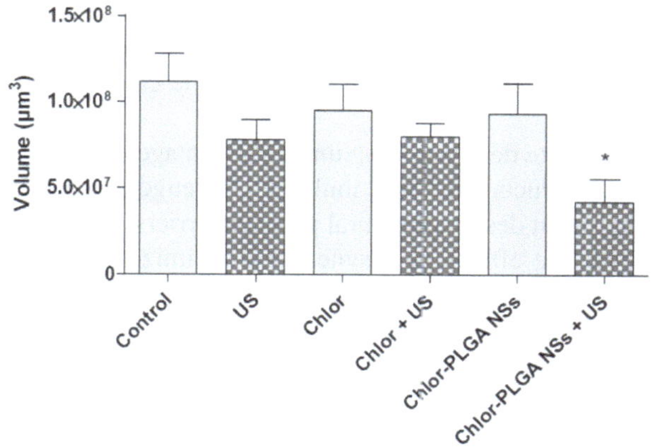

Fig. 4.5 The volumes of tumor spheroids at 24 h after each treatment [20]. (Copyright ©© 2020 by the authors. Licensee MDPI, Basel, Switzerland. Reprinted with permission from Ref. [20])

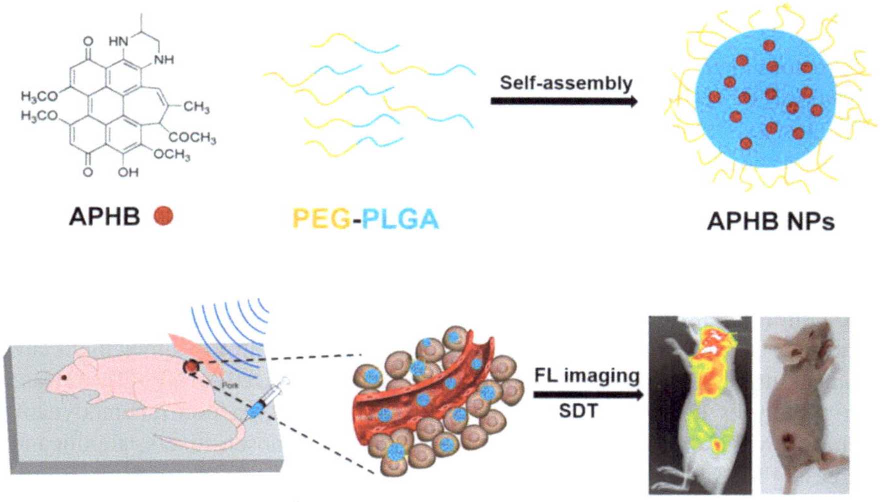

Fig. 4.6 Schematic illustration of the preparation of hypocrellin B derivative-loaded polymer NPs for tumor fluorescence imaging and SDT [21]. (*Copyright © 2019, American Chemical Society.* Reprinted with permission from Ref. [21])

[21]. This approach involved chemical modification of hypocrellin B by introducing 1,2-diaminopropane to form an amphiphilic derivative APHB, followed by self-assembly with PEG-PLGA to form NPs. The resulting APHB NPs demonstrated significantly improved water solubility, NIR fluorescence emission (690 nm), and efficient ROS generation capability under ultrasound stimulation (0.6 W cm^{-2}), along with good biocompatibility, controllable degradability, and tumor-targeting accumulation properties. In vitro experiments showed that APHB NPs were internalized by cancer cells via endocytosis and induced over 90% cell apoptosis upon ultrasound activation, with no significant dark toxicity observed. In vivo studies further confirmed their tumor-targeted accumulation via the enhanced permeability

and retention (EPR) effect. When combined with ultrasound treatment, the NPs significantly inhibited the growth of deeply located tumors (volume reduction through 1-cm-thick tissue barrier) while causing no detectable damage to normal tissues.

Despite demonstrating unique advantages in SDT, the clinical translation of natural products still faces multiple challenges. Future research should prioritize the intelligent design of natural product carriers, spatiotemporal precise control of sonodynamic effects, and synergistic optimization of multimodal therapies to overcome existing limitations and promote the practical application of natural products in SDT [22].

3.2 Inorganic Nano-sonosensitizer for Tumor SDT

1. *Noble Metal Nanomaterials*

Precious metal nanomaterials (e.g., gold, silver, and platinum) exhibit significant advantages in SDT due to their unique physicochemical properties—including surface plasmon resonance effect, catalytic activity, excellent thermal conductivity, surface reactivity, and ease of modification [23, 24].

For example, Lin et al. developed an ultrasound/glutathione dual-responsive Janus gold-manganese oxide nanovesicle (JNP Ve) for synergistic therapy and imaging of deep-seated liver cancer [25]. This nanovesicle is self-assembled from Janus Au-MnO NPs modified with hydrophilic PEG and ROS-sensitive polymers. Upon ultrasound stimulation, it dissociates into smaller NPs, enhancing tumor penetration. The high glutathione (GSH) concentrations within the tumor microenvironment further degrade MnO, releasing Mn^{2+} and small-sized Au NPs. The released Mn^{2+} catalyzes H_2O_2 to produce ·OH via a Fenton-like reaction (chemodynamic therapy), while the Au NPs act as highly efficient sonosensitizers, significantly increasing ROS production (2.2-fold) [25].

In a study by Qu et al. in 2023, an antibacterial nanoparticle based on bovine serum albumin (BSA)—Au@BSA-CuPpIX—was developed to enhance the therapeutic effects of SDT on colorectal cancer (CRC) while reducing phototoxic skin damage (Fig. 4.7) [26]. The synthesis involved biomineralization of Au NPs onto the surface of BSA, followed by EDC/NHS-mediated conjugation of copper protoporphyrin (CuPpIX) to form an ultrasound-responsive nano-sonosensitizer. Ultraviolet-visible spectroscopy and electron paramagnetic resonance (ESR) experiments confirmed that Au@BSA-CuPpIX can efficiently produce ROS under ultrasound excitation, while the incorporated Au NPs significantly suppressed light-induced ROS generation via an energy transfer mechanism, thereby reducing the skin phototoxicity of porphyrin-based sonosensitizers. The results showed that Au@BSA-CuPpIX significantly enhanced the therapeutic efficacy of SDT on orthotopic CRC by eliminating the core pathogenic bacterium *Fusobacterium nucleatum* within tumors, downregulating anti-apoptotic proteins such as Bcl-2, and activating pro-apoptotic proteins such as cleaved caspase 3/9. Animal experiments

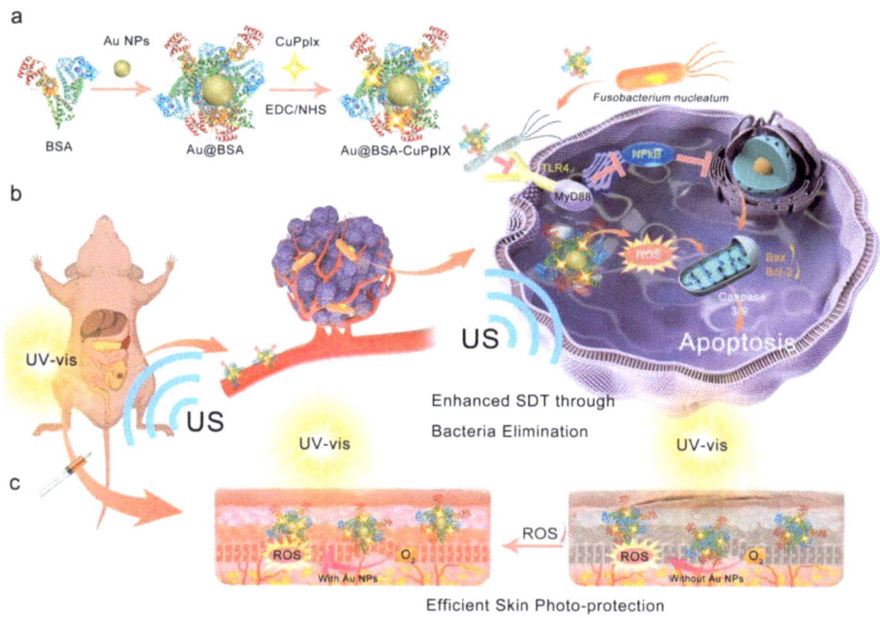

Fig. 4.7 Schematic illustration of the formation of Au@BSA-CuPpIX NPs for cancer SDT [26]. (Copyright © 2023 The Authors. Published by American Chemical Society. Reprinted with permission from Ref. [26])

further demonstrated that these NPs not only inhibited tumor growth (inhibition rate: 94%) but also reduced lung metastasis by regulating the TLR4/MyD88/NF-κB pathway. Additionally, the introduction of Au NPs reduced skin phototoxicity by 76%, mitigating the inflammation and tissue damage typically observed with traditional porphyrin-based photosensitizers in clinical applications.

This study innovatively combined noble metal nanomaterials with antibacterial sonosensitizers, proposing a strategy of "eliminating harmful bacteria synergistically with tumor therapy," thereby providing valuable insights for the development of highly efficient and low-toxicity tumor SDT [26].

2. *Carbon-Based Nanomaterials*

Carbon-based nanomaterials, including carbon quantum dots [27, 28], graphene quantum dots [29, 30], and fullerenes [31], demonstrate significant potential for application in SDT. Benefiting from high specific surface area, tunable electronic structure, and good biocompatibility, these materials can efficiently generate ROS under low-intensity ultrasound activation, rendering them promising for both tumor therapy and antibacterial treatment. However, challenges such as low charge-separation efficiency and poor solubility have limited their performance as sonosensitizers in SDT [32, 33]. Researchers are actively addressing these limitations through techniques like surface modification and the grafting of water-soluble molecules.

In a 2021 study, Yang et al. reported a folic acid (FA)-conjugated nitrogen-doped graphene quantum dot (FA-N-GQDs) for SDT (Fig. 4.10) [35]. The result showed that the pyrrolic nitrogen and pyridinic nitrogen in N-GQDs served as catalytic centers, significantly enhancing the generation efficiency of ROS under ultrasound activation. The quantum yield of singlet oxygen ($^1O_2$) generated by N-GQDs was three to five times higher than that of traditional sonosensitizers (such as porphyrins and metal oxides). The FA-modified FA-N-GQDs exhibited high labeling rates (>96%) across various tumor cell types and induced oxidative stress through activation of the PEX pathway as well as p53 pathway-mediated apoptosis. Both in vitro and in vivo experiments demonstrated tumor suppression efficiency exceeding 90%, with no significant toxicity observed. This study provides an important theoretical basis for the development of metal-free carbon-based sonosensitizers and their application in SDT.

3. *Metal-Containing Semiconductor Quantum Dots*

Quantum dots (QDs) encompass a variety of materials, including metal-containing QDs, graphene QDs, black phosphorus QDs, semiconducting polymer QDs, etc. Metal-containing QDs typically refer to semiconductor nanocrystals composed of elements from group IIB-VIA (e.g., CdS, CdSe, CdTe, ZnS, ZnSe, CdSe/ZnS, etc.) or IIIA-VA (e.g., InP, InAs, etc.) of the periodic table. These materials exhibit excellent fluorescent properties and can function as sonosensitizers for SDT.

For example, in 2020, Li et al. reported the synthesis of a biomimetic nanoplatform based on red blood cell membrane-coated Ag_2S QDs for SDT (Fig. 4.8) [34]. The authors prepared (QD@P)Rs with prolonged blood circulation by modifying Ag_2S QDs with Pluronic F-127 and encapsulating them within red blood cell vesicle membranes. In vitro and in vivo experiments showed that these NPs could catalyze the conversion of H_2O_2 to O_2 within tumor cells, effectively alleviating tumor hypoxia and enhancing the efficacy of SDT. In a murine tumor model, combined with oral administration of the anticancer drug phenethyl isothiocyanate (PEITC), the (QD@P)Rs NPs significantly inhibited tumor growth and prolonged survival time. This study represents a pioneering application Ag_2S QDs as sonosensitizers in SDT, offering a novel strategy for noninvasive tumor therapy.

4. *Ceramic Nanomaterials*

Ceramic nanomaterials are typically crystalline NPs with stable properties [36]. Their resistance to oxidative damage allows them to withstand prolonged ultrasound irradiation, in contrast to organic materials, which are often susceptible to sonobleaching [37]. Owing to their high chemical stability and low phototoxicity, some ceramic nanomaterials—including crystalline titanium dioxide (TiO_2), zinc oxide (ZnO), cuprous oxide (Cu_2O), and manganese oxides (MnO_x)—have gained attention as promising inorganic sonosensitizers [38]. Some of these ceramic nanomaterials also qualify as quantum dots, such as zinc oxide (ZnO) and boron nitride (BN).

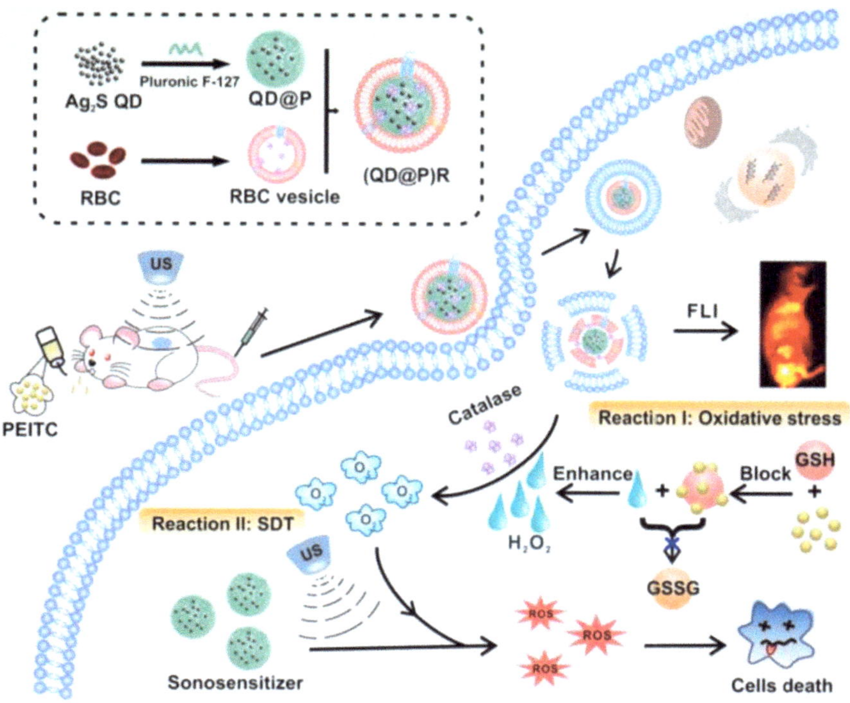

Fig. 4.8 Schematic illustration of the preparation of (QD@P)R NPs for SDT [34]. (Copyright © The author(s). Reprinted with permission from Ref. [34])

Ceramic nanomaterials demonstrate significant therapeutic potential in SDT. For example, TiO_2, as a typical ceramic oxide nanomaterial, exhibits excellent photocatalytic activity and biocompatibility. In 2011, Yamaguchi et al. developed polyethylene glycol (PEG)-coated titanium dioxide (TiO_2/PEG) NPs for SDT [39]. The PEG modification improved the dispersibility and stability of the nanoparticles in aqueous environments, enabling effective adhesion to glioma cell membranes and subsequent cellular uptake. Under ultrasound excitation, the TiO_2/PEG rapidly disrupted cell membrane structures, resulting in the leakage of intracellular contents and subsequent cell death [39].

Although TiO_2 demonstrates preliminary therapeutic efficacy, its wide bandgap (3.2 eV) limits sonoenergy conversion efficiency. Li et al. developed tin sulfide (SnS) NPs with a narrow bandgap (1.18 eV), enabling efficient electron-hole pair separation under ultrasound excitation and generating a substantial amount of ROS (Fig. 4.9) [40]. The surface PEG coating not only enhances biocompatibility but also induces mild photothermal effects (photothermal conversion efficiency: 25.2%), promoting in situ denaturation of the tumor collagen matrix and facilitating deep drug penetration. Experimental results showed that this system significantly enhances antitumor immune responses in triple-negative breast cancer (4 T1) and

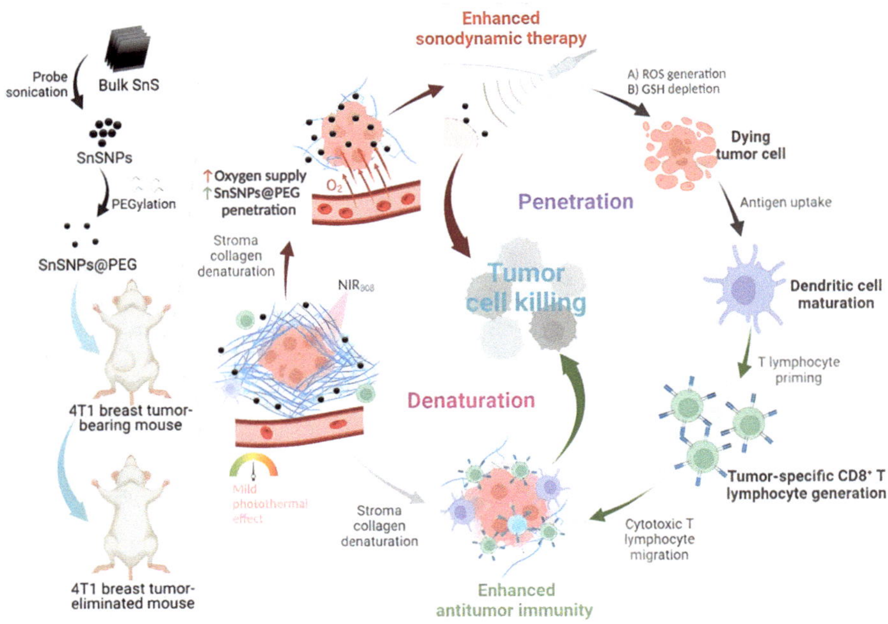

Fig. 4.9 Schematic illustration of preparation of PEG-modified SnS NPs for SDT [40]. (*Copyright © 2023, The Author(s).* Reprinted with permission from Ref. [40])

liver cancer (RIL-175) mouse models, increasing $CD8^+$ T cell infiltration by fivefold and achieving a 100% tumor eradication rate.

In a further research, Cai et al. [41] constructed sulfur vacancy-regulated cobalt sulfide nanocages (Co9S8-x) via defect engineering. By optimizing the sulfur source ratio (TAA/ZIF-67 = 2:1), they reduced the bandgap from 2.06 eV to 1.54 eV while increasing the Co^{2+}/Co^{3+} ratio to 1.97. This system exhibited pH-responsive degradation characteristics in the weakly acidic tumor microenvironment, enabling gradual release of Co^{2+} ions that catalyze a Fenton-like reaction (·OH generation rate constant: 0.096 min^{-1}). Combined with ultrasound-triggered $^1O_2$ generation (rate constant: 0.055 min^{-1}), it achieved synergistic sonodynamic-chemodynamic therapy. After loading the immune adjuvant R837, the system effectively activated dendritic cells ($CD80^+/CD86^+$ ratio increased to 35.9%), significantly inhibiting primary tumor growth and lung metastasis.

Furthermore, Wu et al. [42] discovered that the surface wettability of NPs influences cavitation effects by modulating nanobubble stability. When the contact angle approaches 90°, the desorption energy (ΔG) between mesoporous silica NPs (dMSN-2) and bubbles reaches an extreme value, significantly enhancing cavitation-induced ROS generation (·OH signal intensity increased by 11-fold). This system demonstrated remarkable thrombolytic efficiency showing a 50.3% improvement in a rat thrombosis model.

3.3 Organic-Inorganic Hybrid Nano-sonosensitizers

Organic-inorganic hybrid nano-sonosensitizers represent an emerging type of sonosensitizing material formed by combining organic materials with inorganic nanomaterials through chemical bonding or physical assembly. These hybrids integrate the advantages of organic materials (good biocompatibility and strong designability) with those of inorganic nanomaterials (high stability and catalytic activity). According to present research, organic-inorganic hybrid sonosensitizers mainly include mesoporous structures (e.g., silica-based composite materials which enhance acoustic energy transfer via ordered channels) [43], yolk-shell structures (featuring core-shell separation design to improve cavitation effects) [44], metalloporphyrins (porphyrin ring π-conjugated systems to enhance acoustic responsiveness) [45], and metal-organic frameworks (MOFs) [46].

Nanoscale MOFs are a class of nanomaterials formed by the coordination of metal ions/clusters with organic sonosensitizing ligands. The porous structure of MOFs forms a periodic array that minimizes self-quenching, thereby ensuring rapid diffusion and efficient generation of ROS. In addition, MOFs exhibit a high specific surface area, substantial porosity, and ease of surface functionalization. These advantages enable them to efficiently load various drugs and achieve synergistic enhancement between SDT and other treatment methods, thereby significantly improving therapeutic effects [47].

Pan et al. reported a MOF-derived double-layer hollow manganese silicate nanoparticle (DHMS) using a ZIF-8 template method for multimodal imaging-guided SDT. Its unique double-layer structure not only enhances ultrasound-triggered cavitation effects but also utilizes the redox properties of Mn to promote electron-hole separation, significantly improving the generation efficiency of singlet oxygen ($^1O_2$) and hydroxyl radicals (·OH) ($^1O_2$ levels increased by 339.3%). Additionally, DHMS catalyzes the production of oxygen from H_2O_2 in the tumor microenvironment, effectively alleviating tumor hypoxia. Under ultrasound and magnetic resonance imaging guidance, it achieved a 92.0% tumor suppression rate in in vivo experiments. This study combines the porous structure of MOFs with oxygen generation capability, providing novel strategies for overcoming the hypoxia limitation of SDT [48].

MOF-based SDT is usually combined with other therapies to improve the therapy efficiency. For example, Xu et al. developed a manganese porphyrin-based MOF (Mn-MOF) that alleviates hypoxia by catalyzing H_2O_2-to-O_2 conversion, while Zr ions deplete glutathione (GSH) to reduce GPX4 activity. Under ultrasound irradiation, it generates abundant ROS, synergistically inducing ferroptosis. Experiments demonstrated that Mn-MOF exhibits potent antitumor activity in H22 and 4 T1 tumor models (inhibition rate: 89.2%) and remodels the tumor immune microenvironment by activating CD8+ T cells and mature dendritic cells. This study combines SDT with ferroptosis induction, offering a promising strategy to overcome tumor resistance and metastasis [49].

In another study, Xu et al. [50] constructed Chl-MOF via electrostatic assembly of microalgae *Chlorella vulgaris* with MOF, enabling synergistic PDT and SDT

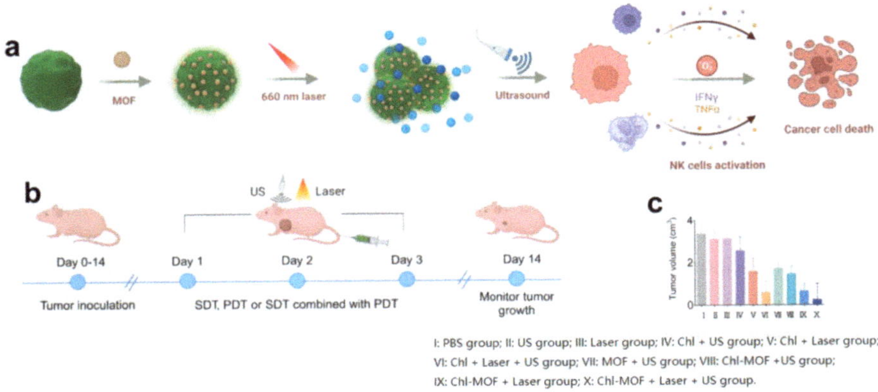

Fig. 4.10 MOF-coated chlorella for photo-sonodynamic therapy and immunotherapy. (**a**) Schematic illustration of the preparation of Chl-MOF and the possible mechanisms of multimodal therapy. (**b**) Schematic illustration of SDT/PDT animal experiments. (**c**) The volumes of tumors after treatments at day 14 in different groups [50]. (*Copyright* © 2024 Acta Materialia Inc. Published by Elsevier Ltd. Reprinted with permission from Ref. [50])

(Fig. 4.10a). Under dual activation with a 660 nm laser and ultrasound, this system utilizes the autonomous motility of microalgae (speed: 100–200 μm/s) to enhance deep tumor penetration, while photosynthesis alleviates the hypoxic tumor microenvironment. In vivo studies revealed an 88.8% reduction in breast tumor volume (Fig. 4.10b, c) and activation of antitumor immune responses by NK cells and DC cells [50].

Furthermore, in the field of antibacterial therapy, MOF-based nanomaterials doped with noble metals have demonstrated excellent efficacy when combining SDT with photothermal therapy (PTT). For example, Guo et al. developed a Janus-structured MOF@Au-DNase I nanomotor which integrates the photothermal effect of Au and the eDNA-degrading capability of DNase I. Under 808 nm laser irradiation, the nanomotor actively penetrates biofilms via thermophoretic effects, while ultrasound-induced ROS synergistically eradicates methicillin-resistant *Staphylococcus aureus* (MRSA). This system achieves rapid biofilm penetration within 15 min and significantly enhances bactericidal efficiency [51].

These studies demonstrate that MOF-based sonosensitizers significantly enhance the therapeutic efficacy of SDT through structural design strategies, including porous frameworks, stimulus responsiveness, and multimodal integration. However, despite considerable progress, challenges remain in areas such as material stability, biocompatibility, and unclear in vivo metabolic mechanisms. Future efforts should focus on optimizing the design of MOF nanocomposites, improving targeting capability, investigating the biosafety of MOF degradation products, integrating immunomodulatory functions to reverse tumor immunosuppressive microenvironment, and combining artificial intelligence technology to precisely regulate sonodynamic therapy parameters—all critical steps toward advancing the clinical translation of MOF-based sonodynamic therapy systems.

4 The Problems and Challenges of SDT

Although nanoparticle-based tumor SDT has demonstrated significant potential, its clinical translation remains challenging. Compared to nano-organic sonosensitizers, nano-inorganic sonosensitizers offer advantages such as high stability, safety, controllable physicochemical properties, and multifunctionality. However, significant obstacles impede their clinical application in SDT. Most inorganic NPs are poorly degradable in vivo, potentially leading to long-term accumulation and toxicity problems. They may also trigger immunogenic reactions, and their long-term toxicity requires further clinical validation, which limits their widespread application in clinical tumor treatment. Developing biodegradable sonosensitizers that can be effectively excreted from the body represents an important future research direction. Additionally, the tumor microenvironment—characterized by hypoxia, low pH, and high concentrations of GSH—may reduce the efficacy of SDT. Many studies have revealed that the EPR effect of NPs is very low. Although targeting ligands such as folic acid, cell membranes, specific antibodies, and proteins can be conjugated to the surface of NPs to enhance their tumor-targeting capabilities, a significant proportion of NPs is still captured by the immune system and prematurely cleared during delivery.

Although it is widely believed that NPs can provide cavitation nucleation sites, lower the cavitation threshold, and promote energy transfer—thereby enhancing the thermal effects of ultrasound—the specific mechanisms underlying the sonodynamic phenomenon remain poorly understood. This lack of clarity hinders more systematic efforts to improve sonodynamic efficiency. Furthermore, due to the complexity of tumor pathology, SDT alone is often insufficient for effective tumor eradication. The combination of multiple treatment methods such as PDT, PTT, immunotherapy, and chemotherapy holds the potential to achieve synergistic effects ("1 + 1 > 2"). However, identifying the optimal combination strategy with SDT requires further systematic exploration.

References

1. Yumita N, Iwase Y, Imaizumi T, Sakurazawa A, Kaya Y, Nishi K, Ikeda T, Umemura S, Chen FS, Momose Y. Sonodynamically-induced anticancer effects by functionalized fullerenes. Anticancer Res. 2013;33(8):3145–51. PMID: 23898072.
2. Liang S, Deng X, Ma P, Cheng Z, Lin J. Recent advances in nanomaterial-assisted combinational sonodynamic cancer therapy. Adv Mater. 2020;32(47):e2003214. https://doi.org/10.1002/adma.202003214. Epub 2020 Oct 16. PMID: 33064322.
3. Zeghimi A, Escoffre JM, Bouakaz A. Role of endocytosis in sonoporation-mediated membrane permeabilization and uptake of small molecules: an electron microscopy study. Phys Biol. 2015;12(6):066007. https://doi.org/10.1088/1478-3975/12/6/066007. PMID: 26599283.
4. Yumita N, Iwase Y, Nishi K, Komatsu H, Takeda K, Onodera K, Fukai T, Ikeda T, Umemura S, Okudaira K, Momose Y. Involvement of reactive oxygen species in Sonodynamically induced apoptosis using a novel porphyrin derivative. Theranostics. 2012;2(9):880–8.

5. Hachimine K, Shibaguchi H, Kuroki M, Yamada H, Kinugasa T, Nakae Y, Asano R, Sakata I, Yamashita Y, Shirakusa T, Kuroki M. Sonodynamic therapy of cancer using a novel porphyrin derivative, DCPH-P-Na(I), which is devoid of photosensitivity. Cancer Sci. 2007;98(6):916–20.
6. Yumita N, Okuyama N, Sasaki K, Umemura S. Sonodynamic therapy on chemically induced mammary tumor: pharmacokinetics, tissue distribution and sonodynamically induced antitumor effect of gallium-porphyrin complex ATX-70. Cancer Chemother Pharmacol. 2007;60(6):891–7.
7. Miyoshi N, Misik V, Riesz P. Sonodynamic toxicity of gallium-porphyrin analogue ATX-70 in human leukemia cells. Radiat Res. 1997;148(1):43–7.
8. Yumita N, Okudaira K, Momose Y, Umemura S. Sonodynamically induced apoptosis and active oxygen generation by gallium-porphyrin complex, ATX-70. Cancer Chemother Pharmacol. 2010;66(6):1071–8.
9. Deng X, Shao Z, Zhao Y. Development of porphyrin and titanium dioxide sonosensitizers for sonodynamic cancer therapy. Biomater Transl. 2021;2(1):72–85.
10. Li SH, Mok GSP, Dai YL. Lipid bilayer-based biological nanoplatforms for sonodynamic cancer therapy. Adv Drug Deliver Rev. 2023;202:115110.
11. Xing XJ, Zhao SJ, Xu T, Huang L, Zhang Y, Lan MH, Lin CW, Zheng XL, Wang PF. Advances and perspectives in organic sonosensitizers for sonodynamic therapy. Coord Chem Rev. 2021;445:214087.
12. Luo YL, Qiao B, Yang C, Zhang P, Xie ZY, Cao J, Yang AY, Xiang QYQ, Ran HT, Wang ZG, Hao L, Cao Y, Zhou ZY, Ren JL. Low intensity focused ultrasound ignited "deep-penetration nanobomb" (DPNB) for tetramodal imaging guided hypoxia-tolerant Sonodynamic therapy against hypoxic tumors. Int J Nanomedicine. 2022;17:4547–65.
13. Zhao PH, Wu YL, Li XY, Feng LL, Zhang L, Zheng BY, Ke MR, Huang JD. Aggregation-enhanced sonodynamic activity of phthalocyanine-artesunate conjugates. Angew Chem Int Edit. 2022;61(5):e202113506.
14. Nakonechny F, Nisnevitch M, Nitzan Y, Nisnevitch M. Sonodynamic excitation of rose Bengal for eradication of gram-positive and gram-negative bacteria. Biomed Res Int. 2013;2013:684930.
15. Nomikou N, Fowley C, Byrne NM, McCaughan B, McHale AP, Callan JF. Microbubble-sonosensitiser conjugates as therapeutics in sonodynamic therapy. Chem Commun. 2012;48(67):8332–4.
16. Cao XL, Li Y, Wang YB, Yu T, Zhu C, Zhang XZ, Guan JL. Curcumin suppresses tumorigenesis by ferroptosis in breast cancer. PLoS One. 2022;17(1):e0261370.
17. Zhu JX, Zhu WT, Hu JH, Yang W, Liu P, Liu QH, Bai YX, Xie R. Curcumin-loaded poly(L-Lactide-co-Glycolide) microbubble- mediated sono-photodynamic therapy in liver cancer cells. Ultrasound Med Biol. 2020;46(8):2030–43.
18. Abrahamse H, Hamblin MR. New photosensitizers for photodynamic therapy. Biochem J. 2016;473:347–64.
19. Foglietta F, Canaparo R, Cossari S, Panzanelli P, Dosio F, Serpe L. Ultrasound triggers hypericin activation leading to multifaceted anticancer activity. Pharmaceutics. 2022;14(5):1102.
20. Bosca F, Foglietta F, Gimenez A, Canaparo R, Durando G, Andreana I, Barge A, Peira E, Arpicco S, Serpe L, Stella B. Exploiting lipid and polymer nanocarriers to improve the anticancer sonodynamic activity of chlorophyll. Pharmaceutics. 2020;12(7):605.
21. Zheng XL, Liu WM, Ge JC, Jia QY, Nan FC, Ding Y, Wu JS, Zhang WJ, Lee CS, Wang PF. Biodegradable natural product-based NPs for near-infrared fluorescence imaging-guided sonodynamic therapy. ACS Appl Mater Interfaces. 2019;11(20):18178–85.
22. Yang L, Yuan M, Ma PA, Chen XR, Cheng ZY, Lin J. Assembling AgAuSe quantum dots with peptidoglycan and neutrophils to realize enhanced tumor targeting, NIR (II) imaging, and sonodynamic therapy. Small Methods. 2023;7(7):2201706.
23. Wang DD, Wu HH, Phua SZF, Yang GB, Lim WQ, Gu L, Qian C, Wang HB, Guo Z, Chen HZ, Zhao YL. Self-assembled single-atom nanozyme for enhanced photodynamic therapy treatment of tumor. Nat Commun. 2020;11(1):357.

24. Wang Y, Dai JY, Wang M, Qi FG, Jin XX, Zhang LX. Enhanced toluene oxidation by photothermal synergetic catalysis on manganese oxide embedded with Pt single-atoms. J Colloid Interface Sci. 2023;636:577–87.
25. Lin XH, Liu SY, Zhang X, Zhu R, Chen S, Chen XY, Song JB, Yang HH. An ultrasound activated vesicle of Janus Au-MnO NPs for promoted tumor penetration and sono-chemodynamic therapy of orthotopic liver cancer. Angew Chem Int Edit. 2020;59(4):1682–8.
26. Qu X, Yin F, Pei M, Chen Q, Zhang Y, Lu S, Zhang X, Liu Z, Li X, Chen H, Zhang Y, Qin H. Modulation of intratumoral fusobacterium nucleatum to enhance sonodynamic therapy for colorectal cancer with reduced phototoxic skin injury. ACS Nano. 2023;17(12):11466–80.
27. Yang HL, Bai LF, Geng ZR, Chen H, Xu LT, Xie YC, Wang DJ, Gu HW, Wang XM. Carbon quantum dots: preparation, optical properties, and biomedical applications. Mater Today Adv. 2023;18:100376.
28. Wu YY, Li C, van der Mei HC, Busscher HJ, Ren YJ. Carbon quantum dots derived from different carbon sources for antibacterial applications. Antibiotics (Basel). 2021;10(6):623.
29. Ponomarenko LA, Schedin F, Katsnelson MI, Yang R, Hill EW, Novoselov KS, Geim AK. Chaotic dirac billiard in graphene quantum dots. Science. 2008;320(5874):356–8.
30. Fan HY, Yu XH, Wang K, Yin YJ, Tang YJ, Tang YL, Liang XH. Graphene quantum dots (GQDs)-based nanomaterials for improving photodynamic therapy in cancer treatment. Eur J Med Chem. 2019;182:111620.
31. Bensasson RV, Berberan-Santos M, Brettreich M, Frederiksen J, Göttinger H, Hirsch A, Land EJ, Leach S, McGarvey D, Schönberger H. Triplet state properties of malonic acid C 60 derivatives C 60 [C (COOR) 2] n; R= H, Et; n= 1–6. Phys Chem Chem Phys. 2001;3(21):4679–83.
32. Ge JC, Lan MH, Zhou BJ, Liu WM, Guo L, Wang H, Jia QY, Niu GL, Huang X, Zhou HY, Meng XM, Wang PF, Lee CS, Zhang WJ, Han XD. A graphene quantum dot photodynamic therapy agent with high singlet oxygen generation. Nat Commun. 2014;5:4596.
33. Robinson JT, Tabakman SM, Liang YY, Wang HL, Casalongue HS, Vinh D, Dai HJ. Ultrasmall reduced graphene oxide with high near-infrared absorbance for photothermal therapy. J Am Chem Soc. 2011;133(17):6825–31.
34. Li C, Yang XQ, An J, Cheng K, Hou XL, Zhang XS, Hu YG, Liu B, Zhao YD. Red blood cell membrane-enveloped O self-supplementing biomimetic NPs for tumor imaging-guided enhanced sonodynamic therapy. Theranostics. 2020;10(2):867–79.
35. Yang SW, Wang XL, He P, Xu AL, Wang G, Duan JL, Shi YQ, Ding GQ. Graphene quantum dots with pyrrole N and pyridine N: superior reactive oxygen species generation efficiency for metal-free sonodynamic tumor therapy. Small. 2021;17(10):2004867.
36. Augustine R, Mathew AP, Sosnik A. Metal oxide NPs as versatile therapeutic agents modulating cell Signaling pathways: linking nanotechnology with molecular medicine. Appl Mater Today. 2017;7:91–103.
37. Xu HB, Sun X, Yao JT, Zhang J, Zhang Y, Chen HB, Dan JH, Tian Z, Tian Y. The decomposition of protoporphyrin IX by ultrasound is dependent on the generation of hydroxyl radicals. Ultrason Sonochem. 2015;27:623–30.
38. Hu CT, Hou B, Xie SL. Application of nanosonosensitizer materials in cancer sono-dynamic therapy. RSC Adv. 2022;12(35):22722–47.
39. Yamaguchi S, Kobayashi H, Narita T, Kanehira K, Sonezaki S, Kudo N, Kubota Y, Terasaka S, Houkin K. Sonodynamic therapy using water-dispersed TiO-polyethylene glycol compound on glioma cells: comparison of cytotoxic mechanism with photodynamic therapy. Ultrason Sonochem. 2011;18(5):1197–204.
40. Li YJ, Chen W, Kang Y, Zhen XY, Zhou ZM, Liu C, Chen SY, Huang XA, Liu HJ, Koo S, Kong N, Ji XY, Xie T, Tao W. Nanosensitizer-mediated augmentation of sonodynamic therapy efficacy and antitumor immunity. Nat Commun. 2023;14(1):6973.
41. Cai JM, Shen QW, Wu Y, Hu JY, Pan DY, Wu Y, Geng BJ. Defect engineering of biodegradable sulfide nanocage sonozyme systems enables robust immunotherapy against metastatic cancers. Adv Funct Mater. 2024;34(52):2411064.

42. Wu QY, Zhang FR, Pan XT, Huang ZJ, Zeng ZJ, Wang HY, Jiao J, Xiong XL, Bai LX, Zhou DS, Liu HY. Surface wettability of nanoparticle modulated sonothrombolysis. Adv Mater. 2021;33(25):2007073.
43. Feng QH, Zhang WX, Yang XM, Li YZ, Hao YW, Zhang HL, Hou L, Zhang ZZ. pH/ultrasound dual-responsive gas generator for ultrasound imaging-guided therapeutic inertial cavitation and sonodynamic therapy. Adv Healthc Mater. 2018;7(5):1700957.
44. Xu FY, Hu M, Liu CC, Choi SK. Yolk-structured multifunctional up-conversion NPs for synergistic photodynamic-sonodynamic antibacterial resistance therapy. Biomater Sci. 2017;5(4):678–85.
45. Ma AQ, Chen HQ, Cui YH, Luo ZY, Liang RJ, Wu ZH, Chen Z, Yin T, Ni J, Zheng MB, Cai LT. Metalloporphyrin complex-based nanosonosensitizers for deep-tissue tumor theranostics by noninvasive sonodynamic therapy. Small. 2019;15(5):1804028.
46. Zhuang F, Ma Q, Dong CH, Xiang HJ, Shen YJ, Sun P, Li CX, Chen YX, Lu BL, Chen Y, Huang BJ. Sequential ultrasound-triggered and hypoxia-sensitive nanoprodrug for cascade amplification of sonochemotherapy. ACS Nano. 2022;16(4):5439–53.
47. Jiang Q, Gao XR, Zhang W, Chen ZG. Recent progress in metal-organic framework-based sonosensitizers for sonodynamic tumor therapy. Biomater Sci. 2023;11(13):4452–70.
48. Pan XT, Wang WW, Huang ZJ, Liu S, Guo J, Zhang FR, Yuan HJ, Li X, Liu FY, Liu HY. MOF-derived double-layer hollow nanoparticles with oxygen generation ability for multimodal imaging-guided sonodynamic therapy. Angew Chem Int Edit. 2020;59(32):13557–61.
49. Xu QB, Zhan GT, Zhang ZL, Yong TY, Yang XL, Gan L. Manganese porphyrin-based metal-organic framework for synergistic sonodynamic therapy and ferroptosis in hypoxic tumors. Theranostics. 2021;11(4):1937–52.
50. Xu CL, Dong JH, Shi XZ, Rui JX, Chen M, Lu W, Zhang AH, Wang SJ, Teng ZG, Ye XH. Engineered microalgae for photo-sonodynamic synergistic therapy in breast cancer treatment. Acta Biomater. 2025;193:531–44.
51. Guo W, Wang YM, Zhang K, Dai XG, Qiao ZZ, Liu ZW, Yu BR, Zhao NA, Xu FJ. Near-infrared light-propelled MOF@au nanomotors for enhanced penetration and sonodynamic therapy of bacterial biofilms. Chem Mater. 2023;35(17):6853–64.

Nanomaterial-Based Cancer Chemodynamic Therapy

5

Zinan Zhao, Tianxiao Mei, Ran Cui, Muladili Mutailipu, Yongkun Wang, Chunxiu Dong, Bingdi Chen, Yihui Hu, Chen Bo, and Wenjun Le

Abstract

Nanotechnology is reshaping tumor therapy, with nanoenzymes—anoenzymesogybiocatalytic nanomaterialsanomaterialsg tumor therapy. with nanoenzymeslationol of Medicine, strates with low or negligible toxicity into cytotoxic species (e.g., reactive free radicals), nanoenzymes can drive tumor cells toward apoptosis or necrosis while reducing the side effects of conventional treatments. Compared with natural enzymes, nanoenzymes offer enhanced stability and versatility, thereby improving both the effectiveness and safety of existing therapeutic approaches. Under mild physiological in vivo conditions, nanoenzymes

Tianxiao Mei and Ran Cui contributed equally with all other contributors.

Z. Zhao · M. Mutailipu
Department of Hepatopancreatobiliary Surgery, Shanghai East Hospital, School of Medicine, Tongji University, Shanghai, China

Institute for Regenerative Medicine, Shanghai East Hospital, School of Medicine, Tongji University, Shanghai, China

T. Mei · Y. Hu (✉) · W. Le (✉)
Institute for Regenerative Medicine, Shanghai East Hospital, School of Medicine, Tongji University, Shanghai, China

Shanghai Institute of Stem Cell Research and Clinical Translation, Shanghai, China
e-mail: yihuihu@tongji.edu.cn; wenjunle@tongji.edu.cn

R. Cui · Y. Wang · C. Dong · C. Bo (✉)
Department of Hepatopancreatobiliary Surgery, Shanghai East Hospital, School of Medicine, Tongji University, Shanghai, China
e-mail: chenbo7349@tongji.edu.cn

B. Chen
Institute for Regenerative Medicine, Shanghai East Hospital, School of Medicine, Tongji University, Shanghai, China

M. Chu (ed.), *Modern Nanomaterials and Nanotechnology for Cancer Therapy*,
https://doi.org/10.1007/978-981-95-6849-9_5

catalyze reactions efficiently and potentiate modalities such as immunotherapy, photodynamic therapy, and sonodynamic therapy, underscoring their unique value in tumor treatment. This review summarizes recent advances in nanoenzyme-based combination therapies and outlines how nanoenzymes enhance antitumor efficacy through multiple mechanisms. It also discusses the challenges and opportunities of integrating nanoenzymes with conventional therapies and looks ahead to the development of next-generation nanoenzymes and the expansion of their biomedical applications.

Keywords

Chemodynamic therapy (CDT) · Nanomaterials · Reactive oxygen species (ROS) · Tumor microenvironment · Cancer therapy

1 Introduction

Cancer is a significant health problem on a global scale. Approximately 2,000,000 new instances of cancer and 600,000 cancer-related deaths are projected to occur in the United States in 2024 [1]. Although conventional treatments such as surgery, chemotherapy, and radiotherapy, along with their combination regimens, are widely used, they still face challenges such as postoperative recurrence, side effects on normal tissues, and treatment resistance. These challenges are major factors contributing to the ongoing rise in cancer mortality [2, 3].

Nanotechnology's swift advancement has presented fresh prospects for cancer treatment. By harnessing the strong reactivity of nanomaterials, scientists have created various innovative treatment methods, such as chemodynamic therapy (CDT), photodynamic therapy (PDT), photothermal therapy (PTT), sonodynamic therapy (SDT), and starvation therapy (ST), among others. These therapeutic techniques demonstrate efficacy in eradicating tumor cells while minimizing hazardous side effects [4]. Nevertheless, the actual implementation of these novel techniques has been hindered by the spreading of cancer cells and the development of therapeutic resistance. Hence, it is imperative to devise novel approaches for cancer treatment in order to enhance its efficacy against tumors.

Enzymes function as potent biocatalysts, exhibiting remarkable selectivity and catalytic effectiveness for biochemical events occurring both within living organisms (in vivo) and in laboratory settings (in vitro) [5]. Nevertheless, the extensive utilization of enzymes is constrained by several drawbacks: enzymes are prone to degradation, challenging to store and transport, and expensive to make and purify. Additionally, the catalytic activity of enzymes is very sensitive to environmental circumstances [6]. In order to tackle these problems, researchers have started investigating a promising strategy: the advancement of artificial enzymes. Specifically, nanoenzymes possessing nanoscale dimensions and enzyme-like capabilities have emerged as a highly intriguing technology [7].

Nanomaterials that mimic the catalytic activity of natural enzymes in the complex biological environment of the human body are called nanoenzymes. Similar to natural enzymes, nanoenzymes possess high catalytic activity [8] and are characterized by good stability, low production costs, and simple preparation steps. So far, numerous nanomaterials have exhibited characteristics similar to those of natural enzymes. These include metal nanoparticles, metal oxide nanoparticles, carbon-based nanomaterials, and metal-organic frameworks [9, 10]. Furthermore, these nanoenzymes exhibit notable benefits in the field of biological applications, particularly in the realm of cancer treatment [11, 12]. For instance, nanoenzymes can effectively utilize the tumor microenvironment (TME) to obtain precise and focused therapeutic outcomes. TME includes not only extracellular matrix components such as microvessels, fibroblasts, immune cells, and lymphocytes but also diverse biological characteristics such as hypoxia, acidic pH, elevated levels of H_2O_2, and high concentrations of glutathione (GSH) in tumor tissues. The strategic utilization of nanoenzymes in cancer treatment has the potential to significantly alter the landscape of oncology in the next few years.

This paper provides a concise overview of the categorization of nanoenzymes. It then explores the utilization of nanoenzymes in both antitumor therapy and complementary antitumor therapy, with a specific emphasis on the combined therapeutic approaches of nanoenzymes with PDT and PTT. Lastly, it briefly outlines the biosafety considerations and challenges associated with nanoenzymes.

2 Different Enzymatic Activities

Nanoenzymes can be divided into two main groups based on their catalytic properties. The first group is the oxidoreductase family, which includes enzymes such as oxidase (OXD), peroxidase (POD), catalase (CAT), superoxide dismutase (SOD), and nitrate reductase. The second group is the hydrolase family, which includes enzymes like nuclease, esterase, phosphatase, protease, and silylates [13]. The major catalytic features of nanoenzymes that are relevant for biomedical applications include POD, CAT, SOD, and OXD enzyme activity.

In 2007, Gao's team made the initial discovery of the intrinsic catalytic activity of Fe_3O_4 nanoparticles toward classical POD substrates, thereby introducing a new realm of nanomaterials with properties similar to POD [14]. Nanoenzymes with characteristics similar to POD enzymes have the ability to facilitate the production of hydroxyl ions from hydrogen peroxide (H_2O_2). The effectiveness of their catalytic activity is influenced by the concentration of H_2O_2 and the affinity of the catalyst toward H_2O_2.Fan et al. modified Fe3O4 nanoparticles by including histidine residues on their surface to replicate the enzyme-enhancing microenvironment found in natural POD enzymes [15]. By altering just one amino acid, the Fe_3O_4 nanoenzymes showed a significant increase in their affinity for the substrate H_2O_2, with a more than tenfold increase in apparent affinity (KM). Additionally, the catalytic efficiency of these nanoenzymes increased by 20-fold. Xie et al. successfully designed Ru nanoenzymes that exhibited exceptionally high POD-like activity [16].

These modified nanoenzymes were used to develop an immunoassay for human alpha-fetoprotein. Compared to the traditional assay, the modified nanoenzymes showed a remarkable 140-fold increase in sensitivity for detection. This study presents a practical approach to creating nanoenzymes that exhibit a high level of specific activity. Additionally, it helps to reduce the disparity in catalytic activity between nanoenzymes and natural enzymes to some degree.

Extensive research has been conducted on POD-like nanoenzymes, focusing on their structural interactions, catalytic processes, and possible uses in disease diagnostics and therapy. At present, the performance of POD nanoenzymes, such as single-atom nanoenzymes, is similar to that of natural peroxidases [17]. Besides POD nanoenzymes that facilitate the production of reactive oxygen species (ROS) within living organisms, other nanoenzymes possessing SOD, CAT, and glutathione peroxidase (GPx) class functions are employed to counteract ROS for the purpose of safeguarding cells, reducing inflammation, or treating tumors [18, 19].

CAT is a widely studied biocatalyst due to its capacity to facilitate the breakdown of H_2O_2 into O_2 and H_2O, which helps safeguard cells against oxidative stress and alleviate hypoxia [20]. Typical CAT-like enzymes consist of iron oxide, cobalt oxide, ruthenium oxide, carbon quantum dots, and Au NPs. A wide range of metal-containing and carbon-based nanomaterials have been synthesized with the purpose of emulating the catalytic function of CAT enzymes [21–23]. Nanoenzymes with catalase-like activity, known as CAT-like nanoenzymes, provide several advantages over natural catalase. These include lower cost, higher catalytic activity and stability, and the ability to easily modify their surface. As a result, CAT-like nanoenzymes have found extensive applications in tumor therapy and other sectors [24].

SOD facilitates the conversion of superoxide radicals into peroxide and oxygen through a process called disproportionation. Presently, the precise catalytic mechanism of SOD-like nanoenzymes remains uncertain. Gao et al. recently published a study on a novel type of nanoenzymes called carbon-dot (C-dot) SOD [25]. The C-dots derived from activated charcoal exhibited an activity level of up to 11,000 U/g, which is similar to that of the natural enzyme. This similarity can be attributed to the distinctive structure of the raw material. Furthermore, the carbon-dot SOD nanoenzymes demonstrated an inherent capacity to selectively target cells that have been harmed by oxidation. The study conducted by the scientists elucidates the correlation between the structure and function of carbon-dotted SOD nanoenzymes, highlighting their potential for treating disorders associated with oxidative stress.

3 Nanozymes in Cancer Therapy

3.1 Nanozyme-Predominant Cancer Therapy

Common cancer treatments like chemotherapy and radiotherapy have made significant strides in the fight against cancer. However, they still encounter significant obstacles in overcoming drug resistance. Enhancing the effectiveness of nanoenzyme-based catalytic therapy has been a focus of researchers in the field of cancer

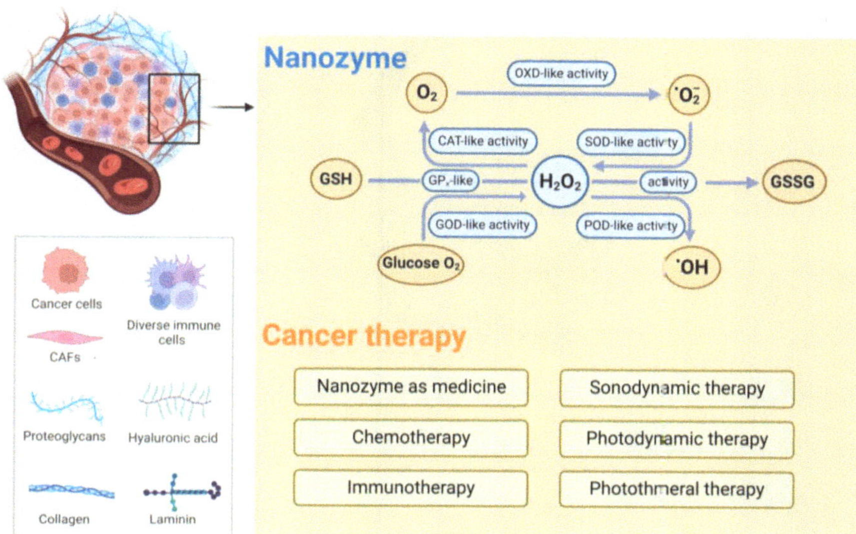

Fig. 5.1 Cancer therapy by nanozymes. (Created with BioRender.com)

treatment. They have devised many techniques to improve the efficacy of tumor death generated by ROS, especially in cases of intratumor hypoxia and hydrogen peroxide deficit (Fig. 5.1).

The low affinity between nanozymes and H_2O_2, along with the low concentration levels of H_2O_2 in the tumor microenvironment, will limit the ability of nanozymes to catalyze the production of highly toxic hydroxyl radicals to kill tumor cells. In response to this issue, Meng and his team have created a pyrite peroxidase nanozyme that has an extremely strong attraction to H_2O_2. This nanozyme demonstrates a catalytic ability that is 4144 times greater than the traditional Fe_3O_4 nanozyme and 3086 times greater than natural horseradish peroxidase [26]. In addition, the unique nanozyme created by Wu and colleagues not only has the above powers but also hinders the growth of tumors by combining the effects of ferroptosis and apoptosis in a synergistic manner [27].

Within living organisms, numerous biological processes are facilitated by a sequence of intracellular enzymes, working in a cascade, to achieve a variety of biochemical reactions. In order to address the inherent constraints of natural enzymes, scientists have developed cascade catalytic systems by combining several nanoenzymes onto a single platform. This integration significantly enhances the efficiency of catalytic reactions. Chen et al. described a method using ultrasound to enhance a cascade catalytic therapy (Fig. 5.2). This therapy is based on nanocrystals of $LaFeO_3$ (LFO), which act as a nano-enzymatic system. The LFO nanocrystals exhibit activities similar to OXD and POD enzymes, which promote the generation of ROS. They also have CAT-like activity, which helps reverse hypoxia, and GPx-like activity, which involves the consumption of GSH. The aforementioned reaction, which involves multiple enzymes, necessitates a substantial amount of

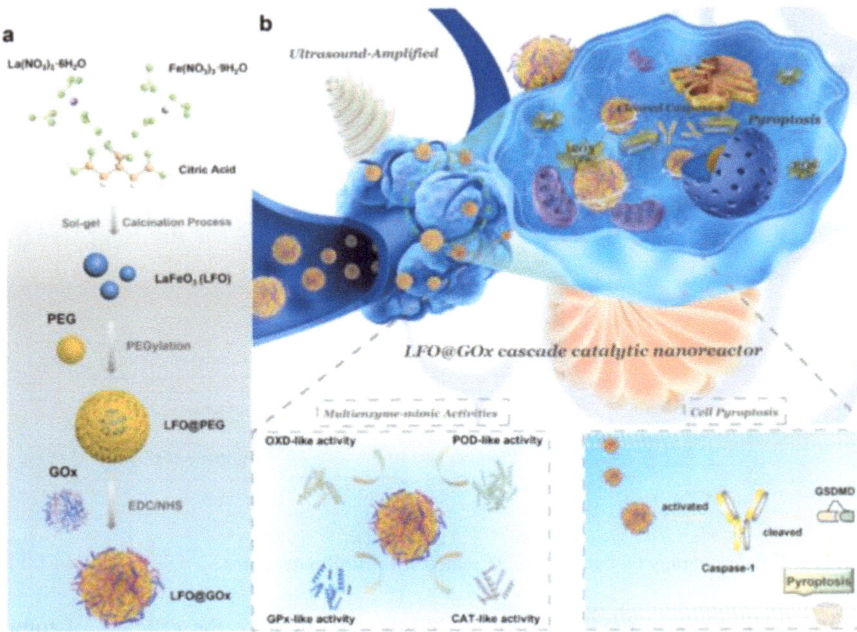

Fig. 5.2 The antitumor effects of nanoenzymes and ferroptosis. (Copyright © 2023 WILEY-VCH. Reprinted with permission from Ref. [28])

H_2O_2. However, the concentration of H_2O_2 is typically low in the TME. To address this, glucose oxidase (GOx) is introduced as a reactant in the enzyme kinetic reaction to generate H_2O_2 and fulfill the requirements of the reaction. The advantages of cell death induced by the LFO nano-system have been validated by both in vitro and in vivo anticancer outcomes.

ROS-induced apoptosis systems rely heavily on the availability of oxygen or external triggers to produce ROS, which significantly hampers their effectiveness as a treatment, particularly in hypoxic tumors. To address this limitation, Wang et al. created a bionic nanoflower called MnO_2@PtCo using nano-enzymatic self-assembly [29]. In this system, PtCo functions as an oxidase mimic, and MnO_2 acts as a catalase mimic. This innovative design allows the system to not only relieve the hypoxia condition at the tumor site but also effectively trigger programmed cell death and specifically inhibit tumor growth through a mechanism involving ROS. Ling et al. have created a new biomimetic nanomaterial that generates H_2O_2 and catalyzes the production of ROS within cells under acidic conditions without the need for external triggers [30]. These studies pave the way for the development of nano-enzymatic systems with multiple enzymatic activities.

3.2 Nanozyme-Assistant Cancer Therapy

3.2.1 Nanozymes in Synergistic Chemotherapy

Chemotherapy is extensively employed in the medical management of cancer, with typically utilized chemotherapeutic drugs including cisplatin [31], methotrexate [32], and paclitaxel [33], among others. Nevertheless, the drawback of using these drugs in treatment is their limited specificity and susceptibility to inducing systemic adverse responses [34].

Recently, the use of nanoenzyme-activated prodrug techniques has opened up new possibilities for targeted therapeutics. These strategies involve converting non-toxic prodrugs into toxic drugs through the activation of nanoenzymes. This process can induce cell death and simultaneously minimize the systemic toxicity response. The little off-target impact of enzyme prodrug treatment (EPT) enhances therapeutic effectiveness. This, along with the high catalytic activity and stability of nanoenzymes, demonstrates significant promise in the realm of targeted medicine [35]. Zhang et al. examined a new approach involving nanoenzymes that do not contain metal to activate prodrugs [36]. In previous studies, the combination of natural peroxidase and indole-3-acetic acid (IAA) demonstrated potential as an effective enzyme-prodrug combination. Nevertheless, the effectiveness of its ability to inhibit tumor growth is greatly hindered by a lack of enough natural peroxidase activity. Therefore, they created a nanoenzyme called porous hollow carbon sphere nanoenzymes (PNCNzymes), which utilizes nanoenzyme-IAA to activate IAA and produce a large amount of ROS, leading to apoptosis. Cao et al. discovered that a nanoenzyme composed of copper-silver alloy nanoparticles (Cu-Ag NPs), which resembles cytochrome c oxidase, can be employed for nanocatalytic cancer treatment [37]. The researchers successfully accomplished the first instance of simultaneous starvation, iron death, and chemotherapy by utilizing nanoenzymes that incorporate a prodrug. This approach substantially enhanced the survival rate of the experimental mice.

Chemotherapeutic drugs can promote the development of drug resistance in tumors, which reduces the effectiveness of treatment [38]. To address this issue, various nanomedicines based on platinum (Pt) have been created for targeted administration and combination therapy. Li et al. have recently developed a novel type of enzyme called nuclease-mimicking Pt nanomimetic enzymes (NMPNs) [39]. These enzymes are capable of specifically targeting the nucleus of tumor cells and modifying the DNA structure necessary for nucleotide excision repair (NER). This modification helps to overcome resistance to platinum (Pt) drugs. The nanomedicine exhibits the ability to release Pt particles and also possesses OXD-like and POD-like actions. As a result, it generates ROS and collaboratively triggers the oxidative cleavage of double-stranded DNA with Pt ions. This discovery paves the way for a novel approach to clinical DNA damage chemotherapy.

Hypoxic TME is a major cause of cancer metastasis, chemotherapy resistance, and tumor recurrence. Guo et al. created mesoporous $CaSiO_3$ nanoparticles that were loaded with MnO_2 and co-modified with PEG [40]. Manganese dioxide nanoparticles (MnO_2 NPs) mitigate the TME condition by interacting with naturally occurring H_2O_2 to generate O2 at the site. Similarly, Wu et al. utilized the naturally occurring carrier ferritin (HFn) for the delivery of MnO_2. This system was able to specifically transport MnO_2 and doxorubicin (Dox) to the tumor site [41].

3.2.2 Nanozyme in Synergistic Chemodynamic Therapy

CDT, as an emerging therapeutic modality, has received much attention due to its unique ROS generation mode. The process relies on the Fenton/Fenton-like reaction, which transforms excessive intracellular H_2O_2 into extremely harmful hydroxyl radicals within the oxygen-deprived TME, resulting in significant oxidative harm [42] (Fig. 5.3). Wang and colleagues engineered copper-amino acid nanoparticles (Cu-Cys NPs) for in situ glutathione activation and enhanced chemodynamic therapy. Utilizing the high concentrations of GSH and H_2O_2 in tumor cells, these nanoparticles sequentially trigger redox reactions. Cu-Cys NPs exhibit relatively higher cytotoxicity toward cancer cells while effectively preventing oxidative damage to normal tissues [43].

Nevertheless, the intracellular pH, H_2O_2 concentration, and reducing chemicals have a direct impact on the effectiveness of the Fenton reaction and, consequently, on the therapeutic outcome of CDT [44]. (i) The pH of the mildly acidic conditions in the TME frequently exceeds the ideal pH for the Fenton reaction. The nanocatalyst (FePt@FeOx@TAM-PEG) developed by Zhao et al. was employed to manipulate the acidic conditions at the tumor site in order to augment the CDT [45]. This was achieved by leveraging the pH-responsive properties of the tamoxifen (TAM) medication, which activates the catalytic function of FePt@FeOx specifically in the acidic microenvironment. The pH-responsive nature of TAM enables the catalytic activity of FePt@FeOx to be activated in acidic conditions and suppressed in neutral conditions. Additionally, the release of TAM from cancer cells leads to an increase in intracellular H+ levels, which in turn enhances the efficiency of Fenton catalysis. (ii) Elevating the concentration of H_2O_2 at the tumor location enhances the effectiveness of the Fenton reaction [46]. Sang et al. created a disruptor that maintains the balance of H_2O_2 in order to increase its concentration in tumor tissues. This disruptor promotes the generation of H_2O_2 and inhibits its elimination, resulting in improved chemodynamic therapy [47]. (iii) High glutathione levels are another characteristic of the TME. Extensive research has been conducted on strategies that aim to decrease the levels of GSH in the TME in order to improve the effectiveness of CDT [48]. Engineered iron oxide nanoparticles (IMOP), designed by Li et al., can exhibit both high catalase and glutathione peroxidase activities. In TME, IMOP acted as a CDT agent to catalyze the decomposition of H_2O_2 into hydroxyl radicals, and in addition the glutathione peroxidase-like activity reduced the concentration of intratumoral GSH, which further reduced the depletion of reactive ROS during the treatment process and achieved synergistic CDT [49].

3.2.3 Nanozymes in Synergistic Photodynamic Therapy

PDT is a minimally invasive procedure that utilizes photosensitizers or photosensitizing medications to produce reactive oxygen species when exposed to light, leading to programmed apoptosis or necrosis [50]. PDT is gaining more recognition as

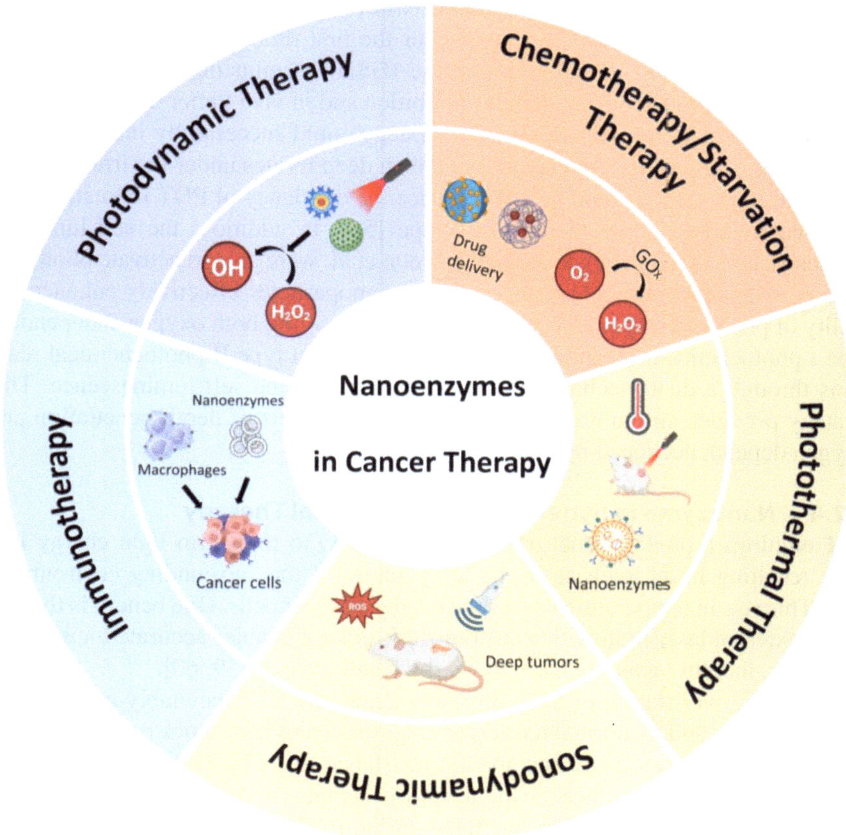

Fig. 5.3 Nanozymes in cancer therapy

a potent cancer treatment due to its exceptional selectivity, absence of cumulative toxicity, lack of developed resistance mechanisms, and noninvasive nature [51].

Zhang et al. have utilized MOFs to perform PDT on cancer patients by including photosensitizers [52]. The incorporation of platinum nanoenzymes onto photosensitizer-integrated MOFs allows for precise control over the release rate of photosensitizers. Both intratumoral and intravenous injections have demonstrated excellent performance for in vivo PDT. Liang et al. developed perfluorocarbon@porphyrin nanoparticles that serve as dual imaging contrast agents, enhancing both fluorescence and CT imaging [53]. These nanoparticles enable visualization of the aggregation of nanoparticles at the tumor site and facilitate precise PDT guidance.

Many researchers have studied methods to improve the effectiveness of PDT by using CAT-like nanoenzymes to break down the overexpression of H_2O_2 at the tumor location, resulting in the production of O_2 [54]. Nevertheless, the application of PDT in the treatment of deep and large-sized tumors has been restricted to alternative therapeutic approaches for disease relapse, tumor size reduction, and palliative care [55]. Pan et al. designed for the first time bimetallic ion-modified MOFs with CAT and glutathione oxidase (GSHOx) mimetic activities, and the authors demonstrated in vitro cellular inhibition and in vivo tumor xenograft evaluation that this nanofold-enhanced PDT strategy could successfully inhibit the differentiation and proliferation of tumor cells in deep tissues under the irradiation of a 660 nm laser, which could help to enhance the efficiency of PDT in treating cancers and to broaden the therapeutic scope [56]. In addition, the self-luminous nanoparticles (CRET-NPs) developed by Jeon et al. were able to activate photosensitizers as endogenous light sources. These nanoparticles effectively enhance the ability of photosensitizers to generate ROS by promoting both oxygen-independent type I photochemical reactions and oxygen-dependent type II photochemical reactions through a dual mechanism of electron transfer and self-luminescence. This strategy provides an innovative solution to the problems of depth penetration and oxygen dependence faced by conventional PDT [57].

3.2.4 Nanozyme in Synergistic Photothermal Therapy

PTT employs a photothermal transfer agent (PTA) to transform light energy into heat, resulting in an increase in the temperature of the surrounding environment [58]. This rise in temperature causes the death of cancer cells. One benefit is that the use of external laser light with a customizable dosage enables accurate focusing on the tumor, thereby reducing harm to nearby healthy tissue [59, 60].

However, overheating of the tumor area treated by PTT inevitably causes damage to the surrounding normal tissues; overexpression of heat shock proteins (HSPs) in some cancers may cause therapeutic resistance to PTT; inefficient delivery of PTA to the lesion may reduce the therapeutic efficiency; and a hypoxic tumor microenvironment and inflammation generated during PTT treatment may lead to tumor recurrence and therapeutic resistance [61]. Thanks to the development of nanoenzyme technology, more and more studies have been conducted to solve the problems during PTT treatment by utilizing nanoenzymes to achieve synergistically enhanced PTT treatment. Wang et al. developed a near-infrared (NIR) photoresponsive nanoreactor (P@DW/BC), which was shown to be able to stimulate the decomposition of bicarbonate in the vicinity while generating heat by photothermal conversion, thereby promoting the generation of CO, which not only achieves highly efficient thermal ablation of tumors but also inhibits PTT-induced inflammation at the same time [62]. Li et al. developed platinum-doped Prussian blue (PtPB) nanoenzymes that have adjustable spectral absorption, high photothermal conversion efficiency, and strong antioxidant activity (Fig. 5.4). These nanoenzymes were found to effectively reduce inflammation caused by hyperthermia. This discovery offers a new approach for creating safe and efficient photothermal treatments for complex oncological diseases.

3.2.5 Nanozymes in Synergistic Sonodynamic Therapy

SDT employs the concurrent application of low-intensity ultrasound and acoustic sensitizers to induce the production of cytotoxic ROS for the purposes of cancer

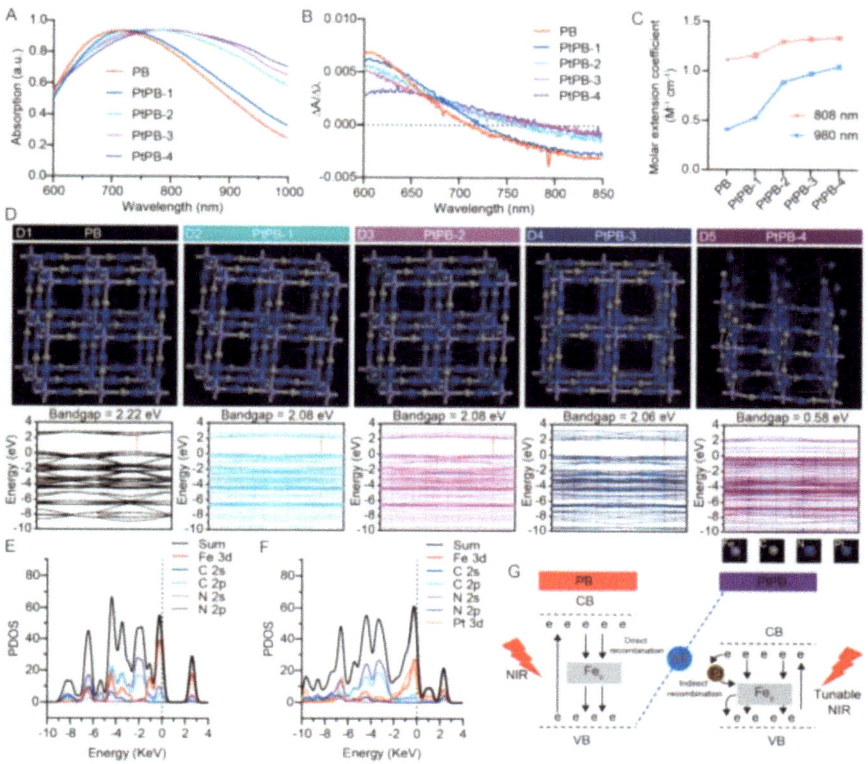

Fig. 5.4 Photothermal enhancement mechanism of PtPB nanozymes. (Copyright © 2021 American Chemical Society. Reprinted with permission from Ref. [63])

treatment and antibiotic therapy [64]. Ultrasound has the ability to penetrate tissues to a depth of tens of centimeters and concentrate acoustic energy on deep tissues without the need for invasive procedures. This makes SDT a highly promising approach for treating cancer that is located deep within the body [65].

However, the efficacy of SDT is severely hampered by the low specificity of acoustic sensitizers and the TME, such as hypoxia and GSH overexpression [66]. Wang et al. reported that $PtCu_3$ nanocages act as acoustic sensitizers with highly efficient ROS-generating ability and can catalyze the decomposition of H_2O_2 like horseradish peroxidase analogs under acidic conditions [67]. What's more, PtCu3 nanocages can also mimic glutathione peroxidase (GSH-Px), accelerating GSH depletion through oxidized molecules and further weakening the ability of tumor cells to scavenge ROS via GSH. Sun et al. encapsulated the acoustic sensitizer and catalase in novel organosilicon nanoparticles, which protected the catalase activity and increased the stability of the acoustic sensitizer and the enzyme, ultimately achieving the goal of alleviating tumor hypoxia [68].

3.2.6 Nanozymes in Synergistic Starvation Therapy

Starvation therapy aims to inhibit tumor cell proliferation and death by cutting down on energy sources, interfering with tumor angiogenesis, reducing glucose levels using Gox, or limiting the action of glucose transporters on the surface of tumor cells [69, 70]. Nevertheless, the oxygen-deficient conditions within the tumor hinder the functioning of O_2-dependent GOx and its analogs. In order to address this constraint, oxygen-donating enzymes, or nanoenzymes, are commonly employed in starvation therapy. Liu et al. designed a dual nano-enzymatic porphyrin metal-organic skeleton (PCN) in which platinum nanoparticles (Pt NPs) mimicking catalase were sandwiched in the PCN, and then ultrasmall gold nanoparticles (Au NPs) mimicking glucose oxidase were embedded inside the shell. The CAT-like Pt NPs catalyzed the production of O_2, which facilitated the oxidation of glucose by the GOx-like Au nanoparticles, which were used for starvation therapy [71].

The combination of starvation therapy with other therapeutic approaches also helps to overcome the drawbacks of some drugs, such as low targeting and the induction of drug resistance [72]. For example, Wang et al. used GOx to reduce the expression of heat shock proteins and enhance the thermal sensitivity of tumor cells, and then combined with AuPtAg nanoenzymes with catalase-like activity, they could improve tumor hypoxia and significantly increase GOx activity. The combined strategy based on AuPtAg-GOx effectively integrates starvation therapy, mild photothermal therapy, and immunotherapy and offers a highly effective way of eradicating cancer cells, providing an efficient pathway (Fig. 5.5).

3.2.7 Nanozymes in Synergistic Immunotherapy

Immunotherapy successfully eliminates tumors by stimulating the body's natural antitumor immune system and triggering lasting immunological surveillance in the body. It has emerged as the most efficacious way for treating tumors, surpassing surgery, radiation, and chemotherapy [74]. Nevertheless, despite the abundance of antitumor T cells, the immune system frequently fails to mount a response against tumor cells. TME is a critical factor in the suppression of the immune system in tumors. It is characterized by low pH, inadequate blood flow, excessive formation of peroxides, and hypoxia.

Nanoenzymes have the ability to enhance the immunosuppressive milieu, trigger immunogenic death in tumor cells, and facilitate the development of immune cells. For instance, substances that mimic catalase can help reduce the lack of oxygen in tumors and promote the activation and specialization of immune cells that fight against tumors, such as T cells and macrophages. Similarly, substances that mimic oxidase or peroxidase can disturb the balance of redox and metabolism within the tumor, leading to the induction of immunogenic apoptosis or the localized death of cancer cells [75]. Wang et al. developed an inorganic nanoenzyme that can auto-oxidize and degrade. The nanoenzymes have a core-shell structure, with a core made of CaO_2/DOX and a shell made of degradable SiO_2/DOX-MnO_2. This delivery system effectively reduces hypoxia and reverses the immunosuppressive TME, thus enhancing the effectiveness of immunotherapy by improving cytotoxic T lymphocyte-associated antigen 4 (CTLA-4) blockade [76].

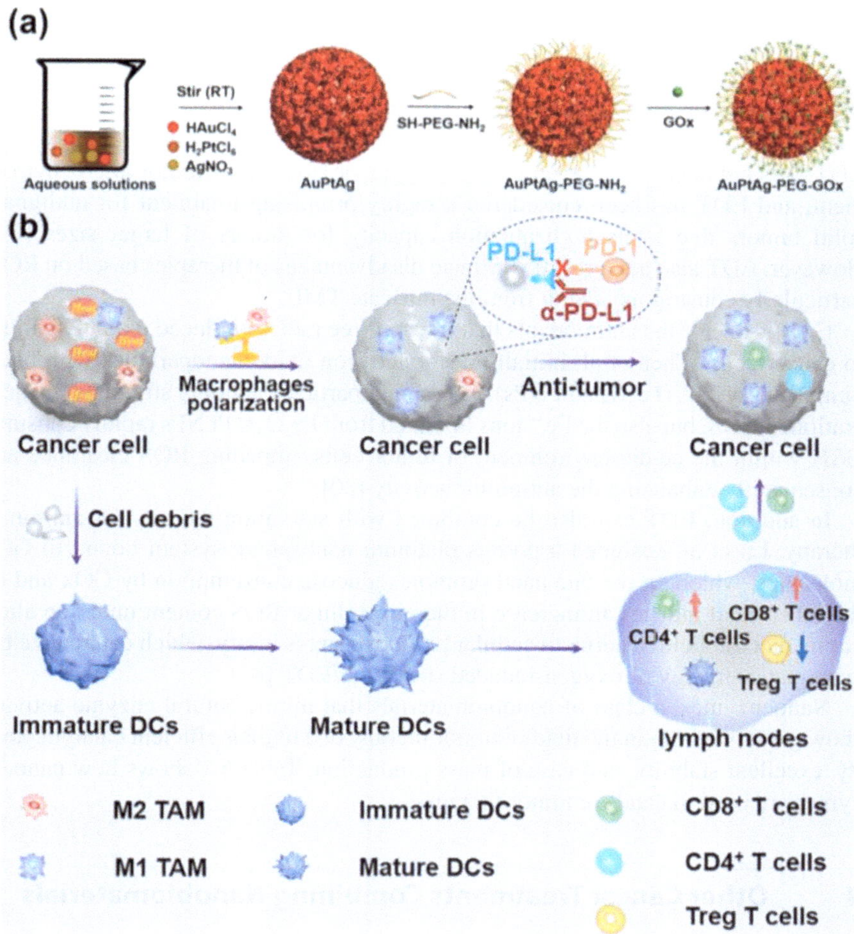

Fig. 5.5 The mechanism of tumor immunotherapy generated by fasting therapy-augmented mild photothermal therapy. (Copyright © 2022 WILEY-VCH. Reprinted with permission from Ref. [73])

Although there have been notable advancements in immunotherapy, it is still difficult to stimulate the immune response and counteract the suppressive environment within tumors. Sun et al. used upconversion nanoparticles (UCNPs) equipped with a double silica layer encapsulation as a nanocarrier and modified with an endoplasmic reticulum (ER)-targeting molecule, and this precise endoplasmic reticulum-targeted PDT could generate ROS in situ under 980 nm laser irradiation, which not only induced cell death directly but also induced endoplasmic reticulum-stress-based immunogenic cell death (ICD) to further enhance the antitumor effect [77].

3.2.8 Nanozymes in Synergistic Electrodynamic Therapy

The principle of EDT is to utilize the electrically driven catalytic reaction of platinum nanoparticles in the presence of an electric field to induce the production of ROS by triggering the decomposition of water molecules, thus enabling the treatment of cancer [78]. Unlike other therapeutic modalities, the production of ROS in EDT does not depend on the amount of oxygen or H_2O_2 in the tumor microenvironment, and EDT has been considered a highly promising treatment for malignant solid tumors due to its high ablation capacity for tumors of larger sizes [79]. However, EDT also acquires the intrinsic disadvantages of therapies based on ROS, particularly constraints arising from the intricate TME.

GSH is one of the antioxidants that prevent free radical-induced oxidative injury to cancer cells. Chen et al. initially decorated iron oxide nanoparticles with platinum nanocrystals (Fe_3O_4@Pt NPs). These nanoparticles not only stimulate the generation of ROS, but also the Fe^{3+} ions liberated from Fe_3O_4@Pt NPs rapidly consume GSH within the acidic environment of tumor cells, impeding ROS clearance and consequently enhancing the antitumor activity [80].

In addition, EDT can also be combined with starvation therapy for antitumor therapy. Lu et al. designed a porous platinum nanosphere system bound to GOx molecules, which on the one hand promotes glucose consumption by GOx and on the other hand induces an increase in the intracellular ROS content under an alternating electric field, leading to cellular oxidative stress injury, which can realize the synergistic strategy of oxygen-induced starvation/EDT [81].

Nanoenzymes, a class of nanobiomaterials that mimic natural enzyme activity, show great potential in the field of tumor therapy due to their efficient catalytic ability, excellent stability, and ease of mass production. Table 5.1 shows how nanoenzymes are used to catalyze tumor therapy.

4 Other Cancer Treatments Combining Nanobiomaterials

4.1 Traditional Chinese Medicine and Nanobiomaterials

Traditional Chinese medicine (TCM) is essential for disease prevention and treatment. Their popularity has grown in recent years and helped China regulate COVID-19 [112, 113]. TCMs and their natural basic ingredients have broad clinical applications, but their weak targeting impact, low solubility, poor stability, quick metabolism, and rapid elimination have limited their use [114].

Fortunately, nanotechnology offers a fresh way to modernize and develop TCM. Nanocarriers can enhance TCM targeting through passive or active targeting and regulate TCM release to boost antitumor efficacy [115]. Liposomal nanocarriers are extensively utilized in the investigation of nano-delivery systems for herbal medicine due to their minimal toxicity and substantial loading capacity [116]. Scholars have synthesized curcumin-loaded liposomal gold nanoparticles for photothermal therapy. The gold coating allows the nanoparticles to absorb specific wavelengths of light and convert it into heat. Due to the increase in temperature, the

Table 5.1 The application of nanoenzymes in cancer treatment

Nanozyme	Activity	Substrate	Application	References
PtFe@Fe_3O_4	POD	H_2O_2,O_2	Synergistic tumor catalytic therapy and PTT	2019 [82]
AuNCs	POD	H_2O_2	Disease monitoring	2019 [83]
TiO_1+x NRs	POD	H_2O_2	Enhancing dual chemotherapy combined with SDT	2020 [84]
IMSN-PEG	POD	H_2O_2,O_2	Enhanced immune therapy	2020 [85]
DMHSs-FeS_2	POD	H_2O_2,O_2	Rapid determination of H_2O_2 and glutathione	2020 [86]
PEG@DMSN	POD	H_2O_2,GSH	Synergistic tumor catalytic therapy and PTT	2020 [87]
FeMSN	POD	H_2O_2	Enhanced CDT efficiency	2021 [88]
PCF-a NEs	POD	H_2O_2,GSH	GSH depletion for enhanced ultrasound/NIR-promoted CDT	2021 [89]
FePc	POD	H_2O_2	Enhanced PDT and PTT efficiency	2021 [90]
Fe_3O_4	POD	H_2O_2	Enhanced CDT and RT efficiency	2021 [91]
TMPAs	POD	TMB	Nanozyme-amplified NIR-II PTT	2021 [92]
RuCu NPs	POD	H_2O_2	Synergistic tumor catalytic therapy and RT	2022 [93]
Pd–C SAzymes	POD	H_2O_2	H_2O_2-supply catalytic therapy	2022 [94]
PdPtAu MNPs	POD	H_2O_2,GSH	Enhanced immune therapy	2023 [95]
AuNC@Fe_3O_4	POD	H_2O_2	ROS production and lysosomal impairment	2023 [96]
RuCu	POD	H_2O_2,GSH	GSH depletion, enhanced CDT efficiency	2023 [97]
MnZ@Au	CAT	H_2O_2,Glu	Synergistic tumor catalytic therapy	2023 [98]
Ru@CeO_2	CAT	H_2O_2	Enhanced dual chemotherapy and PTT	2020 [99]
Au NC-ICG	CAT	H_2O_2	Modulating tumor hypoxia and augmenting PDT	2020 [100]
HIONCs-GOD	CAT	H_2O_2	Drive starvation-CDT synergistic therapy	2020 [101]
Au_2Pt-PEG-Ce_6	CAT	H_2O_2	Enhanced CDT and PDT efficiency	2020 [102]
Pd@Pt-T790	CAT	H_2O_2	Enhanced SDT efficiency, antibacterial infection	2020 [103]
Au@Rh-ICG-CM	CAT	H_2O_2	Bimodal imaging and enhanced PDT efficiency	2020 [104]
Pt@HP@M	SOD	O_2^-	Improve antitumor and SDT efficiency	2023 [105]
Ru-Te	SOD	H_2O_2, O_2^-	Enhanced PTT and PDT efficiency	2020 [106]
FePOs	SOD	H_2O_2, O_2^-	Enhanced CDT efficiency	2021 [107]
Cu-Pd@MIL	SOD	H_2O_2, O_2^-	GSH depletion, enhanced CDT efficiency	2021 [108]
CDs/LDHzyme	SOD	H_2O_2, O_2^-	Enhanced CDT efficiency	2023 [109]
COF-909-Cu/Fe/Ni	SOD	H_2O_2, O_2^-	GSH depletion, enhanced CDT efficiency	2021 [110]
DC@Pt	SOD	H_2O_2, O_2^-	Enhanced SDT and CDT efficiency	2023 [111]

liposomal core becomes unstable, subsequently releasing the encapsulated curcumin. This achieves a synergistic antitumor effect combining nanomedicine with photothermal therapy [117, 118].

5 Conclusions and Perspectives

Biological macromolecules such as DNA, proteins, enzymes, and hemoglobin naturally operate at the nanoscale in living systems, underscoring the promise of nanomaterials for biomedical applications. Within this context, nanoenzymes have attracted sustained interest owing to their exceptional catalytic activity, favorable stability, and advantageous physicochemical properties, as well as their ability to operate under mild reaction conditions.

Presently, the primary emphasis in nanoenzyme research lies in enhancing their catalytic efficiency. However, there is limited knowledge regarding their catalytic mechanism and the structure of the active region, particularly the interactions between multiple enzyme activities and their regulation. Therefore, further exploration is required to gain a comprehensive understanding of these aspects. Understanding the tumor microenvironment and the catalytic efficiency of nanoenzymes, as well as the management of their activities in the biological environment, is crucial in tumor therapy.

Finally, the biosafety of nanoenzymes needs to be further explored. Due to the complexity of the in vivo microenvironment, the catalytic pathways of nanoenzymes in vivo are often uncontrollable, and the potential biocatalytic reactions in vivo and the mechanisms at the biological interface are difficult to predict. Therefore, the in vivo stability, pharmacokinetics, and in vivo toxicity of nanoenzymes should be systematically studied and analyzed in the next phase. Therefore, this review paper is anticipated to encourage the advancement of the nanozyme field.

References

1. Siegel RL, Giaquinto AN, Jemal A. Cancer statistics, 2024. Ca-Cancer J Clin. 2024;74(1):12–49.
2. Nichetti F, Rota S, Ambrosini P, Pircher C, Gusmaroli E, Busset MDD, Pusceddu S, Sposito C, Coppa J, Morano F, Pietrantonio F, Di Bartolomeo M, Mariani L, Mazzaferro V, De Braud F, Niger M. NALIRIFOX, FOLFIRINOX, and gemcitabine with nab-paclitaxel as first-line chemotherapy for metastatic pancreatic cancer: a systematic review and meta-analysis. JAMA Netw Open. 2024;7(1):e2350756.
3. Xue L, Thatte AS, Mai D, Haley RM, Gong N, Han X, Wang K, Sheppard NC, June CH, Mitchell MJ. Responsive biomaterials: optimizing control of cancer immunotherapy. Nat Rev Mater. 2023;9(2):100–18.
4. Pei Z, Lei H, Cheng L. Bioactive inorganic nanomaterials for cancer theranostics. Chem Soc Rev. 2023;52(6):2031–81.
5. Shan JG, Du L, Wang XG, Zhang SD, Li YP, Xue S, Tang QY, Liu PF. Ultrasound trigger Ce-based MOF nanoenzyme for efficient thrombolytic therapy. Adv Sci. 2024;11(20):12.

6. Zhang X, Chen X, Zhao Y. Nanozymes: versatile platforms for cancer diagnosis and therapy. Nano Micro Lett. 2022;14(1):95.
7. Wang Y, Zhang R, Zang PY, Zhao RX, Wu LZ, Zhu YL, Yang D, Gai SL, Yang PP. Synergizing pyroelectric catalysis and enzyme catalysis: establishing a reciprocal and synergistic model to enhance anti-tumor activity. Adv Mater. 2024;36(24):13.
8. Yang L, Dong SM, Gai SL, Yang D, Ding H, Feng LL, Yang GX, Rehman Z, Yang PP. Deep insight of design, mechanism, and cancer theranostic strategy of nanozymes. Nano Micro Lett. 2024;16(1):53.
9. Lou-Franco J, Das B, Elliott C, Cao C. Gold nanozymes: from concept to biomedical applications. Nano Micro Lett. 2021;13(1):10.
10. Liu Q, Zhang A, Wang R, Zhang Q, Cui D. A review on metal- and metal oxide-based nanozymes: properties, mechanisms, and applications. Nano Micro Lett. 2021;13(1):154.
11. Dong S, Dong Y, Liu B, Liu J, Liu S, Zhao Z, Li W, Tian B, Zhao R, He F, Gai S, Xie Y, Yang P, Zhao Y. Guiding transition metal-doped hollow cerium tandem nanozymes with elaborately regulated multi-enzymatic activities for intensive chemodynamic therapy. Adv Mater. 2022;34(7):2107054.
12. Zhang W, Liu J, Li X, Zheng Y, Chen L, Wang D, Foda MF, Ma Z, Zhao Y, Han H. Precise chemodynamic therapy of cancer by trifunctional bacterium-based nanozymes. ACS Nano. 2021;15(12):19321–33.
13. Ju YK, Liu XJ, Ye XZ, Dai MH, Fang BR, Shen XK, Liu LL. Nanozyme-based remodeling of disease microenvironments for disease prevention and treatment: a review. ACS Appl Nano Mater. 2023;6(15):13792–823.
14. Gao L, Zhuang J, Nie L, Zhang J, Zhang Y, Gu N, Wang T, Feng J, Yang D, Perrett S, Yan X. Intrinsic peroxidase-like activity of ferromagnetic nanoparticles. Nat Nanotechnol. 2007;2(9):577–83.
15. Fan K, Wang H, Xi J, Liu Q, Meng X, Duan D, Gao L, Yan X. Optimization of Fe_3O_4 nanozyme activity via single amino acid modification mimicking an enzyme active site. Chem Commun. 2017;53(2):424–7.
16. Fan H, Zheng J, Xie J, Liu J, Gao X, Yan X, Fan K, Gao L. Surface ligand engineering ruthenium Nanozyme superior to horseradish peroxidase for enhanced immunoassay. Adv Mater. 2024;36(10):2300387.
17. Ji S, Jiang B, Hao H, Chen Y, Dong J, Mao Y, Zhang Z, Gao R, Chen W, Zhang R, Liang Q, Li H, Liu S, Wang Y, Zhang Q, Gu L, Duan D, Liang M, Wang D, Yan X, Li Y. Matching the kinetics of natural enzymes with a single-atom iron nanozyme. Nat Catal. 2021;4(5):407–17.
18. Singh N, Savanur MA, Srivastava S, D'Silva P, Mugesh G. A redox modulatory Mn_3O_4 nanozyme with multi-enzyme activity provides efficient cytoprotection to human cells in a Parkinson's disease model. Angew Chem Int Ed. 2017;56(45):14267–71.
19. Miao Z, Jiang S, Ding M, Sun S, Ma Y, Younis MR, He G, Wang J, Lin J, Cao Z, Huang P, Zha Z. Ultrasmall rhodium nanozyme with RONS scavenging and photothermal activities for anti-inflammation and antitumor theranostics of colon diseases. Nano Lett. 2020;20(5):3079–89.
20. Glorieux C, Calderon PB. Catalase, a remarkable enzyme: targeting the oldest antioxidant enzyme to find a new cancer treatment approach. Biol Chem. 2017;398(10):1095–108.
21. Li Q, Gao Y, Shen J, Mu X, Wang J, Ouyang L, Chen K, He H, Pei J, Ren Q, Sun S, Liu H, Zhou L, Sun Y, Long W, Zhang J, Zhang X-D. Catalase-like quantum dots of l-lysine polymerization as free radical scavengers for hypoxic brain injury. Mater Today Commun. 2021;27:102286.
22. Wang C, Li Y, Yang W, Zhou L, Wei S. Nanozyme with robust catalase activity by multiple mechanisms and its application for hypoxic tumor treatment. Adv Healthc Mater. 2021;10(19):2100601.
23. Zhang R, Chen L, Liang Q, Xi J, Zhao H, Jin Y, Gao X, Yan X, Gao L, Fan K. Unveiling the active sites on ferrihydrite with apparent catalase-like activity for potentiating radiotherapy. Nano Today. 2021;41:101317.
24. Xu D, Wu L, Yao H, Zhao L. Catalase-like nanozymes: classification, catalytic mechanisms, and their applications. Small. 2022;18(37):2203400.

25. Gao W, He J, Chen L, Meng X, Ma Y, Cheng L, Tu K, Gao X, Liu C, Zhang M, Fan K, Pang D-W, Yan X. Deciphering the catalytic mechanism of superoxide dismutase activity of carbon dot nanozyme. Nat Commun. 2023;14(1):160.
26. Meng X, Li D, Chen L, He H, Wang Q, Hong C, He J, Gao X, Yang Y, Jiang B, Nie G, Yan X, Gao L, Fan K. High-performance self-cascade pyrite nanozymes for apoptosis–ferroptosis synergistic tumor therapy. ACS Nano. 2021;15(3):5735–51.
27. Wu A, Han M, Ding H, Rao H, Lu Z, Sun M, Wang Y, Chen Y, Zhang Y, Wang X, Chen D. Fe3S4 nanozyme inhibits tumor growth by synergistic effects of ferroptosis and apoptosis. Chem Eng J. 2023;474:145920.
28. Chang MQ, Wang ZY, Dong CH, Zhou RR, Chen L, Huang H, Feng W, Wang ZL, Wang Y, Chen Y. Ultrasound-amplified enzyodynamic tumor therapy by perovskite nanoenzyme-enabled cell pyroptosis and cascade catalysis. Adv Mater. 2023;35(7):12.
29. Wang Z, Zhang Y, Ju E, Liu Z, Cao F, Chen Z, Ren J, Qu X. Biomimetic nanoflowers by self-assembly of nanozymes to induce intracellular oxidative damage against hypoxic tumors. Nat Commun. 2018;9(1):3334.
30. Ling P, Yang P, Gao X, Sun X, Gao F. ROS generation strategy based on biomimetic nanosheets by self-assembly of nanozymes. J Mater Chem B. 2022;10(46):9607–12.
31. Bruno PM, Liu Y, Park GY, Murai J, Koch CE, Eisen TJ, Pritchard JR, Pommier Y, Lippard SJ, Hemann MT. A subset of platinum-containing chemotherapeutic agents kills cells by inducing ribosome biogenesis stress. Nat Med. 2017;23(4):461–71.
32. Li W, Cao Z, Yu L, Huang Q, Zhu D, Lu C, Lu A, Liu Y. Hierarchical drug release designed Au @PDA-PEG-MTX NPs for targeted delivery to breast cancer with combined photothermal-chemotherapy. J Nanobiotechnol. 2021;19(1):143.
33. Shu Y, Yin H, Rajabi M, Li H, Vieweger M, Guo S, Shu D, Guo P. RNA-based micelles: a novel platform for paclitaxel loading and delivery. JCR. 2018;276:17–29.
34. Li ZX, Pei Q, Zhao M, Xie ZG, Zheng M. Self-carrier nanoparticles for delivery of paclitaxel and IDO inhibitor to boost antitumor chemo-immunotherapy. Adv Funct Mater. 2024;34(30):12.
35. Wu Y, Chen W, Wang C, Xing D. Nanozyme-activating prodrug therapies: a review. CCL. 2024;35(2):109096.
36. Liang Q, Xi J, Gao XJ, Zhang R, Yang Y, Gao X, Yan X, Gao L, Fan K. A metal-free nanozyme-activated prodrug strategy for targeted tumor catalytic therapy. Nano Today. 2020;35:100935.
37. Cao C, Yang N, Su Y, Zhang Z, Wang C, Song X, Chen P, Wang W, Dong X. Starvation, ferroptosis, and prodrug therapy synergistically enabled by a cytochrome c oxidase like nanozyme. Adv Mater. 2022;34(29):2203236.
38. Li J, Zhu L, Kwok HF. Nanotechnology-based approaches overcome lung cancer drug resistance through diagnosis and treatment. Drug Resist Updat. 2023;66:100904.
39. Li F, Sun H, Ren J, Zhang B, Hu X, Fang C, Lee J, Gu H, Ling D. A nuclease-mimetic platinum nanozyme induces concurrent DNA platination and oxidative cleavage to overcome cancer drug resistance. Nat Commun. 2022;13(1):7361.
40. Guo SY, Sun D, Ni DL, Yu MR, Qian K, Zhang W, Yang YW, Song S, Li Y, Xi ZY, Wang J. Smart tumor microenvironment-responsive nanotheranostic agent for effective cancer therapy. Adv Funct Mater. 2020;30(17):2000486.
41. Veroniaina H, Wu Z, Qi X. Innate tumor-targeted nanozyme overcoming tumor hypoxia for cancer theranostic use. J Adv Res. 2021;33:201–13.
42. Liu CH, Wang DD, Zhang SY, Cheng YR, Yang F, Xing Y, Xu TL, Dong HF, Zhang XJ. Biodegradable biomimic copper/manganese silicate nanospheres for chemodynamic/photodynamic synergistic therapy with simultaneous glutathione depletion and hypoxia relief. ACS Nano. 2019;13(4):4267–77.
43. Ma B, Wang S, Liu F, Zhang S, Duan J, Li Z, Kong Y, Sang Y, Liu H, Bu W, Li L. Self-assembled Copper–Amino acid nanoparticles for in situ glutathione “AND” H_2O_2 sequentially triggered chemodynamic therapy. J Am Chem Soc. 2019;141(2):849–57.
44. Li SL, Jiang P, Jiang FL, Liu Y. Recent advances in nanomaterial-based nanoplatforms for chemodynamic cancer therapy. Adv Funct Mater. 2021;31(22):2100243.

45. Shi L, Wang Y, Zhang C, Zhao Y, Lu C, Yin B, Yang Y, Gong X, Teng L, Liu Y, Zhang X, Song G. An acidity-unlocked magnetic nanoplatform enables self-boosting ROS generation through upregulation of lactate for imaging-guided highly specific chemodynamic therapy. Angew Chem Int Ed. 2021;60(17):9562–72.
46. Yang K, Yu G, Yang Z, Yue L, Zhang X, Sun C, Wei J, Rao L, Chen X, Wang R. Supramolecular polymerization-induced nanoassemblies for self-augmented cascade chemotherapy and chemodynamic therapy of tumor. Angew Chem Int Ed. 2021;60(32):17570–8.
47. Sang Y, Cao F, Li W, Zhang L, You Y, Deng Q, Dong K, Ren J, Qu X. Bioinspired construction of a nanozyme-based H_2O_2 homeostasis disruptor for intensive chemodynamic therapy. J Am Chem Soc. 2020;142(11):5177–83.
48. Feng L, Liu B, Xie R, Wang D, Qian C, Zhou W, Liu J, Jana D, Yang P, Zhao Y. An Ultrasmall $SnFe_2O_4$ nanozyme with endogenous oxygen generation and glutathione depletion for synergistic cancer therapy. Adv Funct Mater. 2021;31(5):2006216.
49. Li M, Huo L, Zeng J, Zhu G, Shi S, Liu X, Zhu X, Huang G, Qiu D, Jia J, Ni K, Zhao Z. Surface engineered iron oxide nanozyme for synergistic chemodynamic/photodynamic therapy with glutathione depletion and hypoxia relief. Chem Eng J. 2022;440:135966.
50. Lan M, Zhao S, Liu W, Lee CS, Zhang W, Wang P. Photosensitizers for photodynamic therapy. Adv Healthc Mater. 2019;8(13):1900132.
51. Zheng Y, Ye J, Li Z, Chen H, Gao Y. Recent progress in sono-photodynamic cancer therapy: from developed new sensitizers to nanotechnology-based efficacy-enhancing strategies. APSB. 2021;11(8):2197–219.
52. Zhang Y, Wang F, Liu C, Wang Z, Kang L, Huang Y, Dong K, Ren J, Qu X. Nanozyme decorated metal–organic frameworks for enhanced photodynamic therapy. ACS Nano. 2018;12(1):651–61.
53. Liang X, Chen M, Bhattarai P, Hameed S, Dai Z. Perfluorocarbon@porphyrin nanoparticles for tumor hypoxia relief to enhance photodynamic therapy against liver metastasis of colon cancer. ACS Nano. 2020;14(10):13569–83.
54. You Q, Zhang KY, Liu JY, Liu CL, Wang HY, Wang MT, Ye SY, Gao HQ, Lv LT, Wang C, Zhu L, Yang YL. Persistent regulation of tumor hypoxia microenvironment via a bioinspired Pt-based oxygen nanogenerator for multimodal imaging-guided synergistic phototherapy. Adv Sci. 2020;7(17):19.
55. Wang DD, Wu HH, Phua SZF, Yang GB, Lim WQ, Gu L, Qian C, Wang HB, Guo Z, Chen HZ, Zhao YL. Self-assembled single-atom nanozyme for enhanced photodynamic therapy treatment of tumor. Nat Commun. 2020;11(1):13. Sun B, Bte Rahmat JN, Zhang Y. Advanced techniques for performing photodynamic therapy in deep-seated tissues. Biomaterials. 2022;291:121875.
56. Pan MM, Li P, Yu YP, Jiang M, Yang X, Zhang P, Nie J, Hu J, Yu X, Xu L. Bimetallic ions functionalized metal–organic-framework nanozyme for tumor microenvironment regulating and enhanced photodynamic therapy for hypoxic tumor. Adv Healthc Mater. 2023;12(26):2300821.
57. Jeon J, You DG, Um W, Lee J, Kim CH, Shin S, Kwon S, Park JH. Chemiluminescence resonance energy transfer–based nanoparticles for quantum yield–enhanced cancer phototheranostics. Sci Adv. 2020;6(21):eaaz8400.
58. Liu Y, Bhattarai P, Dai Z, Chen X. Photothermal therapy and photoacoustic imaging *via* nanotheranostics in fighting cancer. Chem Soc Rev. 2019;48(7):2053–108.
59. Gao YJ, Liu Y, Li XL, Wang H, Yang YL, Luo Y, Wan YP, Lee CS, Li SL, Zhang XH. A stable open-shell conjugated diradical polymer with ultra-high photothermal conversion efficiency for NIR-II photo-immunotherapy of metastatic tumor. Nano Micro Lett. 2024;16(1):14.
60. Li D, Chen XH, Dai WB, Jin Q, Wang D, Ji J, Tang BZ. Photo-triggered cascade therapy: a nir-ii aie luminogen collaborating with nitric oxide facilitates efficient collagen depletion for boosting pancreatic cancer phototheranostics. Adv Mater. 2024;36(13):13.
61. Tsai M-F, Chang S-HG, Cheng F-Y, Shanmugam V, Cheng Y-S, Su C-H, Yeh C-S. Au Nanorod design as light-absorber in the first and second biological near-infrared windows for *in vivo* photothermal therapy. ACS Nano. 2013;7(6):5330–42.

62. Wang S-B, Zhang C, Ye J-J, Zou M-Z, Liu C-J, Zhang X-Z. Near-infrared light responsive nanoreactor for simultaneous tumor photothermal therapy and carbon monoxide-mediated anti-inflammation. ACS Cent Sci. 2020;6(4):555–65.
63. Li ZH, Chen Y, Sun YX, Zhang XZ. Platinum-doped prussian blue nanozymes for multiwavelength bioimaging guided photothermal therapy of tumor and anti-inflammation. ACS Nano. 2021;15(3):5189–200.
64. Tang YK, Ge L, Jiang L, Jiang XQ. Pore-enhanced reactive oxygen species generation by using covalent organic frameworks for improving sonodynamic therapy of cancer. Nano Today. 2024;55:11.
65. Maleki A, Seyedhamzeh M, Yuan M, Agarwal T, Sharifi I, Mohammadi A, Kelicen-Uğur P, Hamidi M, Malaki M, Al Kheraif AA, Cheng Z, Lin J. Titanium-based nanoarchitectures for sonodynamic therapy-involved multimodal treatments. Small. 2023;19(12):2206253.
66. Liu S, Zhang W, Chen Q, Hou J, Wang J, Zhong Y, Wang X, Jiang W, Ran H, Guo D. Multifunctional nanozyme for multimodal imaging-guided enhanced sonodynamic therapy by regulating the tumor microenvironment. Nanoscale. 2021;13(33):14049–66.
67. Zhong XY, Wang XW, Cheng L, Tang YA, Zhan GT, Gong F, Zhang R, Hu J, Liu Z, Yang XL. GSH-depleted $PtCu_3$ nanocages for chemodynamic- enhanced sonodynamic cancer therapy. Adv Funct Mater. 2020;30(4):12.
68. Zhou H, Sun J, Wu J, Wei H, Zhou X. Biodegradable nanosonosensitizers with the multiple modulation of tumor microenvironment for enhanced sonodynamic therapy. Int J Nanomed. 2021;16:2633–46.
69. Yang L, Dong S, Gai S, Yang D, Ding H, Feng L, Yang G, Rehman Z, Yang P. Deep insight of design, mechanism, and cancer theranostic strategy of nanozymes. Nano Micro Lett. 2024;16(1):28.
70. Wei C, Liu Y, Zhu X, Chen X, Zhou Y, Yuan G, Gong Y, Liu J. Iridium/ruthenium nanozyme reactors with cascade catalytic ability for synergistic oxidation therapy and starvation therapy in the treatment of breast cancer. Biomaterials. 2020;238:119848.
71. Liu C, Xing J, Akakuru OU, Luo L, Sun S, Zou R, Yu Z, Fang Q, Wu A. Nanozymes-engineered metal–organic frameworks for catalytic cascades-enhanced synergistic cancer therapy. Nano Lett. 2019;19(8):5674–82.
72. Yang X, Yang Y, Gao F, Wei J-J, Qian C-G, Sun M-J. Biomimetic hybrid nanozymes with self-supplied H^+ and accelerated O_2 generation for enhanced starvation and photodynamic therapy against hypoxic tumors. Nano Lett. 2019;19(7):4334–42.
73. Wang M, Chang MY, Zheng P, Sun QQ, Wang GQ, Lin J, Li CX. A noble AuPtAg-GOx nanozyme for synergistic tumor immunotherapy induced by starvation therapy-augmented mild photothermal therapy. Adv Sci. 2022;9(31):10.
74. Wang WJ, Zhang L, Zhang YJ, Liu XM, Song AJ, Ren JS, Qu XG. A self-adaptive pyroptosis inducer: optimizing the catalytic microenvironment of nanozymes by membrane-adhered microbe enables potent cancer immunotherapy. Adv Mater. 2024;36(14):11.
75. Zhang L, Wang H, Qu X. Biosystem-inspired engineering of nanozymes for biomedical applications. Adv Mater. 2024;36(10):2211147.
76. Wang J, Fang L, Li P, Ma L, Na W, Cheng C, Gu Y, Deng D. Inorganic nanozyme with combined self-oxygenation/degradable capabilities for sensitized cancer immunochemotherapy. Nano Micro Lett. 2019;11(1):74.
77. Qian Y, Wang M, Xie Y, Sun Q, Gao M, Li C. Rationally integrated precise ER-targeted and oxygen-compensated photodynamic immunostimulant for immunogenicity-boosted tumor therapy. Adv Healthc Mater. 2023;12(32):2301728.
78. Gu T, Wang Y, Lu Y, Cheng L, Feng L, Zhang H, Li X, Han G, Liu Z. Platinum nanoparticles to enable electrodynamic therapy for effective cancer treatment. Adv Mater. 2019;31(14):1806803.
79. Chen T, Gu T, Cheng L, Li X, Han G, Liu Z. Porous Pt nanoparticles loaded with doxorubicin to enable synergistic chemo−/electrodynamic therapy. Biomaterials. 2020;255:120202.
80. Chen T, Chu Q, Li M, Han G, Li X. Fe3O4@Pt nanoparticles to enable combinational electrodynamic/chemodynamic therapy. J Nanobiotechnol. 2021;19(1):206.

81. Lu Z, Gao J, Fang C, Zhou Y, Li X, Han G. Porous Pt nanospheres incorporated with GOx to enable synergistic oxygen-inductive starvation/electrodynamic tumor therapy. Adv Sci. 2020;7(17):2001223.
82. Li S, Shang L, Xu B, Wang S, Gu K, Wu Q, Sun Y, Zhang Q, Yang H, Zhang F, Gu L, Zhang T, Liu H. A nanozyme with photo-enhanced dual enzyme-like activities for deep pancreatic cancer therapy. Angew Chem Int Ed. 2019;58(36):12624–31.
83. Loynachan CN, Soleimany AP, Dudani JS, Lin Y, Najer A, Bekdemir A, Chen Q, Bhatia SN, Stevens MM. Renal clearable catalytic gold nanoclusters for in vivo disease monitoring. Nat Nanotechnol. 2019;14(9):883–90.
84. Wang X, Zhong X, Bai L, Xu J, Gong F, Dong Z, Yang Z, Zeng Z, Liu Z, Cheng L. Ultrafine titanium monoxide (TiO_{1+x}) nanorods for enhanced sonodynamic therapy. J Am Chem Soc. 2020;142(14):6527–37.
85. Xu B, Cui Y, Wang W, Li S, Lyu C, Wang S, Bao W, Wang H, Qin M, Liu Z, Wei W, Liu H. Immunomodulation-enhanced nanozyme-based tumor catalytic therapy. Adv Mater. 2020;32(33):2003563.
86. Huang X, Xia F, Nan Z. Fabrication of FeS_2/SiO_2 double mesoporous hollow spheres as an artificial peroxidase and rapid determination of H_2O_2 and glutathione. ACS Appl Mater Interfaces. 2020;12(41):46539–48.
87. Dong S, Dong Y, Jia T, Liu S, Liu J, Yang D, He F, Gai S, Yang P, Lin J. GSH-depleted nanozymes with hyperthermia-enhanced dual enzyme-mimic activities for tumor nanocatalytic therapy. Adv Mater. 2020;32(42):2002439.
88. Fan X, Gong X, Zhou F, Chen B, Tan S, Xu H, Pan A, Liang S, He Y. Mesoporous peroxidase nanozyme for synergistic chemodynamic therapy and chemotherapy. Colloids Surf B: Biointerfaces. 2022;216:112603.
89. Jana D, Wang D, Bindra AK, Guo Y, Liu J, Zhao Y. Ultrasmall alloy nanozyme for ultrasound- and near-infrared light-promoted tumor ablation. ACS Nano. 2021;15(4):7774–82.
90. Yang H, Xu B, Li S, Wu Q, Lu M, Han A, Liu H. A photoresponsive nanozyme for synergistic catalytic therapy and dual phototherapy. Small. 2021;17(10):2007090.
91. Zhang C, Wang X, Dong X, Mei L, Wu X, Gu Z, Zhao Y. X-ray-facilitated redox cycling of nanozyme possessing peroxidase-mimicking activity for reactive oxygen species-enhanced cancer therapy. Biomaterials. 2021;276:121023.
92. Ma Z, Foda MF, Liang H, Zhao Y, Han H. In situ nanozyme-amplified NIR-II phototheranostics for tumor-specific imaging and therapy. Adv Funct Mater. 2021;31(37):2103765.
93. Hu B, Xiao X, Chen P, Qian J, Yuan G, Ye Y, Zeng L, Zhong S, Wang X, Qin X, Yang Y, Pan Y, Zhang Y. Enhancing anti-tumor effect of ultrasensitive bimetallic RuCu nanoparticles as radiosensitizers with dual enzyme-like activities. Biomaterials. 2022;290:121811.
94. Zhu D, Chen H, Huang C, Li G, Wang X, Jiang W, Fan K. H_2O_2 self-producing single-atom Nanozyme hydrogels as light-controlled oxidative stress amplifier for enhanced synergistic therapy by transforming "cold" tumors. Adv Funct Mater. 2022;32(16):2110268.
95. Hou M, Su H, Wu Q, Sun W, Zhang P, Jiang Y, Qian K, Zhang C. A self-enhancing nanoreactor reinforces radioimmunotherapy by reprogramming nutrients and redox metabolisms. Adv Funct Mater. 2023;33(16):2212510.
96. Shi X, Liu J, Wang G. A peroxidase-like magneto-gold nanozyme AuNC@Fe3O4 with photothermal effect for induced cell apoptosis of hepatocellular carcinoma cells in vitro. Front Bioeng Biotechnol. 2023;11:1168750.
97. Yang J, Fang L, Jiang R, Qi L, Xiao Y, Wang W, Ismail I, Fang X. RuCu nanosheets with ultrahigh nanozyme activity for chemodynamic therapy. Adv Healthc Mater. 2023;12(23):2300490.
98. Luo T, Yang H, Wang R, Pu Y, Cai Z, Zhao Y, Bi Q, Lu J, Jin R, Nie Y, Shuai X. Bifunctional cascading nanozymes based on carbon dots promotes photodynamic therapy by regulating hypoxia and glycolysis. ACS Nano. 2023;17(17):16715–30.
99. Zhu X, Gong Y, Liu Y, Yang C, Wu S, Yuan G, Guo X, Liu J, Qin X. Ru@CeO2 yolk shell nanozymes: oxygen supply in situ enhanced dual chemotherapy combined with photothermal therapy for orthotopic/subcutaneous colorectal cancer. Biomaterials. 2020;242:119923.

100. Dan Q, Hu D, Ge Y, Zhang S, Li S, Gao D, Luo W, Ma T, Liu X, Zheng H, Li Y, Sheng Z. Ultrasmall theranostic nanozymes to modulate tumor hypoxia for augmenting photodynamic therapy and radiotherapy. Biomater Sci. 2020;8(3):973–87.
101. Ying W, Zhang Y, Gao W, Cai X, Wang G, Wu X, Chen L, Meng Z, Zheng Y, Hu B, Lin X. Hollow magnetic nanocatalysts drive starvation–chemodynamic–hyperthermia synergistic therapy for tumor. ACS Nano. 2020;14(8):9662–74.
102. Wang M, Chang M, Chen Q, Wang D, Li C, Hou Z, Lin J, Jin D, Xing B. Au2Pt-PEG-Ce6 nanoformulation with dual nanozyme activities for synergistic chemodynamic therapy/phototherapy. Biomaterials. 2020;252:120093.
103. Sun D, Pang X, Cheng Y, Ming J, Xiang S, Zhang C, Lv P, Chu C, Chen X, Liu G, Zheng N. Ultrasound-switchable nanozyme augments sonodynamic therapy against multidrug-resistant bacterial infection. ACS Nano. 2020;14(2):2063–76.
104. Wang J, Sun J, Hu W, Wang Y, Chou T, Zhang B, Zhang Q, Ren L, Wang H. A porous Au@Rh bimetallic core–shell nanostructure as an H_2O_2-driven oxygenerator to alleviate tumor hypoxia for simultaneous bimodal imaging and enhanced photodynamic therapy. Adv Mater. 2020;32(22):2001862.
105. Zhang Y, Zhao J, Zhang L, Zhao Y, Zhang Y, Cheng L, Wang D, Liu C, Zhang M, Fan K, Zhang M. A cascade nanoreactor for enhancing sonodynamic therapy on colorectal cancer via synergistic ROS augment and autophagy blockage. Nano Today. 2023;49:101798.
106. Kang S, Gil Y-G, Min D-H, Jang H. Nonrecurring circuit nanozymatic enhancement of hypoxic pancreatic cancer phototherapy using speckled Ru–Te hollow nanorods. ACS Nano. 2020;14(4):4383–94.
107. Wang Z, Li G, Gao Y, Yu Y, Yang P, Li B, Wang X, Liu J, Chen K, Liu J, Wang W. Trienzyme-like iron phosphates-based (FePOs) nanozyme for enhanced anti-tumor efficiency with minimal side effects. Chem Eng J. 2021;404:125574.
108. Yang P, Tao J, Chen F, Chen Y, He J, Shen K, Zhao P, Li Y. Multienzyme-mimic ultrafine alloyed nanoparticles in metal organic frameworks for enhanced chemodynamic therapy. Small. 2021;17(7):2005865.
109. Wang Z, Liu J, Feng M, Song K, Li K, Liu W, Guan S, Lin Y. ROS-activated selective fluorescence imaging of cancer cells via the typical LDH-based nanozyme. Chem Eng J. 2023;470:144020.
110. Zhang L, Yang QC, Wang S, Xiao Y, Wan SC, Deng H, Sun ZJ. Engineering multienzyme-mimicking covalent organic frameworks as pyroptosis inducers for boosting antitumor immunity. Adv Mater. 2022;34(13):2108174.
111. Zhao Y, Xu J, Zhang Y, Wu F, Zhao W, Li R, Yang Y, Zhang M, Zhang Y, Guo C. Biomimetic redox-responsive prodrug micelles with diselenide linkage for platinum nanozymes augmented sonodynamic/chemo combined therapy of colon cancer. Chem Eng J. 2023;472:144911.
112. Zou Q, Chen Y, Qin H, Tang R, Han T, Guo Z, Zhao J, Xu D. The role and mechanism of TCM in the prevention and treatment of infectious diseases. Front Microbiol. 2023;14:1286364.
113. Zhao Z, Li Y, Zhou L, Zhou X, Xie B, Zhang W, Sun J. Prevention and treatment of COVID-19 using traditional Chinese medicine: a review. Phytomedicine. 2021;85:153308.
114. Wei D, Yang H, Zhang Y, Zhang X, Wang J, Wu X, Chang J. Nano-traditional Chinese medicine: a promising strategy and its recent advances. J Mater Chem B. 2022;10(16):2973–94.
115. Souho T, Lamboni L, Xiao L, Yang G. Cancer hallmarks and malignancy features: gateway for improved targeted drug delivery. Biotechnol Adv. 2018;36(7):1928–45.
116. Moradi Kashkooli F, Soltani M, Souri M. Controlled anti-cancer drug release through advanced nano-drug delivery systems: static and dynamic targeting strategies. JCR. 2020;327:316–49.
117. Singh SP, Alvi SB, Pemmaraju DB, Singh AD, Manda SV, Srivastava R, Rengan AK. NIR triggered liposome gold nanoparticles entrapping curcumin as in situ adjuvant for photothermal treatment of skin cancer. Int J Biol Macromol. 2018;110:375–82.
118. Yin HL, Xin Y, Yang JY, Luo Q, Yang M, Sun JC, Wang YT, Wang Q, Kalvakolanu D, Guo BF, Jiang W, Zhang L. Multifunctional nanozymes: promising applications in clinical diagnosis and cancer treatment. Biosens Bioelectron. 2025;279:18.

6 Nanomaterials-Mediated Cancer Microwave Dynamic Therapy

Qiong Wu and Xianwei Meng

Abstract

Cancer treatment has long represented a critical global health challenge. In recent years, the continuous advancement of nanotechnology has driven significant progress in nanoparticle-mediated cancer therapeutic strategies. These nanomedicines, distinguished by their unique pharmacological, biochemical, and physicochemical properties, demonstrate targeted and efficient action at tumor sites, thereby enhancing therapeutic efficacy while minimizing damage to normal cells. Among these approaches, nanoparticle-mediated dynamic therapy constitutes an innovative approach to tumor treatment that leverages the tumor microenvironment or external stimuli to activate nanomedicines, thereby generating cytotoxic reactive oxygen species (ROS) for tumor-specific treatment. Dynamic therapy is highly selective and tumor-specific, with minimal systemic toxicity, and has attracted considerable attention from the international medical community. This chapter highlights recent advancements in nanoparticle-mediated microwave dynamic therapy aimed at developing cutting-edge cancer treatment approaches based on nanotechnology. Nanoparticles can accumulate in tumor regions via either passive or active targeting mechanisms and can induce both microwave thermal and dynamic effects under microwave irradiation. The integration of microwave thermal therapy and microwave dynamic therapy in cancer treatment has been explored. Specifically engineered nanoparticles have demonstrated remarkable therapeutic outcomes across various tumor models, offering promising modalities for clinical cancer treatment.

Q. Wu · X. Meng (✉)
State Key Laboratory of Cryogenics Science and Technology, Technical Institute of Physics and Chemistry, Chinese Academy of Sciences, Beijing, P. R. China
e-mail: mengxw@mail.ipc.ac.cn

M. Chu (ed.), *Modern Nanomaterials and Nanotechnology for Cancer Therapy*,
https://doi.org/10.1007/978-981-95-6849-9_6

Keywords

Microwave Dynamic Therapy (MDT) · Nanomaterials · Microwave Irradiation · Reactive Oxygen Species (ROS) · Cancer Therapy

1 Introduction

The global incidence and mortality rates of cancer continue to rise annually, establishing neoplastic diseases as one of the foremost causes of death worldwide and presenting a major public health challenge [1, 2]. Concurrent with advancements in scientific and technological innovation, therapeutic strategies for oncology have evolved beyond conventional interventions such as surgery [3], chemotherapy [4], and radiotherapy [5] to incorporate novel modalities, including targeted therapy [6], immunotherapy [7], thermal therapy [8], and dynamic therapy [9]. Among these, dynamic therapy operates through the generation of cytotoxic reactive oxygen species (ROS) [10], such as hydroxyl radicals (·OH), singlet oxygen ($^1O_2$), superoxide anion ($\cdot O_2^-$), etc., which mediate oxidative reactions with nearby biomolecules, resulting in cytotoxicity. This process leads to apoptosis and necrosis of tumor cells, damages tumor vasculature, and stimulates the host immune system to inhibit tumor growth. Relative to traditional treatments, dynamic therapy provides enhanced selectivity toward malignant cells, shortened therapeutic duration, diminished systemic toxicity, minimal off-target effects, and a reduced propensity for inducing drug resistance.

Photodynamic therapy (PDT), the most extensively studied form of dynamic therapy, utilizes light to excite a photosensitizer from its ground state to an excited state [11]. During the de-excitation process, energy is transferred to molecular oxygen, generating ROS for tumor eradication. Nonetheless, the limited penetration depth of light in biological tissues remains a critical drawback, hindering the clinical applicability of PDT. In contrast, microwave energy offers superior tissue penetration, is minimally impeded by gaseous or bony structures, and thus represents a promising alternative energy source for dynamic therapy. Microwave thermal therapy (MWTT), a clinically established modality [12], employs the biothermal effects of electromagnetic waves with wavelengths ranging from 0.1 mm to 1 m for physical tumor ablation [13]. Recognized as a "green" thermal treatment modality, MWTT is characterized by deep tissue penetration, efficient heating, broad ablation volumes, and low toxicity, securing its prominence in clinical oncology practice [14]. Consequently, the integration of microwave dynamic therapy (MWDT) with MWTT thus emerges as a highly promising strategy for achieving comprehensive tumor eradication.

Advances in nanotechnology have further potentiated this integration through miniaturization, functional integration, and automation across biomedical applications [15]. Nanomaterials, by virtue of their tunable physicochemical properties and enhanced delivery capabilities, afford innovative platforms for the design of sophisticated cancer therapeutics. Variations in nanomaterial composition, structure, and

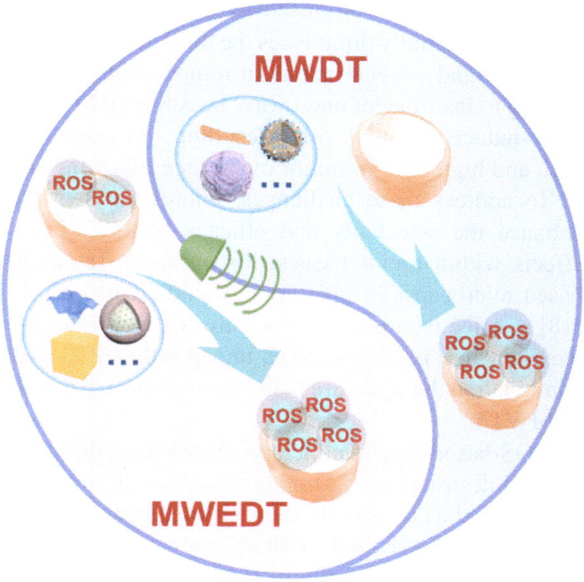

Fig. 6.1 Schematic overview of nanoparticle-mediated MWDT

surface properties can profoundly influence their biological behaviors, offering opportunities to improve tumor selectivity, pharmacokinetic profiles, cellular uptake, cytotoxic potency, physiological stability, and multifunctional capacity for both imaging and treatment. Combining microwave thermal effects, microwave dynamic effects, and functional nanomaterials can enhance the selectivity of microwave thermal effects on tumors, significantly increasing the area of tumor necrosis and improving MWTT efficacy. Moreover, the nonthermal effects of MWDT can effectively complement MWTT by targeting residual malignant cells post-MWTT, thereby reducing tumor recurrence rates, and achieving complete tumor eradication. As illustrated in Fig. 6.1, MWDT can be categorized into two distinct mechanistic classes: microwave-enhanced dynamic therapy (MWEDT), in which microwave irradiation amplifies ROS generation, and pure MWDT, wherein ROS are produced specifically under microwave irradiation rather than increasing the yield of ROS. This review comprehensively summarizes recent progress in nanoparticle-mediated MWDT for more precise tumor targeting and discusses future prospects for its clinical application.

2 Microwave Enhanced Dynamic Therapy (MWEDT)

MWTT is widely regarded as a highly promising clinical treatment modality owing to its operational simplicity, high efficiency, and minimal invasiveness. Nevertheless, MWTT suffers from inherent limitations such as poor tumor selectivity and inadequate conformability to irregular tumor geometries. Tumors often exhibit complex and irregular morphologies, which contrast sharply with the well-defined circular or oval ablation zones produced by conventional MWTT. This morphological mismatch substantially diminishes the therapeutic efficacy of MWTT. Furthermore, the ion types and concentrations in tumors and normal tissues are relatively similar, leading to insufficient microwave targeting of MWTT and a limited range of microwave-induced heating, often resulting in incomplete ablation, residual malignant foci, and high rates of tumor recurrence following standard MWTT.

To address these challenges, microwave sensitizers have been introduced to enhance the selectivity and efficacy of MWTT by promoting targeted thermal effects within tumor tissues. For instance, researchers have developed alginate-based microcapsules [16], ionic liquid (IL)@ZrO_2 [17], and PLGA nanoparticles [18] as effective microwave sensitizers. Nevertheless, the heterogeneous and complex structure of tumors often impedes the effectiveness of such sensitizers. Thus, combination therapies are being explored to achieve improved therapeutic outcomes of MWTT.

ROS-based dynamic therapy represents a promising approach for selective cancer cell destruction and tumor growth inhibition. Typically, PDT has been widely investigated as a spatially targeted treatment for suppressing tumor progression in scientific research and clinical applications. The combination of MWTT with MWDT, which concurrently induces thermal effects and ROS generation under microwave irradiation, offers an attractive synergistic strategy for enhancing antitumor efficiency.

In a recent study, Mn^{2+}-doped Zr-based metal-organic framework nanocubes (Mn-ZrMOF NCs) were employed for the first time to achieve synergistic tumor treatment via combined MWTT and MWDT [19]. The highly porous structure and large surface area of Mn-ZrMOF NCs facilitate enhanced ion adsorption and confinement, thereby improving microwave energy conversion and thermal efficacy. In addition, in vitro studies demonstrated that these nanocubes catalyze and augment the production of hydroxyl radicals under microwave irradiation (Fig. 6.2a, b). The as-developed Mn-ZrMOF NCs not only enhance the conversion of microwave energy to heat energy but also promote the generation of ROS under microwave irradiation. The research results of in vivo animal experiments further confirmed significant tumor growth inhibition attributable to the combined effects of MWTT and MWDT. This work establishes a foundation for the use of nanoplatforms capable of simultaneous microwave-induced thermal and dynamic effects for synergistic tumor therapy of MWTT and MWDT.

The MWDT mechanism described in this work operates through Mn-doped sites in the ZrMOF NCs, which serve as catalytic centers for ·OH generation. Microwave irradiation enhances the peroxidase-like activity of Mn-ZrMOF NCs, while their porous architecture and large surface area expose abundant catalytic active sites, thereby facilitating electron transfer between the nanocubes and substrates and leading to increased ·OH production (Fig. 6.2c). Consequently, this modality is more accurately termed MWEDT. Although Mn-ZrMOF NCs can generate ROS in the absence of microwave irradiation, ROS production is significantly amplified upon microwave irradiation, underscoring the unique capability of Mn-ZrMOF NCs

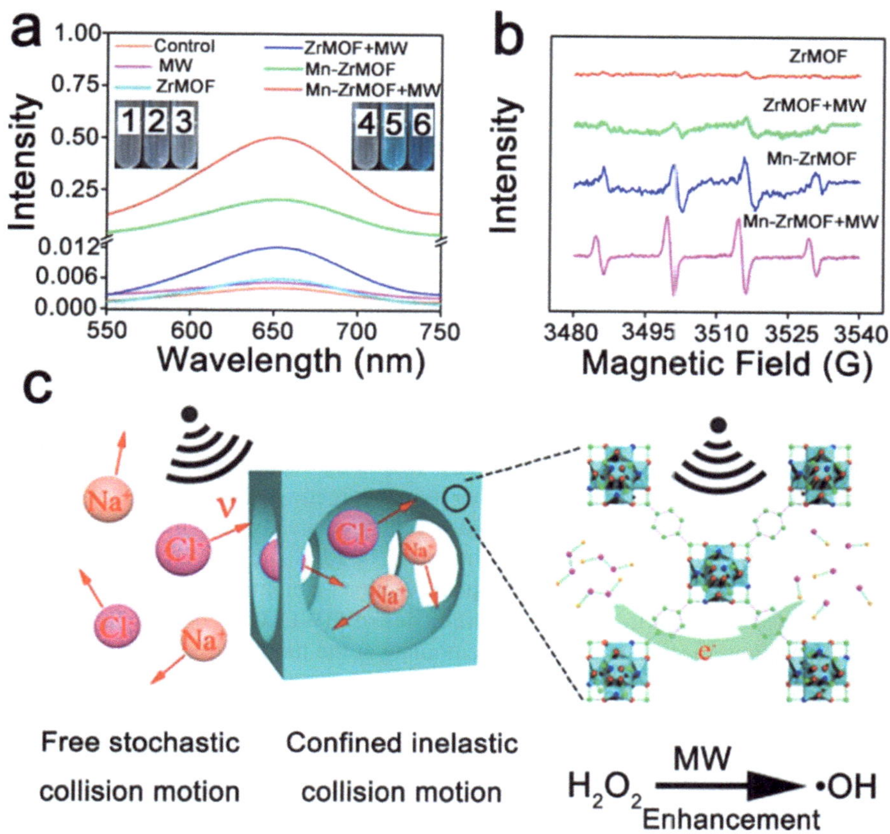

Fig. 6.2 MWEDT of Mn-ZrMOF NCs. (**a**) ROS generation of Mn-ZrMOF and ZrMOF NCs with or without microwave irradiation detected by the TMB method. (**b**) ROS generation of Mn-ZrMOF and ZrMOF NCs with or without microwave irradiation detected by EPR spectra. (**c**) Schematic diagram of ROS generation by Mn-ZrMOF NCs during microwave irradiation. (Copyright © 2018, American Chemical Society)

to enable efficient cancer treatment through the synergistic effect of thermal therapy and dynamic therapy under microwave irradiation for the first time.

In addition to manganese, which exhibits peroxidase-like catalytic activity, iron and copper also serve as excellent candidates for nanozyme design. For instance, the synthesized biodegradable nanoparticles IL@MIL-101(Fe)@BSAAuNCs NPs demonstrate both peroxidase-mimetic catalytic activity and dual-modal imaging capabilities [20]. Within this system, iron-based metal-organic framework nanoparticles (MIL-101(Fe) NPs) function as nanozymes, capable of catalyzing the conversion of hydrogen peroxide into hydroxyl radicals (·OH) within the tumor microenvironment. Moreover, under microwave irradiation, the production of ·OH is significantly accelerated, enabling effective MWEDT. In vivo experimental results confirm that IL@MIL-101(Fe)@BSAAuNCs NPs achieve a tumor suppression rate of up to 96.65%, demonstrating remarkable antitumor efficacy. This nanoplatform thus holds great potential for precise diagnosis and synergistic treatment of tumors through combined MWEDT and MWTT.

In a related study, Mo et al. designed and synthesized a novel microwave-activated copper-doped zirconium-based metal-organic framework (Cu-Zr MOF) to enhance both MWTT and MWEDT [21]. Owing to intense inelastic collisions among ions within its abundant micropores, the Cu-Zr MOF exhibits strong microwave sensitivity and high thermal conversion efficiency, thereby significantly improving MWTT. Furthermore, the presence of Cu^{2+} ions confers Fenton-like activity, enabling the generation of ·OH radicals and enhancing catalytic redox-mediated cytotoxicity. Under microwave irradiation, this platform synergizes MWEDT with MWTT, resulting in significant tumor growth inhibition and maximized therapeutic synergy.

Nevertheless, the therapeutic efficiency of MWDT mediated by cytotoxic ROS is often constrained by several limitations, including inadequate ROS generation, robust angiogenic responses, and poor tumor-targeting efficiency. Studies have revealed that copper-based nanoparticles can generate oxygen under microwave irradiation, thereby promoting the production of ROS such as O_2, ·OH, and $^1O_2$, particularly O_2 [22]. Leveraging this phenomenon, a biomimetic nanoengineering strategy was designed to enhance the efficiency of MWEDT. Following intravenous administration, these nanoparticles accumulate at the tumor site through active targeting mediated by biomimetic modification and continuously produce oxygen under microwave irradiation, elevating the ROS level in the tumor microenvironment. Additionally, concomitant use of the anti-angiogenic agent apatinib downregulates vascular endothelial growth factor (VEGF) expression, effectively suppressing tumor angiogenesis post-MWEDT. The resulting RBC-Zr@APC/C NPs were applied for combined MWTT and MWEDT, achieving an inhibition rate of up to 96.79% in a murine 4T1 breast cancer model.

The therapeutic efficiency of MWEDT mediated by ROS is not only plagued by insufficient ROS generation but also significantly impacted by the high concentration of glutathione (GSH) in the tumor microenvironment. GSH scavenges ROS, thereby reducing the therapeutic efficacy of MWEDT. To address this, a manganese-based nanoplatform was engineered for microwave thermal-dynamic sensitization and GSH-responsive degradation [23]. Through a combination of nanoengineering approaches, including integration with cisplatin (CDDP), encapsulation of phase-change materials (PCM), and coating with tumor cell membrane (TCM) for homologous targeting, the MnMOF-CDDP@MCM-TCM nanomissiles (TMCP Nm) were constructed. These nanostructures effectively deplete GSH via active and passive targeting mechanisms, microwave-triggered and GSH-dependent reactions, and spatially controlled drug release. This process reverses CDDP resistance and amplifies the efficacy of microwave thermal-dynamic therapy combined with chemotherapy, ultimately yielding superior therapeutic outcomes in both in vivo and in vitro drug-resistant tumor models (Fig. 6.3). This strategy, focused on modulating the tumor microenvironment, significantly enhances the therapeutic efficacy of MWTT and MWEDT.

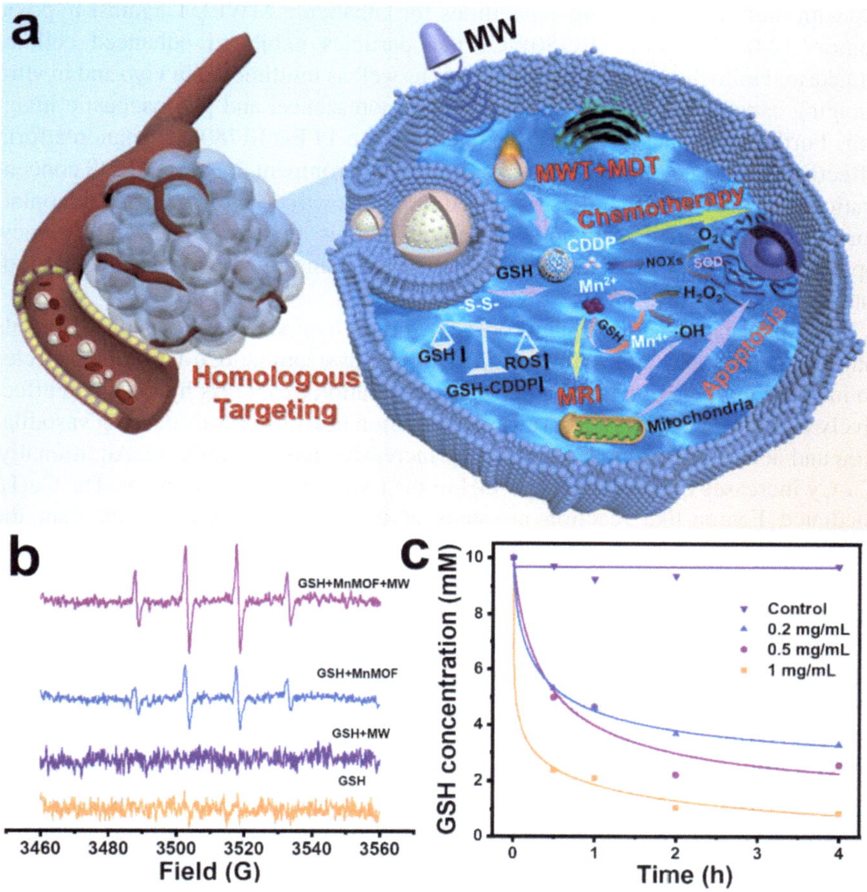

Fig. 6.3 GSH clearance of TMCP Nm enhances MWTT and MWEDT. (**a**) In vivo GSH clearance, MWTT and MWEDT of TMCP Nm. (**b**) Detection of ROS generation by EPR spectra. (**c**) GSH clearance results. (Copyright © 2022, Elsevier)

To address the issue that GSH scavenging ROS compromises the therapeutic efficiency of ROS-mediated MWEDT, Tian et al. prepared Fe-doped Cu bimetallic metal-organic frameworks (FCM) via a hydrothermal method, which not only improved the efficiency of microwave thermal energy conversion and enhanced the therapeutic effect of MWTT but also simultaneously promoted the generation of ROS by consuming GSH and facilitated the Fenton reaction to achieve MWEDT. The experimental results in vitro and in vivo demonstrated that the combined treatment of MWTT and MWEDT effectively induced tumor cell death without causing significant damage to normal tissues.

The tumor microenvironment plays a critical role in determining nanomaterial efficacy. To overcome the limitation of MWEDT in hypoxic tumors, Lu's group developed a liquid perfluorocarbon-based oxygen-rich nanoplatform PFP@IR780@

O_2 with multimodal imaging capabilities for enhancing MWEDT against hypoxic tumors [24]. The PFP@IR780@O_2 nanoparticles exhibited enhanced cellular uptake and mitochondrial targeting ability, as well as multimodal in vivo and in vitro imaging capabilities, including ultrasound, fluorescence, and photoacoustic imaging. Furthermore, the MWEDT mediated by the PFP@IR780@O_2 nanoplatform effectively alleviated the hypoxic tumor microenvironment, increased ROS concentration, and thus significantly suppressed tumor growth. The oxygen-rich nanoplatform with multimodal imaging capability shows higher MWEDT efficiency, opening a new door for treating hypoxic tumors using microwave-triggered nanocarrier-based therapy.

In response to the impediments posed by tumor hypoxia and elevated GSH levels on ROS generation, Yu's team designed copper-cysteamine (Cu-Cy) nanoparticles to increase oxygenation under microwave irradiation to address hypoxia and effectively eliminate GSH [25]. Microwave irradiation heats the tissue, causing vasodilation and accelerating blood flow, thereby increasing the oxygen level. Additionally, Cu-Cy increases the production of ·OH in the tumor microenvironment. The Cu(I)-mediated Fenton-like reaction proceeds at a significantly higher rate than the Fe-based Fenton reaction. Furthermore, excessive GSH in the tumor microenvironment can reduce Cu(II) generated by the Cu(I)-based Fenton reaction, further accelerating the reaction kinetics and amplifying ROS generation. The superior MWEDT performance of Cu-Cy was validated in the HCT15 tumor-bearing mouse model. The study also revealed that microwave-activated Cu-Cy induces ferroptosis in HCT15 cells, providing a new approach to further understand the mechanism of MWEDT in cancer treatment.

Hu's group constructed metal-organic framework (MOF) nanoplates featuring a pronounced lateral size-to-thickness ratio to augment the generation of ROS [26]. In contrast to spherical MOFs, these nanoplates showcase a greater abundance of exposed active sites, facilitating rapid ROS diffusion and thereby enhancing therapeutic potency. The synthesized manganese-doped titanium MOF nanoplates (Mn-Ti MOFs) serve as an avant-garde microwave-sensitizing agent for MWTT and MWEDT in the treatment of liver cancer under the guidance of magnetic resonance imaging (MRI). Owing to their porous structure, Mn-Ti MOFs engender heat through elastic collisions during microwave irradiation via the ion confinement effect. The in situ doping with manganese introduces defects and narrows the bandgap of Ti MOF, thereby escalating the separation efficiency of electron-hole pairs and promoting ROS production. Additionally, manganese confers MRI functionality to the MOF, enabling real-time visualization of nanoplate accumulation within tumors and guiding subsequent microwave irradiation protocols. Both in vitro and in vivo investigations confirmed that PEG-modified Mn-Ti MOFs deliver highly efficacious treatment for liver cancer when subjected to low-power microwave-triggered MWTT and MWEDT, offering novel insights into the research on microwave sensitizers.

To tackle the clinical conundrum of lung cancer bone metastasis, Jin's team engineered a microwave-responsive nanoplatform composed of $MgFe_2O_4$@ZOL nanoparticles (NPs) [27]. These $MgFe_2O_4$@ZOL NPs release Fe^{3+}, Mg^{2+}, and zoledronic acid (ZOL) within the acidic tumor microenvironment. Fe^{3+} is capable of depleting intracellular GSH and catalyzing the conversion of hydrogen peroxide into hydroxyl radicals, thereby initiating chemotherapeutic dynamic therapy (CDT). Microwave irradiation significantly potentiates the generation of ROS, facilitating the effective implementation of MWEDT. Furthermore, Mg^{2+} in conjunction with ZOL promotes osteoclast-like cell differentiation. Additionally, $MgFe_2O_4$@ZOL NPs enable targeted thermal ablation of tumor tissue under microwave irradiation, enhancing the effect of MWTT. Both in vitro and in vivo experiments indicate that the synergistic combination of active targeting, GSH depletion-enhanced CDT, MWEDT, and selective MWTT leads to significant antitumor therapeutic effects and bone repair effects. This multimodal combined therapy provides a promising strategy for treating lung cancer patients with bone metastasis.

Recently, research on nanoparticle-mediated MWEDT has expanded, and combination strategies with MWTT have demonstrated promising therapeutic outcomes across various tumor models [28, 29]. It is noteworthy that in most reported cases, ROS generation originates either from the nanoparticles themselves or from their reaction with the tumor microenvironment, with further amplification under microwave irradiation, thus being classified as MWEDT. This mechanism distinctly differs from MWDT, in which ROS are produced directly under microwave irradiation alone. The following section will focus on introducing advances in MWDT.

3 Microwave Dynamic Therapy (MWDT)

MWDT represents a therapeutic modality for tumor treatment, wherein microwaves act as the exclusive energy source to activate sensitizers for the generation of ROS. The energy of microwaves is approximately 2×10^{-22} J, which is sufficient only to excite molecular rotational transitions and insufficient to directly induce dynamical effects [30]. Drawing upon principles from microwave chemistry, catalysis, and heterogeneous reactions, it is postulated that selective interfacial heating may generate localized "hot spots" within nanoscale pores of materials. This resonant coupling between microwaves and these confined regions leads to the formation of point defects or the weakening of surface bonds, thereby facilitating electron transfer from adsorbed H_2O or O_2 molecules to generate free radicals, such as $\cdot O_2^-$ and $\cdot OH$.

Liquid metals (LMs), such as the eutectic gallium-indium (EGaIn) alloy, which melts near or below room temperature, have attracted significant interest across various fields, including catalysis, flexible electronics, biomedicine, and energy applications [31]. Recent studies have revealed that EGaIn-based ultrafine nanoparticles can function as microwave sensitizers by generating ROS under microwave irradiation, thereby revolutionizing conventional ROS generation methodologies [32]. As illustrated in Fig. 6.4a, EGaIn (composed of 75.5% Ga and 24.5% In) combined with an ionic liquid was uniformly loaded into mesoporous ZrO_2 to form IL-LM-ZrO_2 supernanoparticles (SNPs). These IL-LM-ZrO_2 SNPs exhibited a pronounced microwave dynamic effect: while inert prior to irradiation, they produced

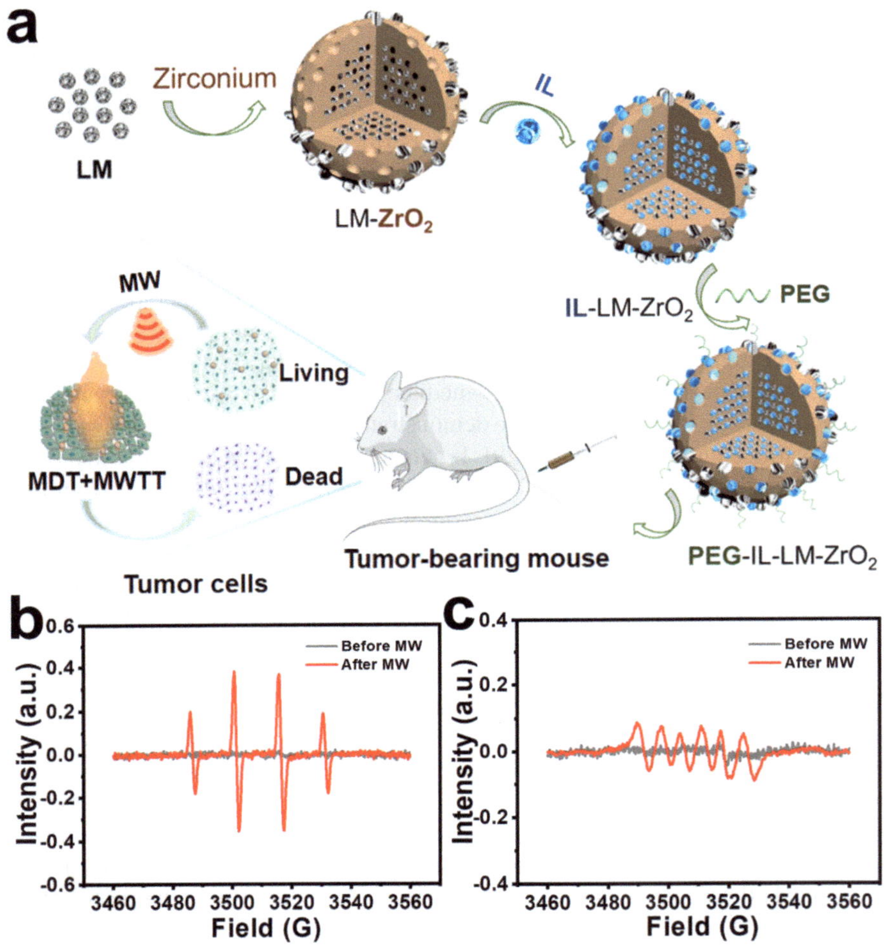

Fig. 6.4 MWDT of IL-LM-ZrO_2 SNPs. (**a**) Schematic diagram of PEG-IL-LM-ZrO_2 SNPs combined with MWDT and MWTT in tumor treatment. (**b**) Detection of ·OH production by EPR spectra. (**c**) Detection of $\cdot O_2$ production by EPR spectra. (Copyright © 2019, American Chemical Society)

substantial ·OH under microwave irradiation (Fig. 6.4b). Similarly, superoxide anion radicals were detected (Fig. 6.4c), collectively promoting tumor cell apoptosis. Owing to the high microwave sensitivity of the ionic liquid, thermal and dynamic effects occurred concurrently. This study constitutes the first conceptual demonstration of microwave dynamic theory, wherein microwave-induced hot spots in mesopores facilitate electron transfer from Ga to adsorbed H_2O and O_2, yielding ·OH and $\cdot O_2^-$. Thus, for the first time, both MWDT and MWTT were simultaneously realized within a single nanoparticle system. This combinatory approach yielded exceptional antitumor efficacy in both subcutaneous and orthotopic hepatocellular carcinoma models, offering a new method for improving the therapeutic efficacy of deep-seated tumors in clinical research.

The introduction of MWDT addresses a key limitation of conventional MWTT, namely, its inability to completely eradicate tumors in clinical practice. However, according to the theory of MWDT, the efficacy of MWDT is constrained by poor delivery of microwave sensitizers and the hypoxic tumor microenvironment. To overcome these challenges, an LM-based nanozyme denoted LM@ZIF@HA (LZH) was engineered, comprising an EGaIn core coated with a bimetallic CoNi-zeolitic imidazolate framework (ZIF) and surface-functionalized with hyaluronic acid (HA) for active targeting [33]. The flexibility of LM and HA-mediated targeting promotes efficient tumor cell internalization, resolving sensitizer delivery issues. Additionally, the catalase-like activity of the nanozyme catalyzes O_2 generation from endogenous H_2O_2, alleviating hypoxia and enhancing microwave-driven ROS production. Experimental validation confirmed that the nanozymes had significant targeting effect, oxygen generation ability, and microwave dynamic effect, which effectively solved the defects of MWDT. Notably, robust therapeutic outcomes were also observed in a patient-derived xenograft (PDX) model, underscoring the clinical translational potential of this platform.

Beyond LM-based microwave sensitizers, the exploration of novel biocompatible nanomaterials with microwave absorption capabilities and innovative mechanisms remains critical for achieving synergistic MWTT/MWDT. Liu's team reported the microwave absorption characteristics of the nonmagnetic CZTS nanocrystals (NCs) for the first time and used them as a new type of microwave sensitizer for tumor treatment [34]. Due to the significant electromagnetic wave absorption characteristics of CZTS NCs within the range of 1–3 GHz, MWTT can be achieved under microwave irradiation. Furthermore, the electromagnetic energy absorbed by CZTS NCs induces electron transfer in the nanocrystals and leads to the generation of singlet oxygen, thereby achieving MWDT. Therefore, the tumor treatment using microwave irradiation of CZTS NCs has a combined mechanism of MWTT and MWDT. The in vivo experimental results confirmed that the MWTT and MWDT produced by CZTS NCs under microwave irradiation have excellent tumor therapeutic effects and can be regarded as ideal drugs for the combined tumor treatment of MWTT and MWDT in vivo.

Nevertheless, metal ion-based sensitizers raise concerns regarding potential toxicity from metal leakage or high sensitizer concentrations. Hence, the development of efficient and biosafe sensitizers is imperative for advancing MWDT. Lumata's group pioneered the use of aggregation-induced emission luminogens (AIEgens) as microwave sensitizers for cancer treatment [35]. Two synthesized AIEgens (TPEPy-I and TPEPy-PF6) were shown to generate ROS, including singlet oxygen, under microwave irradiation, effectively eradicating cancer cells. TPEPy-I nanoaggregates outperformed TPEPy-PF6, with lower IC_{50} values (2.73 μM vs. 3.22 μM), indicating higher potency. AIEgen-mediated MWDT circumvents the limitations of conventional PDT and reduces the required microwave sensitizer dose, thereby minimizing off-target effects. This work establishes a foundation for AIEgen-based microwave therapeutics and expands the repertoire of MWDT sensitizers.

Furthermore, although the energy of microwaves is insufficient to directly activate cytotoxic ROS generation via electron excitation similar to that of theoretical PDT, microwave photons possess low energy (~10^{-3} eV). However, since the surface effect corresponds to different surface energy levels, the surface energy levels can alleviate the difficulty of electron jumping through multilevel electron transitions, which will promote the generation of excited electrons under microwave stimulation and achieve MWDT. Based on this principle, Liu's team designed a CNT-CuHHTP heterojunction with abundant surface states caused by the confinement effect [36]. The material exhibited strong microwave absorption owing to its excellent conductivity, multiple reflections, heterojunction interface polarization, and dipole polarization. Importantly, under microwave irradiation, CNT-CuHHTP generated free electrons through surface energy level transitions, and the built-in electric field formed in the heterojunction accelerated electron transfer, further achieving efficient microwave dynamic characteristics and generating a large amount of ROS, and curing osteomyelitis by reducing the number of bacteria through local MWTT and MWDT treatment. This surface state engineering-guided generation of microwave-excited electrons may bring new insights for the development of microwave energy absorption and conversion materials, providing material reserves for MWTT and MWDT of various deep tissue diseases.

Although the energy of microwave alone is inadequate for direct ROS generation, electron transfer occurs via indirect pathways through the formation of point defects or weak surface bonds caused by local hot spots, or through surface energy level transitions, thereby achieving efficient microwave dynamic characteristics and generating a large amount of ROS, ultimately resulting in MWDT [37, 38]. Currently, there are not many studies on MWDT, and it is still in its infancy. Nevertheless, as a form of dynamic therapy, it further offers opportunities for the application of dynamic therapy in clinical applications.

4 Discussion

The recent advancements and efforts in nanotechnology offer extensive opportunities for designing nanoscale agents for cancer diagnosis and treatment. Nanotechnology-enabled personalized cancer therapeutics is poised to advance cancer management paradigms toward more universal, efficient, and sophisticated approaches. In particular, nanoparticle-mediated MWDT has progressively evolved from initially being unable to overcome electronic transitions and only conducting MWEDT to directly achieving MWDT nowadays. Furthermore, multifunctional nanoparticles have been engineered with multiple functionalities to mitigate the limitations imposed by the tumor microenvironment and fulfill the essential requirements for effective cancer MWDT. These meticulously designed multifunctional and personalized nanoparticle-mediated MWDT have demonstrated exceptional therapeutic outcomes across various types and origins of tumor models, as well as other disease models, and hold considerable promise for further clinical application to enhance the efficacy of cancer treatment in the future. However, the clinical development of these multifunctional personalized nanoparticles necessitates a thorough evaluation of the intricate interactions between nanoparticles, biological systems, and the tumor microenvironment, along with extensive and rigorous preclinical studies. Despite these challenges, nanoparticle-mediated MWDT presents a highly promising therapeutic modality for clinical cancer treatment.

References

1. Bray F, Laversanne M, Sung H, Ferlay J, Siegel RL, Soerjomataram I, Jemal A. Global cancer statistics 2022: GLOBOCAN estimates of incidence and mortality worldwide for 36 cancers in 185 countries. CA Cancer J Clin. 2024;74(3):229–63.
2. Sung H, Ferlay J, Siegel RL, Laversanne M, Soerjomataram I, Jemal A, Bray F. Global cancer statistics 2020: GLOBOCAN estimates of incidence and mortality worldwide for 36 cancers in 185 countries. CA Cancer J Clin. 2021;71(3):209–49.
3. Contrera KJ, Patel MR, Burtness B, Mehra R, Ferris RL. The role of surgery and deescalation for HPV-related oropharyngeal cancer. Cancer. 2025;131(2):e35287.
4. Yan M, Wu S, Wang Y, Liang M, Wang M, Hu W, Yu G, Mao Z, Huang F, Zhou J. Recent progress of supramolecular chemotherapy based on host-guest interactions. Adv Mater. 2024;36(21):2304249.
5. Haque W, Butler EB, Teh BS. Personalized radiation therapy for breast cancer. Curr Oncol. 2024;31(3):1588–99.
6. Wu C, Yao Y, Lu W, Ji Z, Mao Y. SKA3 targeted therapies in cancer precision surgery: a commentary. Int J Surg. 2024;110(8):5151–2.
7. Schuler M, Cuppens K, Plones T, Wiesweg M, Du Pont B, Hegedus B, Koster J, Mairinger F, Darwiche K, Paschen A, Maes B, Vanbockrijck M, Lahnemann D, Zhao F, Hautzel H, Theegarten D, Hartemink K, Reis H, Baas P, Schramm A, Aigner C. Neoadjuvant nivolumab with or without relatlimab in resectable non-small-cell lung cancer: a randomized phase 2 trial. Nat Med. 2024;30(6):1602–11.
8. Li S, Chen Z, Guo W, Tan L, Wu Q, Ren X, Fu C, Wang Q, Huang Z, Meng X. H2S-mediated gas therapy and HSP90 downregulation synergically enhance tumor microwave thermal therapy. Adv Funct Mater. 2024;34(21):2314742.
9. Xiong LH, Yang L, Geng J, Tang BZ, He X. All-in-one alkaline phosphatase-response aggregation-induced emission probe for cancer discriminative imaging and combinational chemodynamic-photodynamic therapy. ACS Nano. 2024;18(27):17837–51.
10. Zhou H, Tang D, Yu Y, Zhang L, Wang B, Karges J, Xiao H. Theranostic imaging and multimodal photodynamic therapy and immunotherapy using the mTOR signaling pathway. Nat Commun. 2023;14(1):5350.
11. Zhang R, Tan W, Ji D-K. DNA aptamer-empowered self-assembly full-API nanodrug for enhanced photodynamic and chemotherapy synergistic cancer therapy. Nano Today. 2024;57:102399.
12. Yu M, Li S, Ren X, Liu N, Guo W, Xue J, Tan L, Fu C, Wu Q, Niu M, Du Y, Meng X. Magnetic bimetallic heterointerface nanomissiles with enhanced microwave absorption for microwave thermal/dynamics therapy of breast cancer. ACS Nano. 2024;18(4):3636–50.
13. Sun Y, Chen Z, Gong H, Li X, Gao Z, Xu S, Han X, Han B, Meng X, Zhang J. Continuous "snowing" thermotherapeutic graphene. Adv Mater. 2020;32(26):2002024.
14. Tang X, Mao X, Ling P, Yu M, Pan H, Wang J, Liu M, Pan H, Qiu W, Che N, Zhang K, Bao F, Peng H, Ding Q, Wang S, Zhou W. Glycolysis inhibition induces anti-tumor central memory CD8(+)T cell differentiation upon combination with microwave ablation therapy. Nat Commun. 2024;15(1):4665.
15. Ni KY, Lan GX, Veroneau SS, Duan XP, Song Y, Lin WB. Nanoscale metal-organic frameworks for mitochondria-targeted radiotherapy-radiodynamic therapy. Nat Commun. 2018;9:4321.

16. Shi H, Liu T, Fu C, Li L, Tan L, Wang J, Ren X, Ren J, Wang J, Meng X. Insights into a microwave susceptible agent for minimally invasive microwave tumor thermal therapy. Biomaterials. 2015;44:91–102.
17. Long D, Liu T, Tan L, Shi H, Liang P, Tang S, Wu Q, Yu J, Dou J, Meng X. Multisynergistic platform for tumor therapy by mild microwave irradiation-activated chemotherapy and enhanced ablation. ACS Nano. 2016;10(10):9516–28.
18. Tang S, Fu C, Tan L, Liu T, Mao J, Ren X, Su H, Long D, Chai Q, Huang Z, Chen X, Wang J, Ren J, Meng X. Imaging-guided synergetic therapy of orthotopic transplantation tumor by superselectively arterial administration of microwave-induced microcapsules. Biomaterials. 2017;133:144–53.
19. Fu C, Zhou H, Tan L, Huang Z, Wu Q, Ren X, Ren J, Meng X. Microwave-activated Mn-doped zirconium metal–organic framework nanocubes for highly effective combination of microwave dynamic and thermal therapies against cancer. ACS Nano. 2017;12(3):2201–10.
20. Ma X, Ren X, Guo X, Fu C, Wu Q, Tan L, Li H, Zhang W, Chen X, Zhong H, Meng X. Multifunctional iron-based metal-organic framework as biodegradable nanozyme for microwave enhancing dynamic therapy. Biomaterials. 2019;214:119223.
21. Feng Y, Chen Q, Jin C, Ruan Y, Chen Q, Lin W, Zhu C, Zhang T, Zhang Y, Gao J, Mo J. Microwave-activated Cu-doped zirconium metal-organic framework for a highly effective combination of microwave dynamic and thermal therapy. J Control Release. 2023;361:102–14.
22. Chen Z, Wu Q, Guo W, Niu M, Tan L, Wen N, Zhao L, Fu C, Yu J, Ren X, Liang P, Meng X. Nanoengineered biomimetic Cu-based nanoparticles for multifunational and efficient tumor treatment. Biomaterials. 2021;276:121016.
23. Wu Q, Du Q, Sun X, Niu M, Tan L, Fu C, Ren X, Zheng Y, Liang T, Zhao J, Lv X, Liang P, Yang D, Meng X, Yu J. MnMOF-based microwave-glutathione dual-responsive nano-missile for enhanced microwave Thermo-dynamic chemotherapy of drug-resistant tumors. Chem Eng J. 2022;439:135582.
24. Yang R, Huang J, Liao M, Huang J, Gao B, Zhang H, Zhou J, Xu J, Lu Q. An oxygen-sufficient nanoplatform for enhanced imaging-guided microwave dynamic therapy against hypoxic tumors. Int J Nanomedicine. 2022;17:5525–45.
25. Zhou H, Liu Z, Zhang Z, Pandey NK, Amador E, Nguyen W, Chudal L, Xiong L, Chen W, Wen Y. Copper-cysteamine nanoparticle-mediated microwave dynamic therapy improves cancer treatment with induction of ferroptosis. Bioact Mater. 2023;24:322–30.
26. Qin Q, Yang M, Shi Y, Cui H, Pan C, Ren W, Wu A, Hu J. Mn-doped Ti-based MOFs for magnetic resonance imaging-guided synergistic microwave thermal and microwave dynamic therapy of liver cancer. Bioact Mater. 2023;27:72–81.
27. Shu M, Wang J, Xu Z, Lu T, He Y, Li R, Zhong G, Yan Y, Zhang Y, Chu X, Ke J. Targeting nanoplatform synergistic glutathione depletion-enhanced chemodynamic, microwave dynamic, and selective-microwave thermal to treat lung cancer bone metastasis. Bioact Mater. 2024;39:544–61.
28. Luo X, Sun HY, Lu SY, Zhou Y, Xu ZQ, Zhong N, Lu YS, Wang SJ, Shi HB, Tian W. Fe-doped Cu-based bimetallic metal-organic frameworks as nanoscale microwave sensitizers for enhancing microwave thermal and dynamic therapy for hepatocellular carcinoma. Nanoscale. 2024;16(23):11069–80.
29. Chen H, Kong X, Yan J, Wang T, Lv Y, Zhang F, Liu D, Qu Y, Fu J, Han W, Shi J. "Butterfly-effect" nano-platforms for cascade immune amplification and tumor clearance through multimodal therapy and immune sensitization. Chem Eng J. 2024;495:153103.
30. Li J, Li Z, Liu X, Li C, Zheng Y, Yeung KWK, Cui Z, Liang Y, Zhu S, Hu W, Qi Y, Zhang T, Wang X, Wu S. Interfacial engineering of Bi(2)S(3)/Ti(3)C(2)T(x) MXene based on work function for rapid photo-excited bacteria-killing. Nat Commun. 2021;12(1):1224.
31. Lu HD, Tang SY, Zhu JY, Huang XM, Forgham H, Li XK, Shen A, Yun GL, Hu JM, Zhang SW, Davis TP, Li WH, Qiao RR. Nanoengineering liquid metal core-shell nanostructures. Adv Funct Mater. 2024;34(6):2311300.

32. Wu Q, Xia N, Long D, Tan L, Rao W, Yu J, Fu C, Ren X, Li H, Gou L, Liang P, Ren J, Li L, Meng X. Dual-functional supernanoparticles with microwave dynamic therapy and microwave thermal therapy. Nano Lett. 2019;19(8):5277–86.
33. Wu Q, Yu Y, Yu X, Du Q, Gou L, Tan L, Fu C, Ren X, Ren J, Xiao K, Meng X. Engineering liquid metal-based nanozyme for enhancing microwave dynamic therapy in breast cancer PDX model. J Nanobiotechnology. 2023;21(1):399.
34. Tang T, Xu X, Wang Z, Tian J, Yang Y, Ou C, Bao H, Liu T. Cu(2)ZnSnS(4) nanocrystals for microwave thermal and microwave dynamic combination tumor therapy. Chem Commun (Camb). 2019;55(87):13148–51.
35. Pandey NK, Xiong W, Wang L, Chen W, Bui B, Yang J, Amador E, Chen M, Xing C, Athavale AA, Hao Y, Feizi W, Lumata L. Aggregation-induced emission luminogens for highly effective microwave dynamic therapy. Bioact Mater. 2022;7:112–25.
36. Qiao Y, Wu S, Zheng Y, Wang C, Li Z, Zhang Y, Zhu S, Jiang H, Cui Z, Liu X. Enhancing microwave dynamic effects via surface states of ultrasmall 2D MOF triggered by interface confinement for antibiotics-free therapy. Adv Sci (Weinh). 2023;10(21):2300084.
37. Chu X, Mao L, Johnson O, Li K, Phan J, Yin Q, Li L, Zhang J, Chen W, Zhang Y. Exploration of TiO(2) nanoparticle mediated microdynamic therapy on cancer treatment. Nanomedicine. 2019;18:272–81.
38. Wang Y, Jin L, Wang C, Liu H, Mao C, Liu X, Li Z, Zhu S, Jiang H, Cui Z, Zheng Y, Wu S. Microwave-excited thermoelectric catalysis of organic-inorganic nanohybrid for highly effective treatment of Staphylococcus aureus-infected osteomyelitis. Small. 2025;21(18):2500915.

Nanomaterials-Mediated Cancer Thermodynamic Therapy

7

Yuliang Guo, Miguel Vladimir Armindo, and Maoquan Chu

Abstract

Thermodynamic therapy (TDT) is a novel cancer therapy modality that has attracted a lot of attention in recent years. In this therapy, TDT reagents react to thermal stimulation (e.g., laser, ultrasound, microwave) to produce highly toxic alkyl radicals, which can kill cancer cells and inhibit tumor growth. These alkyl radicals may also activate an antitumor immune response, preventing distant metastasis. To improve tumor-targeted therapy efficiency, the TDT reagents are usually incorporated into organic or inorganic nanoparticles. This chapter summarizes the different types of TDT reagents, their mechanisms, and TDT reagent-loaded nanoparticles for cancer TDT and combination therapy (for example, TDT and photothermal therapy). The advantages and disadvantages of TDT are also discussed.

Keywords

TDT reagents · Drug-loaded nanoparticles · Thermal sources · Thermodynamic therapy · Combination therapy

1 The Concept of Thermodynamic Therapy

As previously mentioned, photodynamic therapy (PDT) typically requires molecular oxygen to generate singlet oxygen ($^1O_2$) [1, 2]. However, the tumor microenvironment (TME) usually exhibits hypoxia [3], which significantly reduces the

Y. Guo · M. V. Armindo · M. Chu (✉)
School of Life Sciences and Technology, Tongji University, Shanghai, China
e-mail: mqchu98@tongji.edu.cn

M. Chu (ed.), *Modern Nanomaterials and Nanotechnology for Cancer Therapy*,
https://doi.org/10.1007/978-981-95-6849-9_7

efficacy of PDT [4]. Thermodynamic therapy (TDT) is a novel tumor treatment approach that has been intensively investigated in recent years. The primary mechanism of TDT is the decomposition of heat-responsive compounds at elevated temperatures, thereby generating oxygen-independent free radicals with high activity to kill cancer cells [5, 6]. Unlike traditional PDT—in which a photosensitizer converts oxygen into singlet oxygen and other reactive oxygen species under light irradiation [7, 8]—TDT employs heat-responsive reagents to generate alkyl radicals at relatively high temperatures in the absence of oxygen [9]. Therefore, in severely hypoxic TMEs, TDT has the potential to exhibit better efficacy compared to PDT [10, 11].

TDT usually requires an external heat source to trigger the decomposition of the radical precursor (i.e., the TDT reagent). Consequently, a photothermal effect also occurs in TDT. When the TDT reagent is combined with a photothermal therapy (PTT) reagent, laser-triggered TDT/PTT reagents typically exhibit a synergistic effect between TDT and PTT. The photothermal reagent can convert laser light energy into heat under red and/or near-infrared (NIR) irradiation [12, 13]. This process directly kills cancer cells through thermal effects and provides the thermal stimulus necessary for the TDT reagent to produce free radicals. This combination results in both chemical damage and thermal damage to cancer cells, significantly enhancing overall therapeutic efficacy.

TDT reagents are relatively stable at normal body temperature and only become toxic when heated locally. As a result, unlike nanocarriers loaded with chemotherapeutic drugs, nanoparticle-based tumor TDT therapy causes no systemic toxicity or side effects under physiological conditions.

2 Thermal Sources and TDT Reagents

2.1 Thermal Sources

TDT requires an effective heat source to raise the local tumor temperature to a specific level, thereby inducing the TDT reagent to decompose and release alkyl radicals to kill cancer cells. Commonly used heating methods include near-infrared (NIR) laser irradiation, ultrasound, and high-frequency microwaves. Among these, NIR laser irradiation—which causes photothermal materials to elevate the tumor temperature for rapid TDT reagent decomposition—is the most widely used method due to its convenience and ability for precise localized treatment. High-intensity focused ultrasound can also generate controllable high temperatures in a specific area in vivo, offering the advantage of being noninvasive [14]. Microwave heating (such as microwave ablation) causes ions and water molecules in tumor tissue to vibrate and produce heat [15], thereby raising the tumor temperature to trigger the release of free radicals from heat-responsive reagents (see Fig. 7.1). Regardless of the heat source, the key is to elevate the local tumor temperature to ≥42–45 °C to trigger the decomposition of the radical precursor while avoiding thermal damage to surrounding normal tissues [16].

2.2 TDT Reagents

The efficacy of TDT relies on the appropriate choice of heat-responsive reagents (i.e., TDT reagents) that decompose under heat to produce free radicals. A typical example is the azo compound [17], such as AIPH (2,2′-azobis[2-(2-imidazolin-2-yl) propane] dihydrochloride), which can release nitrogen gas and generate two alkyl radicals upon heat treatment (see Fig. 7.1c). These alkyl radicals are highly reactive, capable of damaging biomacromolecules in cancer cells, causing effects such as membrane lipid peroxidation and DNA damage, thereby inducing tumor cell apoptosis or necrosis. AIPH is relatively stable at 37 °C, but once the temperature rises to approximately 44 °C, its decomposition rate increases significantly with rising temperature, releasing a large amount of alkyl radicals [18]. Although many azo compounds can decompose under heat to release alkyl radicals, AIPH's suitable decomposition temperature makes it the preferred TDT reagent in most studies.

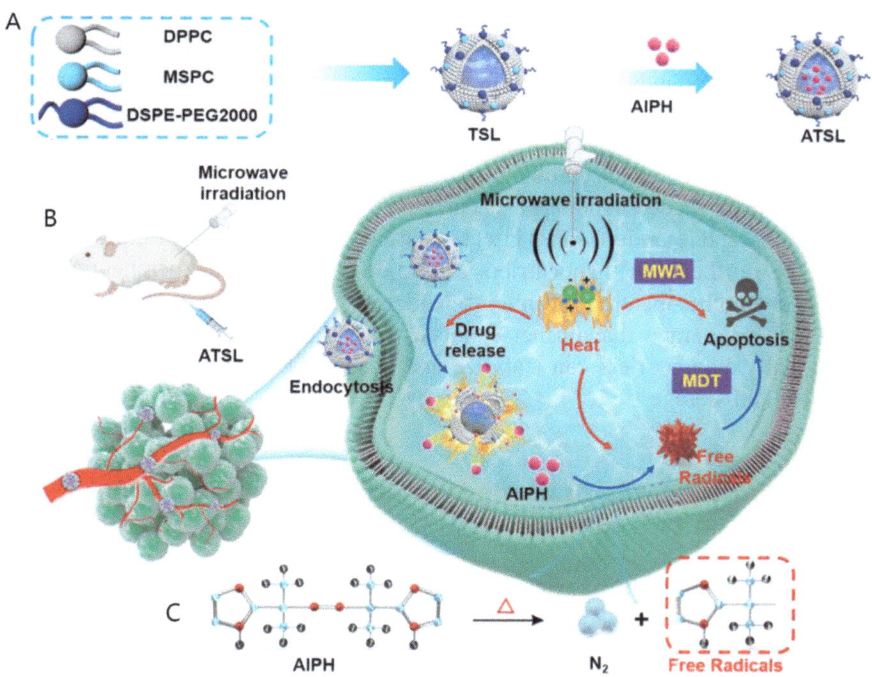

Fig. 7.1 A typical study example of nanoparticles used in cancer thermodynamic therapy, which generally includes the following: (**a**) the synthesis route of TDT reagent (most commonly AIPH)-loaded nanoparticles; (**b**) the drug administration method (in this case, intravenous injection (I.V)), the heating method (here, microwave irradiation), and the killing mechanism (usually due to the release of oxygen-independent free radicals from thermal-triggered decomposition of AIPH, as shown in (**c**)) [15]. (Copyright © 2023 by Wiley-VCH GmbH. Reprinted with permission from Reference [15])

When using AIPH as the TDT reagent, raising the local tumor temperature to around 45 °C is necessary to ensure potent radical generation.

In addition to AIPH, some peroxides or compounds containing peroxide bonds can also be used as heat-activated radical sources. For example, artesunate (ARS), an artemisinin derivative, contains a peroxide bridge that can cleave upon heating to release reactive oxygen species (ROS) radicals [19], which have been utilized in PTT/TDT combination therapy.

In summary, the azo compound AIPH—due to its optimal decomposition temperature—is currently the most used TDT reagent. Its advantages include stability and safety at normal body temperature, as well as its ability to release alkyl radicals at elevated temperatures, allowing for precise targeting of tumor tissue.

3 Nanomaterials for Tumor Thermodynamic Therapy

3.1 The Need for Nanoformulation of TDT Reagents

As previously stated, most TDT reagents that produce alkyl radicals are water-soluble azo compounds. These small molecules, when administered intravenously, exhibit poor tumor targeting, and TDT has limited tumor inhibition efficiency when used alone. However, by incorporating TDT reagents into nanocarriers, the concentration of these reagents accumulating in the tumor is significantly increased. This is because drug-loaded nanoparticles can target tumor tissue through the enhanced permeability and retention (EPR) effect and transcytosis of tumor vascular endothelium [20]. Furthermore, nanocarriers can co-deliver additional therapeutic drugs, such as chemotherapeutics and photosensitizers, resulting in dual-mode or multimodal cancer therapy that further enhances tumor suppression efficacy [21, 22]. Additionally, nanocarriers can stabilize and protect TDT reagents during in vivo treatment and storage.

3.2 Polymer Systems Loaded with TDT Reagents for Tumor Treatment

3.2.1 Polymer Nanoparticle Carriers

Polymer carriers can create PTT/TDT synergistic nanosystems by simultaneously loading photothermal reagents and TDT reagents. Albumin nanoparticles are a common example of polymer carriers. For instance, Lee et al. [23] developed a bovine serum albumin (BSA) nanoparticle loaded with indocyanine green (ICG) and AIPH (named BSA-ICG-AIPH nanoparticles, abbreviated as BIA NPs), leveraging albumin's excellent biocompatibility and the enhanced permeability and retention (EPR) effect for drug delivery. Under 808 nm laser irradiation, the ICG within BIA NPs produces a photothermal effect that kills cancer cells while raising the local tumor temperature to approximately 50 °C, inducing AIPH decomposition and release of alkyl radicals, thereby producing the TDT effect. Cell experiments showed that

AIPH-loaded BIA NPs under laser stimulation exhibited significant cytotoxicity to cancer cells and remained effective even in hypoxic conditions; in contrast, the conventional photodynamic therapy (PDT) photosensitizer protoporphyrin IX almost completely lost its cytotoxic activity in an oxygen-deprived environment. In a mouse tumor model, BIA NPs were administered intravenously and subjected to laser irradiation to increase the local tumor temperature by approximately 25 °C, sufficient to cleave the thermosensitive reagent and release radicals. As a result of laser irradiation, the BIA NPs completely ablated solid tumors, whereas the control group receiving ICG alone (without AIPH) only partially inhibited tumor growth. This study demonstrates that polymeric nanocarriers can effectively combine PTT and PDT in vivo, leading to high therapeutic efficacy without oxygen dependence. In addition to natural polymers such as albumin, synthetic polymer nanoparticles have been used to load TDT reagents for tumor therapy. For example, [24], a study developed PPy-AIPH@LA nanoparticles by attaching AIPH onto the surface of polypyrrole (PPy) nanoparticles and then coating them with lauric acid (LA) as a thermally responsive encapsulation layer. Under 808 nm laser irradiation, PPy converts light into heat (photothermal conversion), enabling PTT treatment at the tumor site and facilitating the decomposition of AIPH to produce oxygen-independent alkyl radicals, thereby achieving the TDT therapeutic effect (see Fig. 7.2).

3.2.2 Polymer Hydrogel Carriers with Nanoporous Structure

Hydrogel materials have gained popularity for localized tumor therapy due to their excellent biocompatibility and injectability. TDT and PTT reagents can be loaded into a temperature-responsive hydrogel to create an implantable or injectable system for combined PTT and TDT therapy. Sun et al. [25] used an NIR-activated, thermo-responsive, and biodegradable hydrogel to treat hypoxic tumors. This hydrogel was created by combining hydrophobic poly(N-isopropylacrylamide) (PNIPAM) and hydrophilic hyaluronic acid-glycidyl methacrylate (HA-GMA) to form a three-dimensional network. The network was then loaded with AIPH and gold nanorods (AuNRs) simultaneously. After being injected into the tumor, the precursor solution cross-links in place to form a gel. Upon 808 nm laser irradiation, the AuNRs generated heat, elevating the local temperature to the gel's phase transition point. The heated gel undergoes volume contraction (PNIPAM transitions from a hydrophilic swollen state to a hydrophobic shrunken state), which accelerates the release of AIPH from the gel and promotes AIPH decomposition under heat. This method enabled the sustained release of alkyl radicals at the tumor site, resulting in a significant antitumor effect even under severe hypoxia.

The hydrogel system can be used as a long-acting implant with a single injection providing continuous therapeutic effects in vivo. Furthermore, due to its high biocompatibility, this system minimizes systemic toxicity. In animal studies, injection of this hydrogel followed by NIR laser irradiation resulted in a significant reduction in tumor growth, which was superior to PTT alone. This strategy maximizes the hydrogel's local retention and thermal responsiveness, converting TDT from a *nanoparticle roaming through the body* model to an in situ *slow-release factory* model that improves solid tumor inhibition while minimizing systemic toxicity.

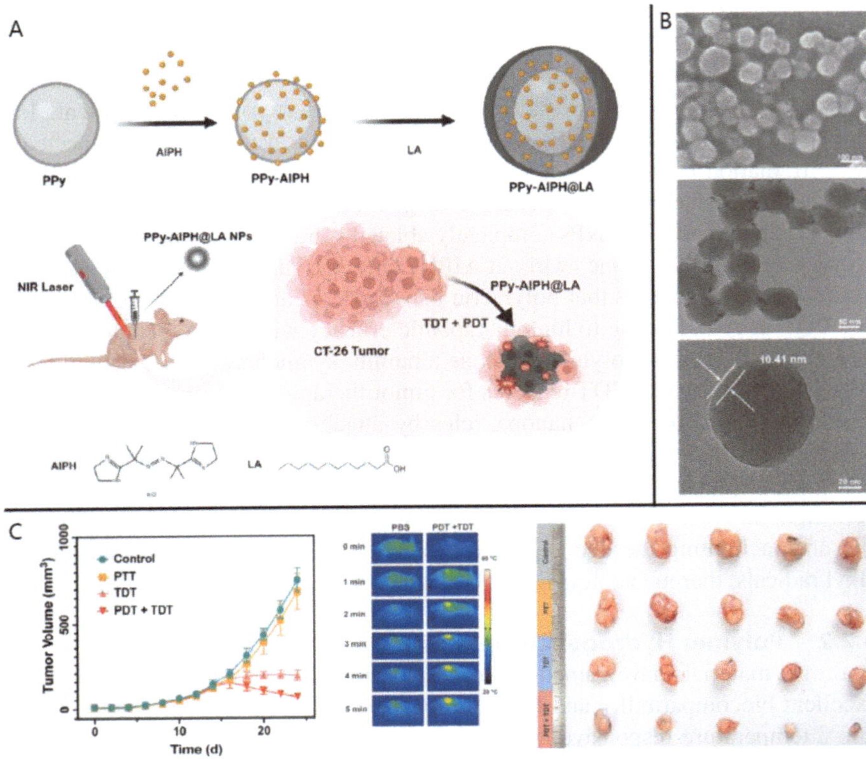

Fig. 7.2 (**a**) The synthesis process of PPy-AIPH@LA, including ROS generation, tumor TDT, and PDT effects. (**b**) SEM and TEM images of the PPy-AIPH@LA nanoparticles. (**c**) Tumor therapy effects of PDT + TDT and PPy-AIPH@LA under 808 nm laser irradiation [24]. (Copyright © 2025 by Wiley-VCH GmbH. Reprinted with permission from Reference [24])

3.3 Inorganic Nanoparticle Carriers Loaded with TDT Reagents

Inorganic nanomaterials (such as metals, metal oxides, or silica-based materials) are often used to construct multifunctional TDT nanoplatforms due to their stable physicochemical properties. Some inorganic nanocarriers possess both photothermal conversion capability and catalytic activity, enabling them to serve as the heat source for TDT and participate in chemical reactions.

3.3.1 Mesoporous Silica Nanocarriers

Mesoporous silica nanoparticles (MSNs) are notable in drug loading and delivery due to their unique pore structures, large surface area, and pore volume. These features enable high payload loading, controlled release, and environmental responsiveness. Their chemical stability, biocompatibility, and scalable synthesis further enhance safety and potential for clinical translation. Therefore, MSNs have been selected to construct nanomaterials that possess both tumor detection (TDT) and photothermal therapy (PTT) functions by loading appropriate agents.

For example, Wang et al. [26] reported a mitochondria-targeted nanoparticle in which natural sepia melanin was used to coat the mesoporous silica core, and triphenylphosphonium (TPP) was decorated onto the silica surface to provide mitochondrial targeting functionality. AIPH was then loaded into the composite, resulting in the final nanoparticles named AIPH@MCTPP. After intratumoral injection, these nanoparticles could accumulate in the tumor and specifically target the mitochondria of cancer cells. Upon 1064 nm laser irradiation, the melanin within the particles generated a strong photothermal effect, causing the local temperature to rise sufficiently to cleave AIPH and release alkyl radicals, thereby delivering a dual killing effect to the cancer cells. It should be noted that the combination of mitochondrial damage and abundant intracellular radical generation induced immunogenic cell death, characterized by the release of danger-associated molecular patterns (DAMPs) and other signals, which activated an antitumor immune response. In mouse models of orthotopic and metastatic breast cancer, treatment with AIPH@MCTPP nanoparticles plus NIR laser not only effectively ablated the primary tumors but also significantly inhibited the growth of distant metastatic tumors. In contrast, the control group using photothermal nanoparticles without AIPH achieved only limited therapeutic efficacy. Furthermore, in the AIPH@MCTPP group, the levels of immune cytokines such as tumor necrosis factor-alpha (TNF-α) in the tumor microenvironment were significantly higher than those in the control group, indicating that this combined PTT/TDT therapy induced a robust antitumor immune response.

For another example, Lin et al. [27] developed a virus-inspired nanogenerator for treating hypoxic tumors. In Lin's experiments, hollow mesoporous silica nanoparticles were used as carriers to load AIPH. The silica surface was then coated with tannic acid–Fe^{3+} coordination, forming a layer of iron–tannic acid metal framework (which provides a photothermal effect), followed by modification with a natural killer (NK) cell membrane for biomimetic camouflage (see Fig. 7.3a, b). The nanoparticle (named AsHMTA/Fe^{III}@NK) has the following functions: Fe^{3+} ions in the coating shell are reduced to Fe^{2+} in the acidic tumor microenvironment, and the generated Fe^{2+} ions then react with endogenous hydrogen peroxide to produce highly toxic hydroxyl radicals (·OH), delivering a chemodynamic therapy (CDT) effect. Meanwhile, the iron–tannic framework shell can convert 808 nm laser energy into heat, causing AIPH decomposition and the release of alkyl radicals to produce a thermodynamic therapy (TDT) effect (see Fig. 7.3c). Additionally, the photothermal effect induces a structural change in the nanoparticle, similar to virus infection, resulting in the release of more AIPH and further enhancing the TDT effect. Due to the NK cell membrane camouflage, these nanoparticles can evade immune clearance, leading to a longer circulation time in vivo and improved tumor accumulation (see Fig. 7.3d). Experimental results showed that, without laser irradiation, the drug-loaded nanoparticles were nontoxic to cells; however, under 808 nm laser exposure, they efficiently killed HepG2 liver cancer cells. These laser-triggered drug-loaded nanoparticles exhibited similar cytotoxic efficacy under both normoxic

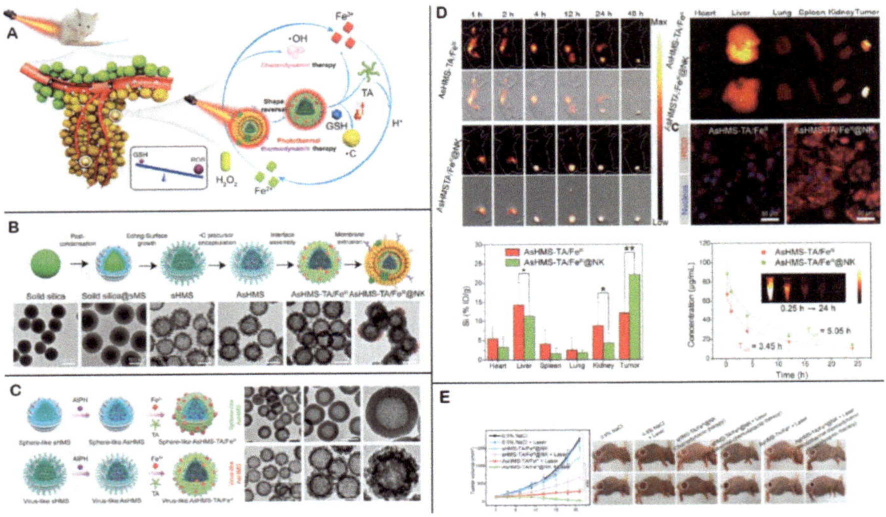

Fig. 7.3 Shows an NK membrane-cloaked, virus-mimicking nanogenerator for NIR-triggered TDT. (**a**) Schematic representation of an AsHMTA/FeIII@NK nanogenerator for NIR/GSH-induced shape reversal and •C/•OH release for PTT + TDT therapy. (**b**) Synthesis of AsHMTA/Fe^{III}@NK. (**c**) Schematic of the synthesis process with TEM images of sphere- and virus-like nanogenerators. (**d**) NK cloaking enables nanoparticles to target and accumulate more effectively at the tumor site in vivo. (**e**) In vivo tumor treatment demonstrating the PTT + TDT effect of AsHMS-TA/Fe^{III}@NK [27]. (Copyright © 2021 by The Authors. Reprinted with permission from Reference [27])

and hypoxic conditions, demonstrating that their killing effect was oxygen-independent. In animal studies, the combination of photothermal effect and free radical generation from this nanoplatform under laser irradiation destroyed tumor tissue (see Fig. 7.3e). This work demonstrated that the combination of photothermal therapy (PTT), thermodynamic therapy (TDT), and chemodynamic therapy (CDT) could significantly inhibit tumor growth in a hypoxic tumor model.

3.3.2 Other Inorganic Nanocarriers

In addition to the previously mentioned examples, TDT research has also investigated other inorganic nanomaterials, including magnetic oxides (e.g., ZrO_2 [14], Fe_3O_4 [29]) and metal sulfides (e.g., CuS [28]). For example, Si et al. reported that they loaded AIPH into hollow mesoporous CuS nanoparticles, followed by coating the nanoparticle surface with BSA. The photothermal effect of laser-triggered CuS activated AIPH molecules in the nanocomposites, resulting in the release of alkyl radicals. Copper ions (Cu^{2+}/Cu^{+}) released from CuS can catalyze the conversion of hydrogen peroxide into hydroxyl radicals (·OH), thereby enabling free radical therapy and magnetic resonance imaging. This type of inorganic nanosystem is stable and allows for the integration of imaging diagnostics—such as photoacoustic imaging and magnetic resonance imaging—with therapeutic functions.

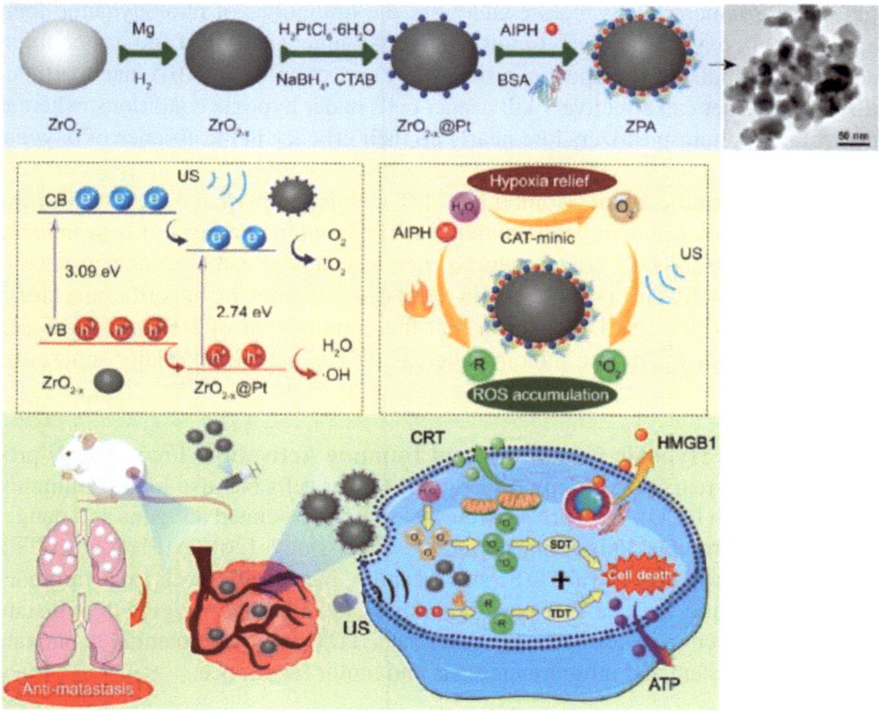

Fig. 7.4 The formulation procedure and a TEM image of the ZPA nanocomplex, along with its mechanism of action for enhancing intracellular oxidative stress under ultrasound (US) irradiation for antitumoral applications [14]. (Copyright © 2022 by Wiley-VCH GmbH. Reprinted with permission from Reference [14])

Li et al. [14] described an AIPH-loaded zirconia-platinum nanocomplex (ZPA) that enables ultrasound-activated sonodynamic and thermodynamic cancer therapy (see Fig. 7.4). The system increases reactive oxygen species (ROS) and alkyl radical production under ultrasound by combining ZrO_{2-x} with Pt nanoparticles and loading AIPH, inducing significant intracellular oxidative stress. This strategy effectively suppresses tumor growth in vivo, exhibits good biosafety, and induces immunogenic cell death.

4 Advantages and Limitations of TDT

4.1 Advantages

Oxygen Independence TDT uses heat-triggered radical precursors to generate alkyl radicals without the need for molecular oxygen, making it effective in hypoxic tumor environments. This feature addresses the limitations of photodynamic therapy (PDT) in oxygen-deprived tumor microenvironments (TMEs) and can counteract hypoxia-induced therapeutic resistance. For example, the BIA nanoparticles discussed earlier can effectively kill cancer cells under hypoxic conditions, whereas control PDT photosensitizers lose nearly all their efficacy in the absence of oxygen.

Synergistic Enhancement Combining TDT with photothermal therapy (PTT) can significantly enhance therapeutic outcomes. The combined effects of heat-induced cellular damage and free radical-induced chemical damage can overcome protective mechanisms, such as heat shock protein-mediated activation by hyperthermia alone. Several studies [23, 25] have shown that the combination of PTT and TDT can completely eradicate mouse tumors, whereas PTT alone only partially suppresses tumor growth.

Adaptation to Hypoxic Conditions and Immune Activation Free radicals produced by TDT reagents not only directly kill cancer cells but also induce immunogenic cell death (ICD). This process releases tumor-associated antigens and danger signals that promote antitumor immunity. In a study [26], AIPH@MS-CTPP nanoparticle-mediated PTT/TDT significantly increased levels of immune-stimulating factors such as TNF-α within tumors and inhibited the growth of distant tumors compared to control groups. As a result, TDT has great potential to activate the immune system and inhibit metastasis and tumor recurrence.

Controllability and Safety TDT reagents are relatively stable at normal body temperature and only produce toxic free radicals when heated at the target site. This ensures that, as long as heating is kept to a minimum in the tumor region, effective tumor tissue ablation can be achieved while minimizing damage to normal tissues.

4.2 Limitations

Thermal Sensitivity Control Precise temperature regulation at the tumor site is essential because the radical initiators used in TDT can only decompose effectively above a certain temperature. If the activation temperature is set too high, there is an increased risk of damaging healthy tissue. For example, some studies [19, 25] have designed dual-responsive systems by encapsulating AIPH in phase-change materials (such as stearic acid or lauric acid), so that the free radicals are only released when the material melts at a specific temperature. This results in a temperature-gated radical release mechanism.

Targeted Delivery TDT nanomaterials, like many other nanomedicines, must accumulate at the tumor site to be effective. If they accumulate in normal tissues instead of the targeted area, the subsequent heating process may cause off-target damage. To address this, various strategies are employed to improve tumor targeting and enrichment of TDT nanoparticles, such as camouflaging them with red blood cell or natural killer (NK) cell membranes or modifying their surfaces with targeting peptides to enhance active targeting. Improving delivery efficiency not only allows for lower drug doses but also enhances the local photothermal heating effect and enables targeted radical release, thereby increasing overall efficacy.

Short Radical Half-Life The free radicals produced by TDT reagents (e.g., alkyl radicals) are extremely reactive and typically have very short lifespans, usually measured in milliseconds or even microseconds—meaning they diffuse only over a limited distance within tissues [30]. An alkyl radical generally affects only the cell in which it was generated and its immediate neighbors and cannot reach distant cancer cells. This very short range limits the depth of tumor destruction in large or invasive tumors, implying that continuous generation of radicals is necessary to effectively treat the entire tumor. Consequently, a key challenge is achieving sustained, high-efficiency radical generation within the tumor—such as via long-circulating carriers or in situ slow-release hydrogels—to ensure comprehensive treatment.

Potential Systemic Toxicity and Safety Concerns Although TDT reagents are nontoxic or of low toxicity at normal body temperature, potential risks must be considered in clinical applications. For example, an excessive burst of free radicals may cause local inflammatory responses, and certain nanocarrier components (such as heavy metal ions or non-biodegradable materials) may accumulate over time and lead to adverse effects. Furthermore, if a patient develops a fever or systemic hyperthermia, it may unintentionally activate heat-sensitive reagents, resulting in unwanted radical release throughout the body and posing safety risks. Currently, most TDT research is in the preclinical stage, using animal models. Successful clinical translation will require additional verification of these nanomaterials' metabolic pathways and long-term safety. All of these issues and challenges must be addressed before TDT can be used in clinical applications.

5 Summary and Discussion

Nanomaterial-mediated tumor thermodynamic therapy (TDT) offers distinct advantages in overcoming tumor hypoxia-induced therapeutic resistance and enhancing treatment specificity. TDT has the potential to transform tumor phototherapy by continuously improving the formulation and delivery of heat-responsive reagents, optimizing the targeting capabilities and controlled-release functions of nanocarriers, and integrating TDT with other treatments such as immunotherapy.

Compared to photodynamic therapy (PDT), photothermal therapy (PTT), and chemodynamic therapy (CDT), the options for reagents in TDT are currently very limited. Expanding these options will be an important area for future research. Additionally, it is necessary to thoroughly evaluate the safety of TDT and standardize treatment protocols to facilitate its transition from laboratory research to clinical application. The emerging trend and goal of nanomaterial-mediated TDT research is to provide cancer patients with a new, more effective, and safer treatment option.

References

1. Agostinis P, Berg K, Cengel KA, Foster TH, Girotti AW, Gollnick SO, et al. Photodynamic therapy of cancer: an update. CA Cancer J Clin. 2011;61:250–81.
2. Jiang W, Liang M, Lei Q, Li G, Wu S. The current status of photodynamic therapy in cancer treatment. Cancers (Basel). 2023;15:585.
3. Shen Z, Ma Q, Zhou X, et al. Strategies to improve photodynamic therapy efficacy by relieving the tumor hypoxia environment. NPG Asia Mater. 2021;13:39.
4. Zheng X, Sun W, Ju M, Wu J, Huang H, Shen B. A chemical biology toolbox to overcome the hypoxic tumor microenvironment for photodynamic therapy: a review. Biomater Sci. 2022;10:4681–93.
5. Wang Y, Wang H, Zhang Y, et al. On-demand generation of heat and free radicals for dual cancer therapy using thermal initiator- and gold nanorod-embedded PLGA nanocomplexes. Biomater Res. 2018;22:1–12.
6. Li R, Hu X, Shang F, Wu W, Zhang H, Wang Y, et al. Treatment of triple negative breast cancer by near infrared light triggered mild-temperature photothermal therapy combined with oxygen-independent cytotoxic free radicals. Acta Biomater. 2022;148:218–29.
7. Kessel D. Cellular mechanisms of singlet oxygen in photodynamic therapy. Int J Mol Sci. 2023;24:16890.
8. Robertson CA, Hawkins Evans D, Abrahamse H. Photodynamic therapy (PDT): a short review on cellular mechanisms and cancer research applications for PDT. J Photochem Photobiol B. 2009;96:1–8.
9. Liu Y, Wang Y, Yang F, et al. Tumor microenvironment/NIR-responsive oxygen-irrelevant radical generation for enhanced thermodynamic therapy. Biomaterials. 2022;287:121640.
10. Li X, Liu Y, Fu F, Cheng M, Liu Y, Yu L, et al. Single NIR laser-activated multifunctional nanoparticles for cascaded photothermal and oxygen-independent photodynamic therapy. Nano Micro Lett. 2019;11:68.
11. Shih C-Y, Wang P-T, Su W-C, Teng H, Huang W-L. Nanomedicine-based strategies assisting photodynamic therapy for hypoxic tumors: state-of-the-art approaches and emerging trends. Biomedicine. 2021;9:137.
12. Li J, Zhang W, Ji W, Wang J, Wang N, Wu W, et al. Near-infrared photothermal conversion materials: mechanism, preparation, and applications. J Mater Chem B. 2021;9:7391–410.
13. Zhang Y, et al. Photothermal nanomaterials: a powerful light-to-heat converter. Chem Rev. 2023;123:1234–78.
14. Li Y, Huang J, Yu H, Zhao Y, Xu Z, Kang Y, Xue P. Zirconia-platinum nanohybrids for ultrasound-activated sonodynamic-thermodynamic bimodal therapy by inducing intense intracellular oxidative stress. Small. 2022;18:e2203080.
15. Zhang W, Zhou H, Gong D, Wu H, Huang X, Miao Z, et al. AIPH-encapsulated thermo-sensitive liposomes for synergistic microwave ablation and oxygen-independent dynamic therapy. Adv Healthc Mater. 2023;12:e2202947.
16. Kim K, Park M. Advancing cancer treatment: enhanced combination therapy utilizing photothermal and thermodynamic approaches. Biomedicine. 2024;12:326.
17. Yoshino K, Ohkatsu Y, Tsuruta T. Study on the modes of thermal decomposition of several azo-type initiators. Polym J. 1977;9:275–82.

18. Zhang L, Chu C, Lin X, Sun R, Li Z, Chen S, et al. Tunable nanoparticles with aggregation-induced emission heater for precise synergistic photothermal and thermodynamic therapy. Adv Sci. 2023;10:e2206874.
19. Gunjan S, Singh SK, Sharma T, Dwivedi H. Artemisinin derivatives and synthetic trioxane trigger apoptotic cell death in plasmodium falciparum via reactive oxygen species generation. Front Cell Infect Microbiol. 2018;8:256.
20. Liu Y, Huo Y, Yao L, Xu Y, Meng F, Li H, et al. Transcytosis of nanomedicine for tumor penetration. Nano Lett. 2019;19:8010–20.
21. Zhang Y, et al. Ultralong circulating lollipop-like nanoparticles assembled with gossypol, doxorubicin and polydopamine via π–π stacking for tumor synergistic therapy. Adv Funct Mater. 2018;28:1805582.
22. Zhang P, Zhang Y, Ding X, Shen W, Li M, Wagner E, et al. A multistage cooperative nanoplatform enables intracellular co-delivery of proteins and chemotherapeutics for cancer therapy. Adv Mater. 2020;32:e2000013.
23. Wan Y, Lu G, Zhang J, Wang Z, Li X, Chen R, et al. A biocompatible free radical nanogenerator with real-time monitoring capability for high performance sequential hypoxic tumor therapy. Adv Funct Mater. 2019;29:1903436.
24. Zhang L, Qian J, Zhang X, Lv Y, Zhao J, Wang S, Xu H. Dual-responsive PPy-AIPH@LA nanoplatform for synergistic photothermal and thermodynamic therapy of colorectal cancer. ChemistrySelect. 2025;10(10) https://doi.org/10.1002/slct.202500263.
25. Sun X, Liu D, Xu X, Shen Y, Huang Y, Zeng Z, et al. NIR-triggered thermo-responsive biodegradable hydrogel with combination of photothermal and thermodynamic therapy for hypoxic tumor. Asian J Pharm Sci. 2020;15:713–27.
26. Zhang Y, Wang Q, Ji Y, Fan L, Ding B, Lin J, et al. Mitochondrial-targeted melanin@mSiO_2 yolk-shell nanostructures for NIR-II-driven photo-thermal-dynamic/immunotherapy. Chem Eng J. 2022;435:134869.
27. Lin J, Li Y, Wang P, Wu M, Zhu F, Zhang Y, et al. Natural killer cell membrane-cloaked virus-mimicking nanogenerator with NIR-triggered shape reversal and •C/•OH storm for synergistic thermodynamic–chemodynamic therapy. Adv Sci. 2022;9:2103498.
28. Si P, Yu W, Li C, Chen H, Zhang E, Gu J, et al. Oxygen-independent alkyl radical nanogenerator enhances breast cancer therapy. Nanomedicine. 2023;48:102630.
29. Li W, Li B, Wu B, Tian B, Chen X, Wang C, et al. Free-radical cascade generated by AIPH/Fe_3O_4-coloaded nanoparticles enhances MRI-guided chemo/thermodynamic hypoxic tumor therapy. ACS Appl Mater Interfaces. 2022;14:29563–76.
30. Griendling K, Touyz RM, Zweier JL, Dikalov S, Chilian WM, Harrison DG, et al. Measurement of reactive oxygen species, reactive nitrogen species, and redox-dependent signaling in the cardiovascular system. Circ Res. 2016;119(5):e39–75.

Nanomaterials-Mediated Cancer Magneto-Mechanical Force Therapy

8

Maoquan Chu

Abstract

Magnetic materials are widely used in biomedical research, including cancer diagnosis and therapy. Magneto-mechanical force-mediated cancer therapy, one of several magnetic materials-based cancer therapy methods, has received recent attention. A low-frequency dynamic magnetic field is used to activate nano- or microscale magnetic particles, causing them to vibrate, rotate, or perform other complex motions. When moving magnetic particles come into contact with cells and tissues, they transfer energy to them, which may result in damage. If cancer cells within tumor tissue are damaged by such physical forces, tumor growth may be slowed. This magneto-mechanical force-mediated therapy may also trigger an antitumor immune response.

This chapter explores different types of magnetic materials and magnetic fields used to destroy cancer cells through magneto-mechanical forces, as well as the parameters that influence cell-killing efficiency, along with the benefits and drawbacks of these therapeutic approaches.

Keywords

Magnetic materials · Low-frequency dynamic magnetic field · Magneto-mechanical force · Cancer therapy

M. Chu (✉)
School of Life Sciences and Technology, Tongji University, Shanghai, China
e-mail: mqchu98@tongji.edu.cn

M. Chu (ed.), *Modern Nanomaterials and Nanotechnology for Cancer Therapy*,
https://doi.org/10.1007/978-981-95-6849-9_8

1 Introduction

Cancer is one of the most serious diseases worldwide. Biomaterials, especially nanoscale biomaterials, have been widely used in fundamental research of cancer diagnosis and therapy. However, most of these materials are difficult to use in the human body due to their potential toxicity. For example, semiconductor quantum dots have unique optical properties that have been applied to in vivo tumor-targeted fluorescence imaging and therapy [1–5]. However, these inorganic nanoparticles are composed of heavy metal ions (e.g., Cd^{2+}, Hg^{2+}). In contrast, superparamagnetic iron oxide nanoparticles are less toxic to cells and animals compared to most inorganic nanoparticles, and they have been approved for use in humans [6, 7].

Magnetic materials have been widely used for bioimaging, separation, detection, and therapy [6–13]. Magnetic materials-based cancer therapies include at least seven methods, as follows: (1) magnetic targeted drug delivery, (2) hyperthermia therapy induced by an alternating magnetic field, (3) laser-induced photothermal therapy, (4) chemodynamic therapy, (5) a magnetic switch that causes death receptors to aggregate on the cancer cell membrane, (6) magneto-mechanical force-mediated therapy; and (7) magnetic microspheres for embolizing tumor blood vessels, among other applications.

Magneto-mechanical force for cell destruction and tumor inhibition is a simple and low-cost technique that has gained popularity in recent years [14–52]. In general, magnetic materials that are remotely triggered by a low-frequency dynamic magnetic field exhibit mechanical motions, such as shifting, swaying, rotation, oscillation, and other complex movements. These modes of motion are determined by the properties of dynamic magnetic fields and magnetic materials. When moving magnetic particles come into contact with cancer cells or tumor tissues, mechanical forces act on the cells and tissues. In cells, these mechanical forces are translated into chemical signals, leading to changes in cell structure and functions [53–58]. When the permeability of the cell membrane and/or internal membrane increases significantly following mechanical stimulation, proteins, lactate dehydrogenase (LDH), ions, and other intracellular components leak out, potentially leading to cell death. As a result, magnetic materials-mediated magneto-mechanical force triggered by a magnetic field can be used to kill cancer cells and inhibit tumor growth.

This chapter discusses the use of magnetic materials-mediated magneto-mechanical force as a standalone cancer therapy. However, magneto-mechanical force alone, like many other therapies, is insufficient to completely inhibit tumor growth. Therefore, the disadvantages of this method are also discussed.

2 Magnetic Materials and Low-Frequency Dynamic Magnetic Fields: Their Size and Shape

Since only a small number of nanoparticles can accumulate in tumor tissue following intravenous injection, and because the safety of intravenous injection of most inorganic materials is questionable, intratumoral injection is typically used for cancer therapy. This method of administration is also commonly employed for magnetic materials for magneto-mechanical force-mediated cancer therapy. As a result, magnetic materials do not always need to be reduced to extremely small sizes. Both nanoscale and microscale magnetic materials are effective at killing cancer cells via magneto-mechanical force. These magnetic materials have two types of morphologies (see Fig. 8.1):

- Magnetic particles with smooth or less rough surfaces, such as smooth solid nano- or microspheres (e.g., Refs. [16, 17, 22]), nanocubes (e.g., Ref. [36]), hollow porous microspheres (e.g., Ref. [26]), nanorods (e.g., Refs. [21, 24]), nanowires (e.g., Refs. [25, 30]), nanotubes (e.g., Ref. [18]), and microdiscs (e.g., Ref. [16]).
- Magnetic particles with rough or sharp surfaces, such as nanoflowers (e.g., Ref. [45]) and urchin-like particles (e.g., Refs. [40, 46, 47]). These rough microspheres also have solid and hollow porous structures.

The dynamic magnetic fields used to induce the movement of these magnetic materials are all of low frequency, including low-frequency alternating magnetic

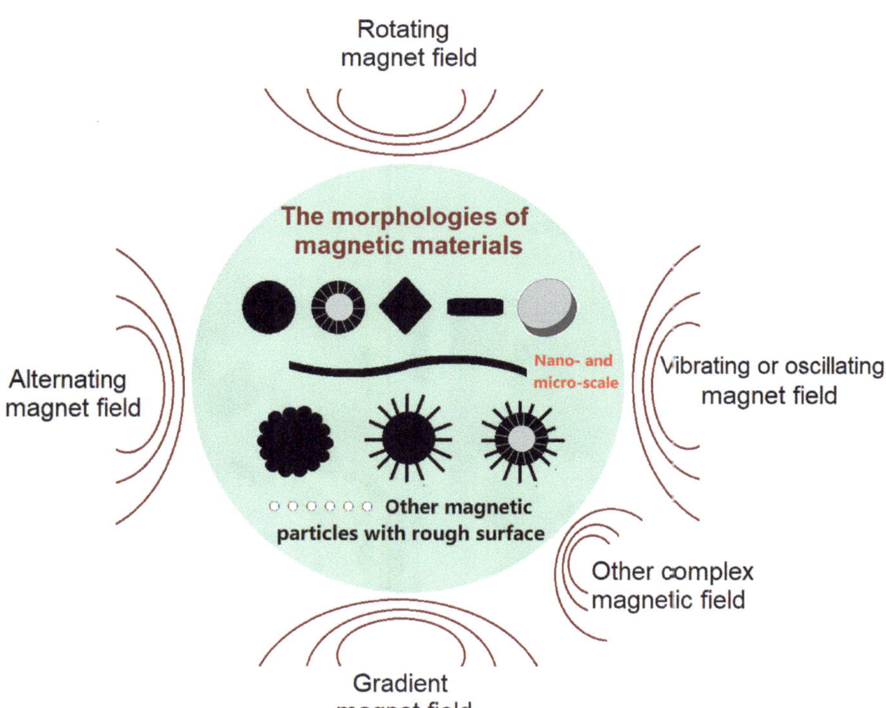

Fig. 8.1 Various morphologies of magnetic materials with different sizes, as well as the types of applied dynamic magnetic fields used in this study

fields (AMF) (e.g., Ref. [16]), rotating magnetic fields (RMF) (e.g., Ref. [17]), vibrating or oscillating magnetic fields (VMF) (e.g., Refs. [19, 46]), gradient magnetic fields (e.g., Ref. [33]), and others (see Fig. 8.1). The frequency of these magnetic fields is no more than 100 Hz and sometimes less than 4 Hz [26, 40, 44–48]. Magnetic materials triggered by low-frequency magnetic fields produce little to no heat. It is known that traditional AMF used for magnetic thermal therapy has a frequency exceeding 100 kHz or MHz. Devices for generating low-frequency magnetic fields are usually simple and low cost. For example, commercial magnetic stirrers [15, 17, 18, 36, 43] can produce a rotating magnetic field. Therefore, there is no need to develop additional apparatuses if magnetic stirrers are available.

3 The Effect of Morphology, Size, and Magnetism of Magnetic Materials on Cell Killing Efficiency Through Magneto-Mechanical Force

There is no doubt that magnetic particles with sharp surfaces exhibit greater destructive power toward objects than particles with smooth surfaces. The sharper the surface is, the more severe the damage is, especially for large-sized magnetic materials. In early 2014, Shi and colleagues reported that rod-shaped magnetic particles were more effective at killing cancer cells than spherical magnetic particles when both were triggered by the same dynamic magnetic field [21]. Shi's experiments produced rod-shaped pure Fe_3O_4 nanoparticles (average length: 200 ± 50 nm; cross-section diameter range: 50–120 nm) and Fe_3O_4 nanospheres (average diameter: 200 ± 50 nm) with strong magnetism (see Fig. 8.2), used for cancer cell killing via

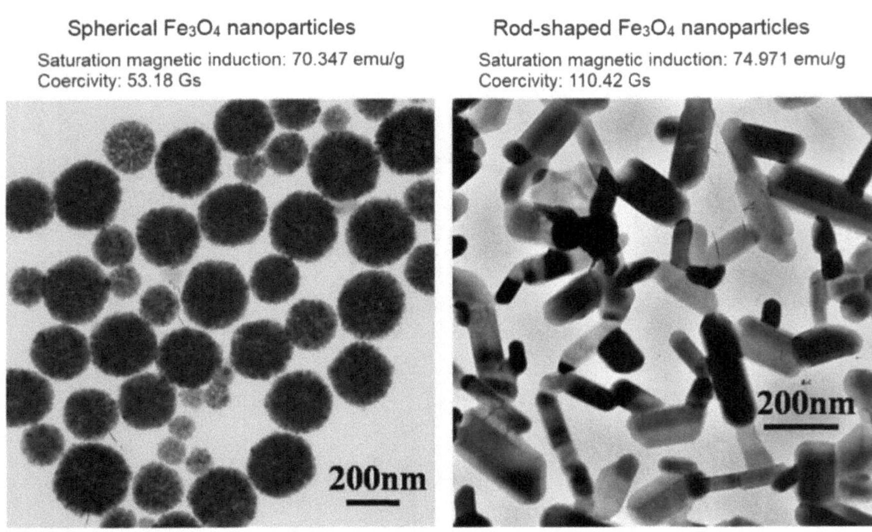

Fig. 8.2 TEM images of two different morphologies of Fe_3O_4 nanoparticles [21]. (Copyright © 2014 by Cheng et al.; licensee Springer. Reprinted with permission from Ref. [21])

magneto-mechanical force. These two types of magnetic nanoparticles exhibited minimal toxicity to human cervical carcinoma cells (i.e., HeLa cells). For example, when 100 μg/mL magnetic nanoparticles were incubated with HeLa cells (initial density: 1×10^4 cells per well in a 96-well plate) for 20 hours, at least 85% of the nanoparticles were taken up by cells. However, cell viability did not decrease significantly. When cells incubated with magnetic nanoparticles were washed with PBS, then digested, suspended, and exposed to an alternating magnetic field (AMF) at 35 kHz and 0.3 W, cell viability decreased significantly with increasing AMF exposure time (see Fig. 8.3a). During AMF exposure, only a small temperature increase was observed. The damage to the cells was primarily caused by the Fe_3O_4-mediated magneto-mechanical force. The cell-killing efficiency of the rod-shaped Fe_3O_4 nanoparticles was much higher than that of the spherical Fe_3O_4 under the same AMF conditions (see Fig. 8.3a). Scanning electron microscopy (SEM) images revealed severe damage to cell membranes following AMF exposure (see Fig. 8.3b). Rod-shaped Fe_3O_4 nanoparticles may possess higher magnetic torque, coercivity, and involved area compared to spherical Fe_3O_4. Furthermore, nanorods have two ends resembling needle tips, making them more likely to puncture cell membranes than spherical nanoparticles. The concentration of Fe_3O_4 nanoparticles (whether nanorods or nanospheres) did not influence cell killing during AMF exposure, whether increased to 500 μg/mL or decreased to 20 μg/mL.

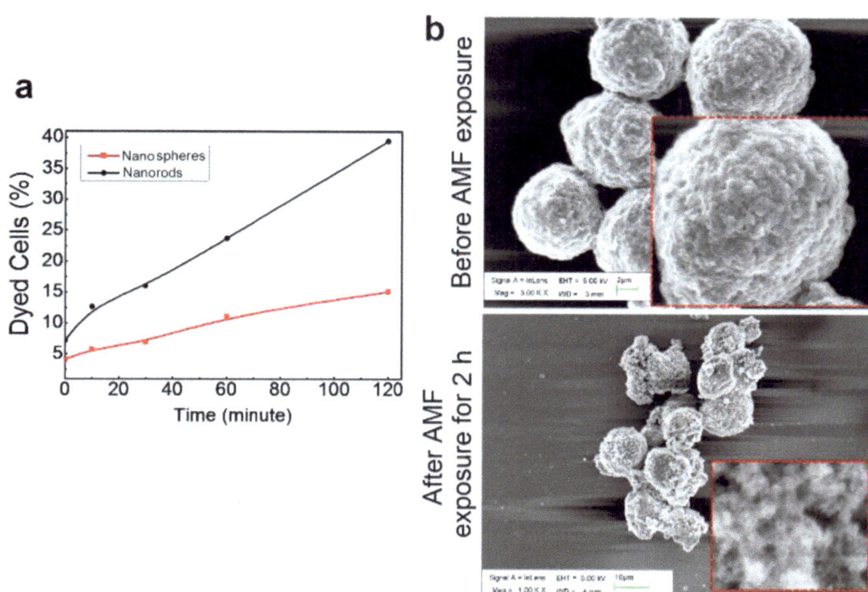

Fig. 8.3 HeLa cancer cells destroyed by Fe_3O_4 nanoparticles-mediated magneto-mechanical force [21]. (**a**) Comparison of the cell-killing ability between Fe_3O_4 nanorods and nanospheres. (**b**) SEM images of cells loaded with Fe_3O_4 nanorods, taken before and after exposure to an alternating magnetic field (AMF). (Copyright © 2014 by Cheng et al., licensee Springer. Reprinted with permission from Ref. [21])

In 2022, Guo et al. further compared the killing efficiency of magnetic particles with different surface roughness [46]. In Guo's report, bovine serum albumin (BSA) microspheres embedded with cubic superparamagnetic Fe_3O_4 nanoparticles (Fe_3O_4/BSA) were prepared (see Fig. 8.4a). The surfaces of these magnetic microspheres were then coated with short rod-like SiO_2 (**S**rSiO_2), medium rod-like SiO_2 (**M**rSiO_2), long rod-like SiO_2 (**L**rSiO_2), and curled rod-like SiO_2 (**C**rSiO_2) shells, respectively (see Fig. 8.4b, samples with long silica rods). The size of these microspheres (i.e., Fe_3O_4/BSA/rSiO_2) ranged from 1 to 4 μm. Fe_3O_4/BSA microspheres without silica shells were used as controls. After incubating Fe_3O_4/BSA/rSiO_2 microspheres (1 mg/mL) with human laryngeal carcinoma cells (TU212) for 1 hour, followed by a low-frequency vibrating magnetic field (VMF, 400 mT and 2 Hz) exposure for 1 hour, and a 4-hour culture, the leakage of lactic dehydrogenase (LDH) from cells was detected. Moreover, the longer the silica rods were—assuming they were straight—the greater the amount of LDH released (see Fig. 8.4c). However, for the longest rods with a circular structure, less LDH was released under the same VMF exposure.

For Fe_3O_4/BSA without silica rod coating, the surface of magnetic spheres appeared smooth. In this case, only a small amount of LDH release was detected under the same treatment (see Fig. 8.4c). The order of LDH leakage, from largest to

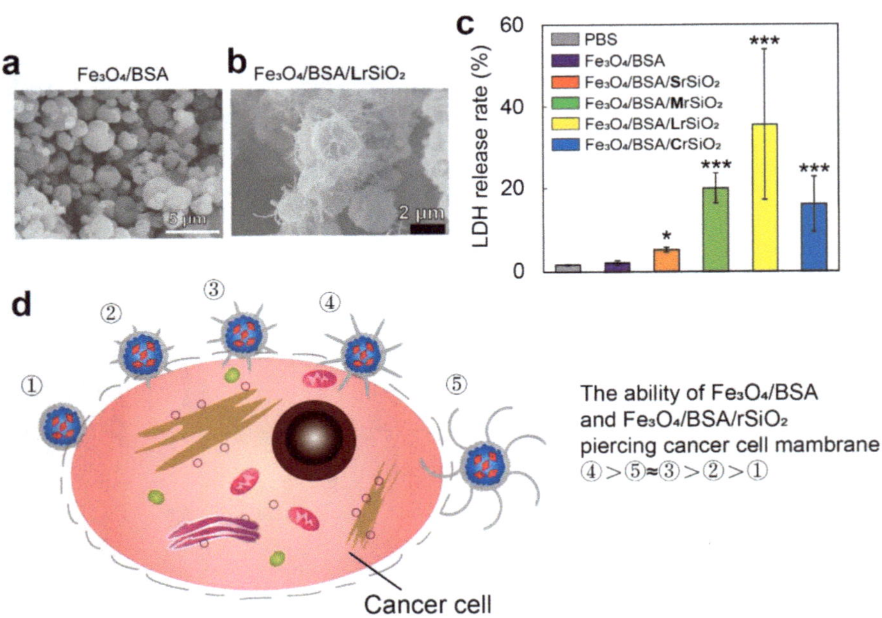

Fig. 8.4 The effect of the surface structure of magnetic microspheres on their cell-destructing ability through magneto-mechanical force. (**a**, **b**) SEM images of Fe_3O_4/BSA and Fe_3O_4/BSA/**L**rSiO_2. (**c**) LDH release in different groups under the same VMF exposure. (**d**) Schematic illustration comparing the cell-destructing ability of different groups under the same VMF exposure [46]. (Copyright © 2022 by the Author(s). Reprinted with permission from Ref. [46])

smallest, is Fe_3O_4/BSA/**L**rSiO$_2$, Fe_3O_4/BSA/**M**rSiO$_2$, Fe_3O_4/BSA/**S**rSiO$_2$, and Fe_3O_4/BSA. There is no statistical difference in the LDH release rate between the Fe_3O_4/BSA/**M**rSiO$_2$ and Fe_3O_4/BSA/**C**rSiO$_2$ groups (see Fig. 8.4c). Since LDH release is associated with cell membrane damage, the VMF-triggered magnetic microspheres exhibit a similar degree of cell damage or capacity to pierce cell membranes, as described above (see Fig. 8.4d). These results indicate that magnetic particles with sharp or rough surfaces are beneficial for killing cancer cells through magneto-mechanical forces.

Recently, Zhu et al. demonstrated that magnetic microspheres with sharp surfaces penetrate cancer cells more easily than those with smooth surfaces [59]. In their experiments, Zhu's team prepared Fe_3O_4/BSA microspheres ranging from 0.5 to 2.5 μm and hedgehog-like magnetic microspheres (HMMs) approximately 1.2 μm in size. The surface of Fe_3O_4/BSA microspheres was smooth compared to the HMMs. After loading equal amounts of doxorubicin (DOX), the drug-loaded microspheres were incubated with TU212 cells for 4 hours. Fluorescence imaging revealed a substantial amount of red fluorescence from DOX within the cells in the HMMs group, whereas only weak red fluorescence was observed in the *Fe_3O_4/BSA group*. Consequently, magnetic particles with sharp features—such as hedgehog-like, chestnut-like, star-like, or other pointed surfaces—are highly damaging to cancer cells when exposed to a dynamic magnetic field.

Larger magnetic microspheres are typically selected for cell destruction via magneto-mechanical force, because the force acting on such particles is proportional to their volume [46]. Therefore, the force generated by a larger magnetic particle exceeds that of a smaller one, and the force produced by an aggregated large particle is greater than that of a single or small-sized aggregate.

In early 2010, Novosad's research team reported that micrometer-scale magnetic discs could be used to destroy cancer cells through magneto-mechanical force [16]. These microdiscs consisted of an iron–nickel core (approximately 1 μm in diameter and 60 nm thick) and a gold (Au) shell (5 nm thick), and they exhibited a spin-vortex ground state. Due to the magnetic spin distribution, vortex-domain magnetic particles do not easily aggregate under a magnetic field, which is advantageous for biomedical applications. When these microdiscs were incubated with an antihuman IL-13α2R antibody and then with N10 glioma cancer cells, followed by exposure to a low-frequency alternating magnetic field (AMF) with a frequency of approximately 10–50 Hz and a magnetic field strength of less than 100 Oe for 10 minutes, they caused severe destruction of the cell membranes (see Fig. 8.5a) and significantly increased the release of lactate dehydrogenase (LDH) from the cells. Atomic force microscopy images confirmed that AMF treatment induced notable changes in cell size and morphology (see Fig. 8.5b). Due to the low frequency, the temperature increase of the microdiscs during AMF exposure was negligible.

In early 2010, Gao's research group reported magneto-fluorescent nanospheres with a small diameter for damaging cancer cells through magneto-mechanical force [17]. These nanospheres are much smaller than those reported by Novosad's research group. In Gao's experiments, the authors synthesized uniform-sized nanospheres in a microemulsion using oleic acid-coated γ-Fe_2O_3 (5 nm) or Fe_3O_4 (15 nm)

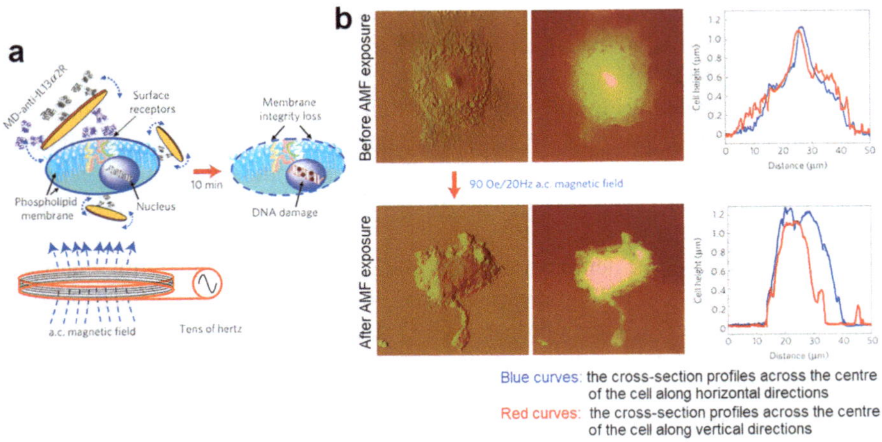

Fig. 8.5 Antibody-conjugated magnetic microdiscs triggered by a low-frequency AMF for targeted cancer cell destruction [16]. (**a**) Schematic illustration of cancer cell killing. (**b**) Changes in cancer cell structure after microdiscs-mediated magneto-mechanical force treatment: on the left, AFM images of cells; on the right, cell size data. (Copyright © 2009 Springer Nature. Reprinted with permission from Ref. [16])

nanoparticles, pyrene (a fluorescent dye), poly(styrene-block-allyl alcohol) (PS_{16}-*b*-PAA_{10}), and polyvinyl alcohol (PVA) as precursors (see Fig. 8.6a). The resulting nanospheres had a biphasic structure, with magnetic nanoparticles located on one side of the polymer nanospheres (see Fig. 8.6b, c). Compared to previous magnetic vortex microdiscs, which are micron-scale in size and used for magnetolytic therapy [16], these small magnetic nanoparticles (approximately 120–220 nm, varying with the weight ratio of magnetic nanoparticles to polymer) (see Fig. 8.6d) possess a high surface area-to-volume ratio. This enhances their ability to bind to cancer cells and penetrate cells more easily. The greater the number of magnetic nanoparticles in the precursors, the larger the magnetic segments in the final nanocomposites. Since the magnetic nanoparticles and fluorescence molecules are separated within the nanocomposites, they can maintain both strong magnetism and fluorescence simultaneously. Most of the cancer cells were killed after exposure to a spinning magnetic field at 50 rpm for 15 minutes (see Fig. 8.7a), as demonstrated by trypan blue staining (see Fig. 8.7b).

The magnetic segments on the biphasic nanospheres do not create a sharply defined surface; instead, their surfaces are smooth or only slightly rough. Nonetheless, these nanospheres may still cause significant mechanical damage to cancer cells for several reasons:

First, the biphasic nanospheres, activated by the magnetic field, can quickly and nonspecifically bind to prostate tumor LNCaP cells. Such firm binding facilitates subsequent magnetolytic therapy because the nanospheres can continuously exert mechanical forces—dragging or pushing and pulling—on the cell membrane, leading to rapid cellular damage.

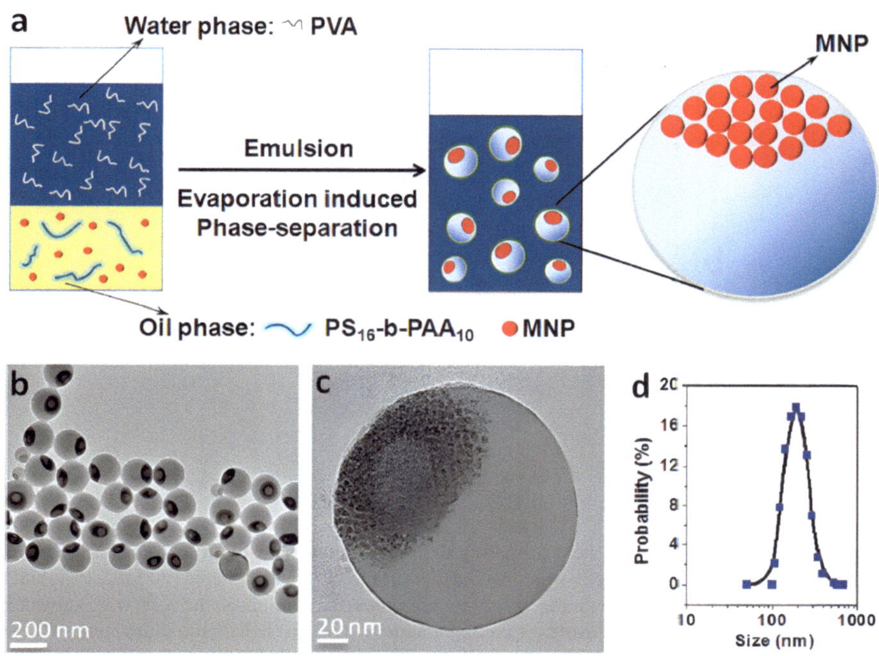

Fig. 8.6 Biphasic magneto-fluorescent nanospheres [17]. (**a**) Schematic illustration of the synthesis of biphasic nanospheres. (**b, c**) TEM images of the biphasic nanospheres at (b) low and (c) high magnification. (**d**) Distribution of biphasic nanosphere sizes (hydrodynamic diameters). (Copyright © 2010 American Chemical Society. Reprinted with permission from Ref. [17])

Secondly, these small nanospheres triggered by the magnetic field may aggregate into larger particles. The magneto-mechanical force generated by a larger aggregate is much stronger than that produced by a single small particle.

Thirdly, for the small-sized nanoparticles without aggregation, each nanosphere is similar to the tip of a needle. The smaller the nanospheres are, the sharper the needle tip.

If the fluorescent dye in the above biphasic nanospheres were replaced with anticancer agents such as chemotherapy drugs, or if both the fluorescent dye and the drugs were incorporated into the nanospheres simultaneously, the resulting product could be more effective at killing cancer cells than mechanical force-mediated cell damage alone.

The magneto-mechanical force is directly proportional to the average magnetic susceptibility of the magnetic particles [46]. The greater the magnetism of the magnetic particles, the greater the mechanical force generated when activated by a dynamic magnetic field.

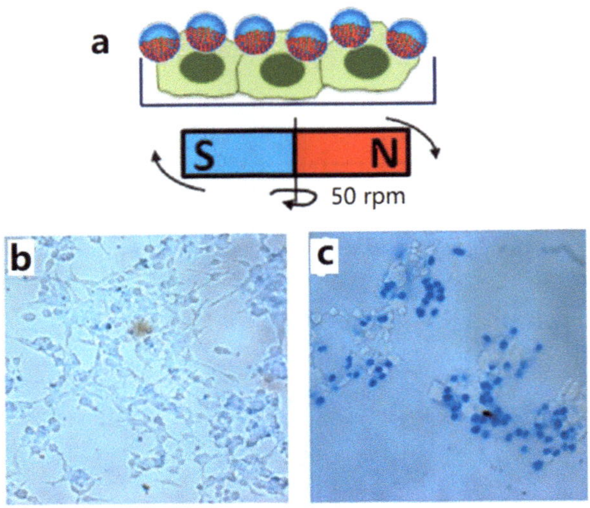

Fig. 8.7 A spinning magnetic field triggering the biphasic magneto-fluorescent nanospheres for killing cancer cells [17]. The cells were incubated with biphasic nanospheres (200 μL, 1%), and a magnet was used to facilitate their attachment to the cells. After 3 minutes, the cells were subjected to a magneto-mechanical force using a spinning magnetic field. (**a**) Schematic illustration of a cell experiment. (**b**) Cells incubated with nanospheres without magnetic field exposure. (**c**) Cells incubated with nanospheres and treated with 15 minutes of magnetic field exposure. (Copyright © 2010 American Chemical Society. Reprinted with permission from Ref. [17])

4 The Effect of Magnetic Field Properties, Exposure Time, and the Concentration of Magnetic Materials on Cell Killing Efficiency Through Magneto-Mechanical Force

As mentioned above, several types of dynamic magnetic fields can be selected to activate magnetic particles for killing cancer cells. Since the motion modes of magnetic particles differ under magnetic fields of different types, the magneto-mechanical force generated also varies. When a specific type of magnetic field is chosen, its properties—including frequency and strength—play crucial roles in determining cell-killing efficiency.

For example, when Fe_3O_4/BSA/$LrSiO_2$ microspheres, as described above, were added to TU212 cells and then exposed to a vortex magnetic field (VMF) of 400 mT for 1 hour, cell viability decreased as the VMF frequency increased from 0.5 to 2 Hz (see Fig. 8.8) [46]. This is because the magnetic microspheres vibrated more frequently as the VMF frequency increased. In other words, the cells were hit more times per minute by magnetic microspheres at higher VMF frequencies. However, cell viability did not further decrease when the VMF frequency increased from 2 to 4 Hz. This may be due to a reduction in the vibration amplitude of the magnetic microspheres at higher VMF frequencies.

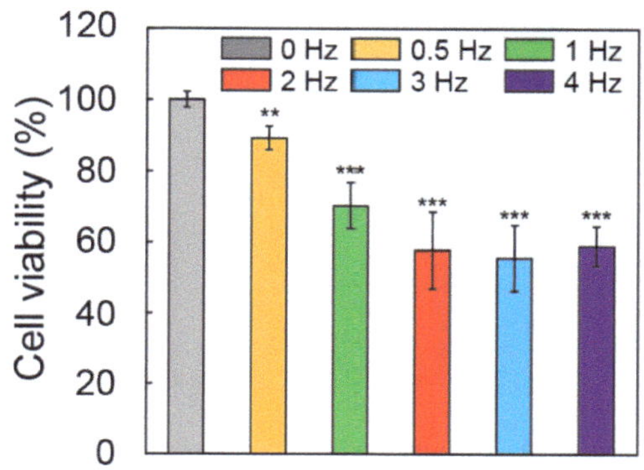

Fig. 8.8 The relationship between cancer cell viability and VMF frequency [46]. TU212 cancer cells were treated with sharp-surfaced magnetic microspheres under VMF exposure. (Copyright © 2022 by the Author(s). Reprinted with permission from Ref. [46])

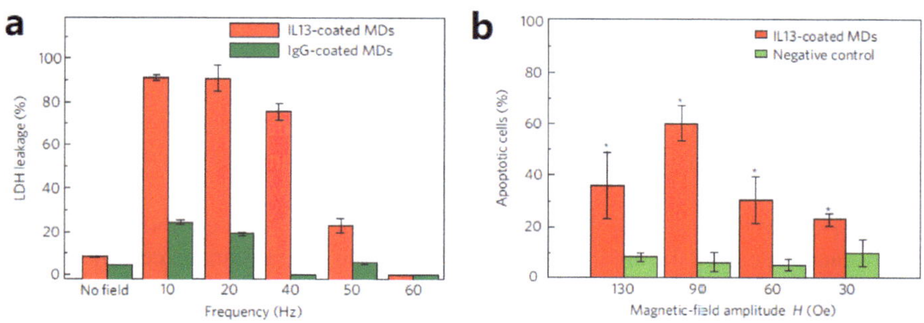

Fig. 8.9 The effect of (**a**) magnetic field frequency and (**b**) strength on cancer cell damage efficiency. N10 glioma cancer cells were treated with magnetic microdiscs conjugated with antihuman IL13α2R antibody or IgG1 under low-frequency alternating magnetic field (AMF) exposure [16]. (Copyright © 2009, Springer Nature. Reprinted with permission from Ref [16])

A similar result was found in other reports. For example, as shown in [16], when N10 glioma cancer cells were incubated with antibody-conjugated magnetic microdiscs (specifically, gold (Au) shell-coated iron–nickel microdiscs conjugated with antihuman-IL13α2R antibodies and mentioned above) and exposed to an alternating magnetic field (AMF) with a magnetic field strength of 90 Oe, the degree of cell damage reached its maximum at lower magnetic field frequencies (e.g., 10 and 20 Hz). When the frequency increased from 20 to 60 Hz, the level of cell damage did not increase; instead, it decreased. Under the 60 Hz condition, the cells were almost undamaged, which may be because the microdiscs could not vibrate effectively at such a high frequency. These results were confirmed by the amount of lactate dehydrogenase (LDH) leakage from the cancer cells (see Fig. 8.9a).

Additionally, in Novosad's experiments, as the magnetic field strength increased from 30 to 90 Oe, the number of apoptotic cells increased with field strength. This was attributed to the larger magneto-mechanical force exerted by the antihuman-IL13α2R antibody-conjugated microdiscs acting on the cells at higher strengths (see Fig. 8.9b). However, when the magnetic field strength was further increased to

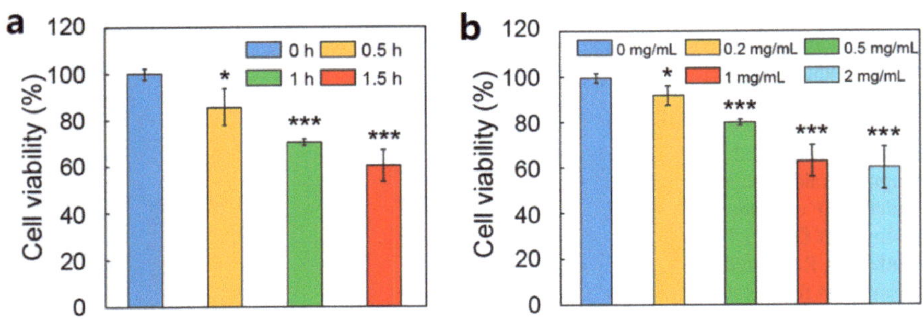

Fig. 8.10 The relationship between cancer cell viability and (**a**) VMF exposure time and (**b**) the concentration of the magnetic microspheres [46]. TU212 cancer cells were treated with magnetic microspheres that had a smooth surface under VMF exposure. (Copyright © 2022 by the Author(s). Reprinted with permission from Ref. [46])

130 Oe, the number of apoptotic cells decreased, which may be due to the limited flexible movement of the microdiscs under stronger magnetic fields.

It should be noted that when the antihuman-IL13α2R antibody on the microdiscs was replaced by IgG1, the N10 glioma cancer cells showed little to no significant damage under the same low-frequency AFM exposure described above. This may be because most of the microdiscs can firmly bind to the cancer cell membrane through the specific recognition between the IL13α2R antibody and IL13α2R overexpressed on glioma cells, allowing the microdiscs to transfer mechanical force to the cell membrane and subsequently trigger intracellular apoptotic pathways.

The surface of the microdiscs used in Novosad's group was smooth (not rough). If the magnetic particles have sharp surfaces, then particles without antibody conjugation can also efficiently kill cancer cells via magneto-mechanical forces.

Other parameters, such as the exposure time of the dynamic magnetic field and the concentration of the magnetic particles, also significantly affect the cell-killing efficiency. For example, as shown in [46], the cell viabilities of TU212 cells treated with the Fe_3O_4/BSA/LrSiO_2 microspheres and a variable magnetic field (VMF) decreased with increasing VMF exposure time (see Fig. 8.10a) and decreased as the microsphere concentration increased (see Fig. 8.10b).

5 Magnetic Materials for Inhibiting Tumor Growth Through Magneto-Mechanical Force

Cancer cells in an animal's body can also be destroyed by magneto-mechanical forces. For example [40], when mice were subcutaneously injected with TU212 cells and then Hedgehog-like gold-coated magnetic microspheres (HGMMS) were injected into the same sites, followed by exposure to a vibrating magnetic field (VMF) (see Fig. 8.11a), the cancer cells struggled to develop into tumors (see Fig. 8.11b, c). HGMMS without VMF exposure could also inhibit cell growth

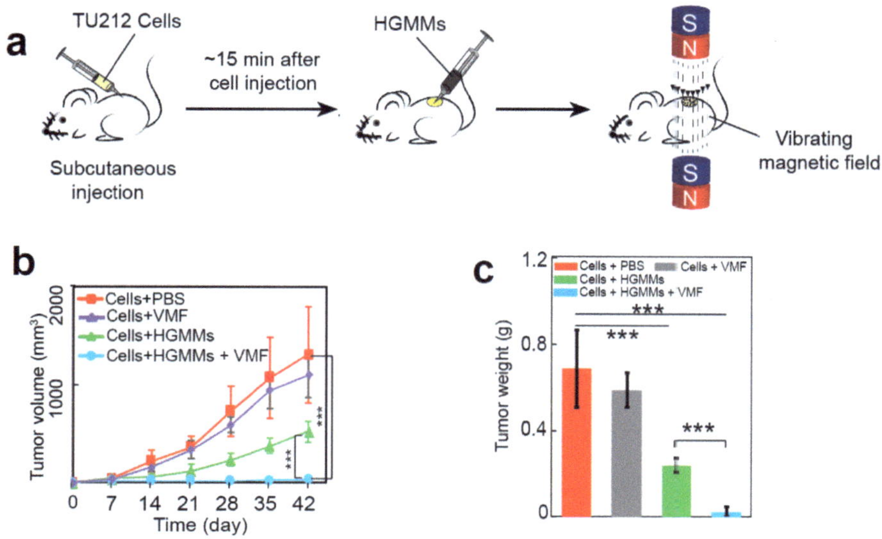

Fig. 8.11 VMF-triggered magnetic microspheres for inhibiting the growth of cancer cells in subcutaneous tissue of mice [40]. (**a**) Schematic illustration of the animal experiment: cancer cells were injected into mice, and approximately 15 minutes later, HGMMs were injected into the same sites, followed by exposure to a VMF. (**b**) The rates of in vivo cancer cell growth leading to tumor formation, and (**c**) the final tumor weights in different groups. (Copyright © 2018 by WILEY-VCH Verlag GmbH & Co. KGaA, Weinheim. Reprinted with permission from Ref. [40])

in vivo but were unable to prevent cancer cells from developing into larger tumors. Furthermore, VMF alone had little effect on cancer cell growth in vivo.

When the cancer cells in a mouse's body grow into larger tumors, the growth of such tumors can also be inhibited by magnetic particles-mediated magneto-mechanical force. For example [45], magnetic microspheres with sharp surfaces (4 mg/mL, 50 μL per mouse) were directly injected into mouse tumors (approximately 4.5–5 mm in size, grown from TU212 cells), followed by exposure to a vibrating magnetic field (VMF) (400 mT, 2 Hz, 2 hours daily) (see Fig. 8.12a). The average tumor volume at 20 days postinjection was about 60% of that in mice injected with PBS alone (the control group). However, for mice intratumorally injected with the same magnetic microspheres without VMF exposure, the average tumor volume at the same time point was not statistically different from that of the control group. Similar results were observed in the average tumor weights at 20 days postinjection (see Fig. 8.12b).

6 Discussion

Magneto-mechanical force can help overcome the drug resistance of cancer cells [59]. The mechanical movement of drug-loaded particles can accelerate drug release [60]. Since the cell membrane can be first damaged by mechanical force,

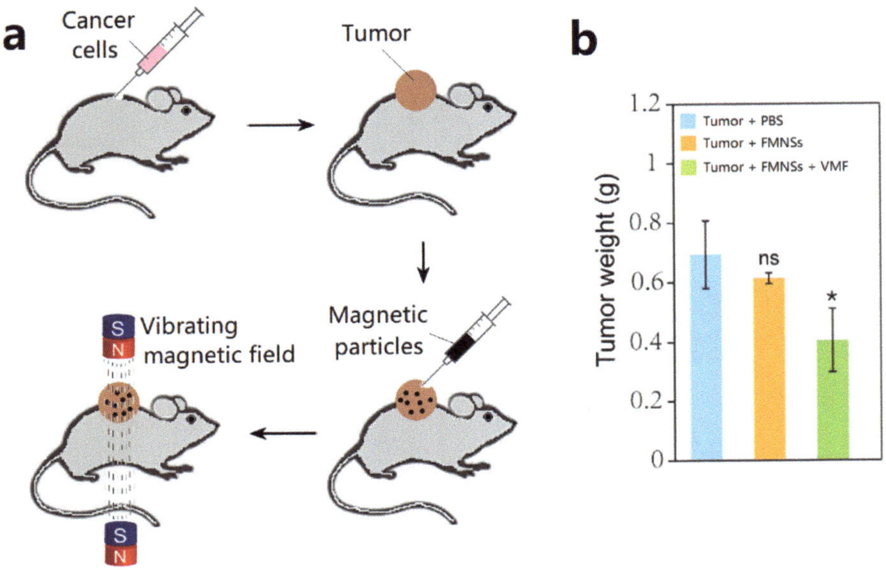

Fig. 8.12 VMF-triggered magnetic microspheres for inhibiting mouse tumor growth [45]. (**a**) Schematic illustration of the animal experiments: flower-like magnetic nanospheres (FMNSs) were intratumorally injected into mice and then exposed to a VMF. (**b**) The average weights of tumors resected from mice 20 days after injection. (See (**b**): Copyright © 2024 by Elsevier Inc. Reprinted with permission from Ref. [45])

extracellular drugs released from carriers have an increased opportunity to enter the cells [59]. When cancer cells in tumor tissue are damaged by magneto-mechanical force, the intracellular contents released from the broken cells may be toxic to nearby cells. Both these released contents and cellular species may also stimulate an antitumor immune response.

As previously mentioned, both the magnetic field apparatus and the therapy procedures are simple, and the magnetic field frequency is very low. The force applied can damage any type of cancer cells, and cancer cells may be stimulated multiple times without developing resistance, as they do not exhibit multicellular drug resistance. Additionally, the magnetic field has excellent tissue penetration capabilities. Magneto-mechanical force-mediated therapy may be suitable for treating tumors in both superficial and deep tissues and could be easily adopted for clinical use.

However, the therapeutic efficiency of this method alone is not high. As the self-repairing ability of mechanically damaged cancer cells diminishes, combining magneto-mechanical force with other therapies—such as photothermal therapy [26, 40, 45, 47], photodynamic therapy [26, 47], chemodynamic therapy [44], chemotherapy [26], and others—often results in enhanced tumor inhibition. Therefore, integrating traditional or novel cancer therapies with mechanical-force-mediated treatment is recommended.

Future research should address key questions, such as whether mechanical force can induce cancer cell metastasis, how magnetic particles can be safely removed from the body, and how magnetic particles can be directly and accurately injected into deep tumors multiple times. These issues warrant further in-depth investigation in the future.

References

1. Wagner AM, Knipe JM, Orive G, Peppas NA. Quantum dots in biomedical applications. Acta Biomater. 2019;94:44–63.
2. Kargozar S, Hoseini SJ, Milan PB, Hooshmand S, Kim H-W, Mozafari M. Quantum dots: a review from concept to clinic. Biotechnol J. 2020;15(12):e2000117.
3. Chu M, Pan X, Zhang D, Qiang W, Peng J, Hai W. The therapeutic efficacy of CdTe and CdSe quantum dots for photothermal cancer therapy. Biomaterials. 2012;33(29):7071–83.
4. Wu Q, Chu M, Shao Y, Wo F, Shi D. Reduced graphene oxide conjugated with $CuInS_2$/ZnS nanocrystals with low toxicity for enhanced photothermal and photodynamic cancer therapies. Carbon. 2016;108:21–37.
5. Babu LT, Paira P. Current application of quantum dots (QD) in cancer therapy: a review. Mini Rev Med Chem. 2017;17(14):1406–15.
6. Mohammed L, Gomaa HG, Ragab D, Zhu J. Magnetic nanoparticles for environmental and biomedical applications: a review. Particuology. 2017;30:1–14.
7. Lima-Tenorio MK, Pineda EA, Ahmad NM, Fessi H, Elaissari A. Magnetic nanoparticles: in vivo cancer diagnosis and therapy. Int J Pharm. 2015;493:313–27.
8. He J, Huang M, Wang D, Zhang Z, Li G. Magnetic separation techniques in sample preparation for biological analysis: a review. J Pharm Biomed Anal. 2014;101:84–101.
9. Farzin A, Etesami SA, Quint J, Memic A, Tamayol A. Magnetic nanoparticles in cancer therapy and diagnosis. Adv Healthc Mater. 2020;9(9):e1901058.
10. Wu K, Su D, Liu J, Saha R, Wang J-P. Magnetic nanoparticles in nanomedicine: a review of recent advances. Nanotechnology. 2019;30(50):502003.
11. Tishin AM, Shtil AA, Pyatakov AP, Zverev VI. Developing antitumor magnetic hyperthermia: principles, materials and devices. Recent Pat Anticancer Drug Discov. 2016;11(4):360–75.
12. Rezaei B, Yari P, Sanders SM, Wang H, Chugh VK, Liang S, Mostufa S, Xu K, Wang J-P, Gómez-Pastora J, Wu K. Magnetic nanoparticles: a review on synthesis, characterization, functionalization, and biomedical applications. Small. 2024;20(5):e2304848.
13. Chu M, Shao Y, Peng J, Dai X, Li H, Qingsheng W, Shi D. Near-infrared laser light mediated cancer therapy by photothermal effect of Fe3O4 magnetic nanoparticles. Biomaterials. 2013;34:4078–88.
14. Hughes S, McBain S, Dobson J, et al. Selective activation of mechanosensitive ion channels using magnetic particles. J R Soc Interface. 2008;5(25):855–63.
15. Fung AO, Kapadia V, Pierstorff E, Ho D, Chen Y. Induction of cell death by magnetic actuation of nickel nanowires internalized by broblasts. J Phys Chem C. 2008;112:15085–8.
16. Kim DH, Rozhkova EA, Ulasov IV, Bader SD, Rajh T, Lesniak MS, Novosad V. Biofunctionalized magnetic-vortex microdiscs for targeted cancer-cell destruction. Nat Mater. 2010;9:165–71.
17. Shang-Hsiu H, Gao X. Nanocomposites with spatially separated functionalities for combined imaging and magnetolytic therapy. J Am Chem Soc. 2010;132:7234–7.
18. Liu D, Wang L, Wang Z, Cuschieri A. Magnetoporation and magnetolysis of cancer cells via carbon nanotubes induced by rotating magnetic felds. Nano Lett. 2012;12:5117–21.
19. Wang BR, Bienvenu C, Mendez-Garza J, Lancon P, Madeira A, Vierling P, Di Giorgio C, Bossis G. Necrosis of HepG2 cancer cells induced by the vibration of magnetic particles. J Magn Magn Mater. 2013;344:193–201.

20. Domenech M, Marrero-Berrios I, Torres-Lugo M, et al. Lysosomal membrane permeabilization by targeted magnetic nanoparticles in alternating magnetic fields. ACS Nano. 2013;7(6):5091–101.
21. Cheng D, Li X, Zhang G, Shi H. Morphological effect of oscillating magnetic nanoparticles in killing tumor cells. Nanoscale Res Lett. 2014;9:195.
22. Zhang EM, Kircher MF, Koch M, Eliasson L, Goldberg SN, Renstrom E. Dynamic magnetic fields remote-control apoptosis *via* nanoparticle rotation. ACS Nano. 2014;8:3192–201.
23. Leulmi S, Chauchet X, Morcrette M, Ortiz G, Joisten H, Sabon P, Livache T, Hou YX, Carriere M, Lequien S, Dieny B. Triggering the apoptosis of targeted human renal cancer cells by the vibration of anisotropic magnetic particles attached to the cell membrane. Nanoscale. 2015;7:15904–14.
24. Kilinc D, Lesniak A, Rashdan SA, et al. Mechanochemical stimulation of MCF7 cells with rod-shaped Fe-Au janus particles induces cell death through paradoxical hyperactivation of ERK. Adv Healthc Mater. 2015;4(3):395–404.
25. Contreras MF, Sougrat R, Zaher A, Ravasi T, Kosel J. Non-chemotoxic induction of cancer cell death using magnetic nanowires. Int J Nanomed. 2015;10:2141–53.
26. Wo FJ, Xu RJ, Shao YX, Zhang ZY, Chu MQ, Shi DL, Liu SP. A multimodal system with synergistic effects of magneto-mechanical, photothermal, photodynamic and chemo therapies of cancer in graphene-quantum dot-coated hollow magnetic nanospheres. Theranostics. 2016;6:485–500.
27. Cheng Y, Muroski ME, Petit DCMC, Mansell R, Vemulkar T, Morshed RA, Han Y, Balyasnikova IV, Horbinski CM, Huang XL, Zhang LJ, Cowburn RP, Lesniak MS. Rotating magnetic field induced oscillation of magnetic particles for *in vivo* mechanical destruction of malignant glioma. J Control Release. 2016;223:75–84.
28. Muroski ME, Morshed RA, Cheng Y, Vemulkar T, Mansell R, Han Y, Zhang LJ, Aboody KS, Cowburn RP, Lesniak MS. Controlled payload release by magnetic field triggered neural stem cell destruction for malignant glioma treatment. PLoS One. 2016;11:e0145129.
29. Master AM, Williams PN, Pothayee N, et al. Remote actuation of magnetic nanoparticles for cancer cell selective treatment through cytoskeletal disruption [J]. Sci Rep. 2016;6:33560.
30. Martínez-Banderas AI, Aires A, Teran FJ, Perez JE, Cadenas JF, Alsharif N, Ravasi T, Cortajarena AL, Kosel J. Functionalized magnetic nanowires for chemical and magneto-mechanical induction of cancer cell death. Sci Rep. 2016;6:1–11.
31. Ju H, Cui Y, Chen Z, Fu Q, Sun M, Zhou Y. Effects of combined delivery of extremely low frequency electromagnetic field and magnetic Fe3O4 nanoparticles on hepatic cell lines. Am J Transl Res. 2016;8:1838–47.
32. Hapuarachchige S, Kato Y, Ngen EJ, Smith B, Delannoy M, Artemov D. Non-temperature induced effects of agnetized iron oxide nanoparticles in alternating magnetic field in cancer cells. PLoS One. 2016;11:1–12.
33. Brossel R, Yahi A, David S, Moreno Velasquez L, Guinebretière J-M. Mechanical signals inhibit growth of a grafted tumor in vivo: proof of concept. PLoS One. 2016;11:e0152885.
34. Vegerhof A, Barnoy EA, Motiei M, Malka D, Danan Y, Zalevsky Z, Popovtzer R. Targeted magnetic nanoparticles for mechanical lysis of tumor cells by lowamplitude alternating magnetic field. Materials. 2016;9:1–12.
35. Wong DW, Gan WL, Liu N, Lew WS. Magnetoactuated cell apoptosis by biaxial pulsed magnetic field. Sci Rep. 2017;7:1–8.
36. Shen Y, Wu C, Uyeda TQP, Plaza GR, Liu B, Han Y, Lesniak MS, Cheng Y. Elongated nanoparticle aggregates in cancer cells for mechanical destruction with low frequency rotating magnetic field. Theranostics. 2017;7:1735–48.
37. Zamay TN, Zamay GS, Belyanina IV, Zamay SS, Denisenko VV, Kolovskaya OS, Ivanchenko TI, Grigorieva VL, Garanzha IV, Veprintsev DV, Glazyrin YE, Shabanov AV, Prinz VY, Seleznev VA, Sokolov AE, Propenko VS, Kim PD, Gargaun A, Berezovski MV, Zamay AS. Noninvasive microsurgery using aptamer-functionalized magnetic microdisks for tumor cell eradication. Nucleic Acid Ther. 2017;27:105–14.

38. Mansell R, Vemulkar T, Petit DCMC, Cheng Y, Murphy J, Lesniak MS, Cowburn RP. Magnetic particles with perpendicular anisotropy for mechanical cancer cell destruction. Sci Rep. 2017;7:1–7.
39. Li W, Liu Y, Qian Z, Yang Y. Evaluation of tumor treatment of magnetic nanoparticles driven by extremely low frequency magnetic field. Sci Rep. 2017;7:1–9.
40. Chen Y, Han P, Wu Y, Zhang ZF, Yue Y, Li WH, Chu MQ. Hedgehog-like gold-coated magnetic microspheres that strongly inhibit tumor growth through magnetomechanical force and photothermal effects. Small. 2018;14:1802799.
41. Chiriac H, Radu E, Ţibu M, Stoian G, Ababei G, Lăbuşcă L, Herea D-D, Lupu N. Fe–Cr–Nb–B ferromagnetic particles with shape anisotropy for cancer cell destruction by magneto-mechanical actuation. Sci Rep. 2018;8:1–9.
42. Lunov O, Uzhytchak M, Smolkova B, Lunova M, Jirsa M, Dempsey NM, Dias AL, Bonfim M, Hof M, Jurkiewicz P, Petrenko Y, Kubinova S, Dejneka A. Remote actuation of apoptosis in liver cancer cells *via* magneto-mechanical modulation of iron oxide nanoparticles. Cancer. 2019;11:1873.
43. Chen M, Wu J, Ning P, Wang J, Ma Z, Huang L, Plaza GR, Shen Y, Xu C, Han Y, Lesniak MS, Liu Z, Cheng Y. Remote control of mechanical forces via mitochondrial-targeted magnetic nanospinners for efficient cancer treatment. Small. 2020;16:1905424.
44. Liu J, Yang W, Huang Y, et al. Oxygen and hydrogen peroxide self-supplying magnetic nanoenzymes for cancer therapy through magneto-mechanical force, force-induced reactive oxygen species, chemodynamic effects and cytotoxicity of Ca^{2+} ions. Nano Res. 2023;16:7134–47.
45. Wang M, Li J, Liu J, Huang Y, Yang L, Zhu C, Zhang Y, Gui X, Peng H, Chu M. Smart nanozymes coupled with dynamic magnet field and laser exposures for cancer therapy. J Colloid Interface Sci. 2024;676:110–26.
46. Guo Y, Yang W, Guangjin P, Zhu C, Zhu Y, Li J, Huang Y, Wang B, Chu M. Low frequency vibrating magnetic field-triggered magnetic microspheres with a nanoflagellum-like surface for cancer therapy. J Nanobiotechnol. 2022;20:316.
47. Wu K, Mohsin A, Zaman WQ, Zhang Z, Guan W, Chu M, Zhuang Y, Guo M. Urchin-like magnetic microspheres for cancer therapy through synergistic effect of mechanical force, photothermal and photodynamic effects. J Nanobiotechnol. 2022;20:224.
48. Li J, Huang Y, Niu W, Chu M. Irregular-shaped Fe3O4 nanoparticles-mediated magneto-mechanical force for killing tumor cells. J Med Biomech. 2023;38(2):238–45.
49. Wang J, Hou Q, Qu J, Huo X, Li H, Feng Y, Wang Q, Le Chang CX. Polyhedral magnetic nanoparticles induce apoptosis in gastric cancer stem cells and suppressing tumor growth through magnetic force generation. J Control Release. 2024;373:370–84.
50. Almosawy W, Alizadeh A'a, Koosha N, Najafi N, Abdi N, Najafi M, Sadeghi Z, Ardalan A. Magnetically guided nanoparticles in cancer treatment: impact of non-Newtonian behavior, interaction force, and wall shear stress. J Drug Delivery Sci Technol. 2024;93:105410.
51. Stavilă C, Minuti AE, Herea DD, Lăbuşcă L, Gherca D, Lupu N, Chiriac H. Synergistic effect of chemotherapy and magnetomechanical actuation of Fe-Cr-Nb-B magnetic particles on cancer cells. ACS Omega. 2024;9(28):30518–33.
52. Mallick S, Sokolich M, Rivas D, Das S. Targeted vibration-induced necrosis in liver cancer cells using paramagnetic microrobots. Int Conf Manip Autom Robot Small Scales. 2023;2023. https://doi.org/10.1109/marss58567.2023.10294136.
53. Wang N, Butler JP, Ingber DE. Mechanotransduction across the cell surface and through the cytoskeleton. Science. 1993;260:1124–7.
54. Martinac B. Mechanosensitive ion channels: molecules of mechanotransduction. J Cell Sci. 2004;117(12):2449–60.
55. Ingber DE. Cellular mechanotransduction: putting all the pieces together again. FASEB J. 2006;20:811–27.
56. Orr AW, Helmke BP, Blackman BR, Schwartz MA. Mechanisms of mechanotransduction. Dev Cell. 2006;10:11–20.
57. Mannix RJ, et al. Nanomagnetic actuation of receptor-mediated signal transduction. Nat Nanotechnol. 2007;3:36–40.

58. Dobson J. Remote control of cellular behaviour with magnetic nanoparticles. Nat Nanotechnol. 2008;3:139–48.
59. Zhu C. Chestnut-like drug-loaded magnetic nanoparticles driven by dynamic magnetic field for multidrug resistance cancer therapy. Master's thesis. 2023.
60. Wo F, Rujiao X, Shao Y, Zhang Z, Chu M, Shi D, Liu S. A multimodal system with synergistic effects of magneto-mechanical, photothermal, photodynamic and chemo therapies of cancer in graphene-quantum dot-coated hollow magnetic nanospheres. Theranostics. 2016;6(4):485–500.

Nanomaterials-Mediated X-Ray-Induced Cancer Photodynamic Therapy

9

Haiyan Wang, Miguel Vladimir Armindo, and Maoquan Chu

Abstract

Cancer is a major disease threatening human health in the twenty-first century, with a rising global incidence. Conventional therapeutic approaches, such as surgery, radiotherapy, and chemotherapy, often fail to achieve curative effects in patients with advanced-stage cancer. Therefore, the development of more precise and effective treatment strategies is essential. X-ray-activated nanoscintillators, an emerging modality, demonstrate significant potential in cancer treatment due to their unique energy conversion capabilities and ability to penetrate deep tissues.

This review outlines the fundamental concepts of nanoscintillators and explains the mechanisms underlying their X-ray-excited luminescence. It further explores their applications in oncology, including direct tumor ablation, X-ray energy transduction for photodynamic therapy (X-PDT), and synergistic combinations with chemotherapy. Furthermore, the advantages (such as high penetration depth and multifunctionality) and disadvantages (such as low energy conversion efficiency and biocompatibility concerns) of X-ray-activated nanoscintillators are critically examined. By optimizing matrix materials, dopants, and energy transfer designs, the therapeutic performance of nanoscintillators can be significantly improved. With ongoing advancements, X-ray-activated nanoscintillators hold promising prospects for broad clinical applications, offering patients additional effective treatment options.

H. Wang · M. V. Armindo · M. Chu (✉)
School of Life Sciences and Technology, Tongji University, Shanghai, China
e-mail: mqchu98@tongji.edu.cn

M. Chu (ed.), *Modern Nanomaterials and Nanotechnology for Cancer Therapy*,
https://doi.org/10.1007/978-981-95-6849-9_9

Keywords

Nanoscintillators · X-ray · Energy transfer · Photodynamic therapy · Tumor

1 Introduction

Cancer has emerged as the leading global cause of mortality in the twenty-first century [1, 2], with its incidence escalating due to industrialization [3], air pollution [4], environmental degradation [5], accelerated lifestyles, and detrimental habits, such as smoking and alcohol abuse [6]. Beyond its impact on individual health, cancer constitutes a critical societal challenge, profoundly affecting families and public health systems [7]. Current clinical interventions include surgical resection, radiotherapy, chemotherapy, molecularly targeted therapies, and minimally invasive techniques [8, 9]. However, advanced-stage malignancies remain resistant to conventional treatments, underscoring an urgent need for innovative therapeutic paradigms with enhanced tissue specificity and efficacy.

As an emerging therapeutic modality, X-ray-activated nanoscintillators demonstrate increasing potential in cancer treatment. Nanoscintillators are luminescent materials capable of absorbing high-energy particles, such as X-rays, and emitting light via radiative recombination [10]. In medical applications, conventional scintillators have been extensively used in diagnostic imaging techniques, such as X-ray radiography [11] and computed tomography (CT) [12]. Recent advancements in nanotechnology have enabled the integration of nanophotonics into scintillator design, resulting in next-generation nanoscintillators with enhanced detection sensitivity and reduced radiation exposure [13]. X-ray-excited nanoscintillators for in vivo imaging and cancer therapy have attracted considerable attention in recent years. By optimizing the material structure and dopant composition of nanoscintillators, their X-ray-induced fluorescence can be significantly improved, leading to higher precision in medical imaging and supporting early cancer diagnosis. In addition to fluorescence, nanoscintillators can generate reactive oxygen species (ROS) upon X-ray irradiation, which have strong therapeutic effects—for example, singlet oxygen ($^1O_2$). X-ray-induced photodynamic therapy (X-PDT) leverages the penetrating ability of X-rays to stimulate nanoscintillators to produce ROS within deep tumor tissues, damaging tumor cells. This approach offers several advantages over light-based photodynamic therapy (PDT), such as laser-induced PDT. Furthermore, by combining nanoscintillators with specific drugs or prodrugs, greater ROS generation can be achieved under X-ray irradiation, enabling precise treatment of localized tumors while reducing damage to surrounding healthy tissues [14]. This method combines the benefits of PDT and radiotherapy (RT), making it more widely applicable.

The most common X-PDT techniques combine inorganic scintillators and organic photosensitizers, with the former converting X-rays into visible light emission and the latter producing singlet oxygen species that kill the tumor site. This chapter summarizes the concept and types of nanoscintillators, the fluorescence emission mechanism of X-PDT, nanoscintillators directly used for tumor X-PDT, the use of nanoscintillators as X-ray energy transmitters for tumor X-PDT, and the combination of X-ray-excited nanoscintillators with other treatments (such as chemotherapy), as well as the strengths and weaknesses of nanoscintillators-based X-PDT.

2 Types of Nanoscintillators

Nanoscintillators, as a class of materials with unique luminescent properties, play an important role in several fields, such as high-energy physics, radiology, bio-imaging, and display technologies [15, 16]. From the perspective of material composition, nanoscintillators can be classified into the following four categories:

1. *Oxide nanoscintillators*: Oxide nanoscintillators are composed of oxide materials. This type of scintillator has stable chemical properties and good luminescence performance [17]. Like many other nanoscintillators, the fluorescence emission mechanism of oxide nanoscintillators generally involves electrons jumping between energy bands, resulting in the emission of light at a specific wavelength under external excitation.
2. *Sulfide nanoscintillators*: Sulfide nanoscintillators are based on sulfide materials. These materials have unique optical and electrical properties, which make sulfide nanoscintillators notable for their fluorescence efficiency and wavelength tunability. Upon X-ray irradiation, sulfide nanoscintillators exhibit high X-ray absorption efficiency, excellent stability, high fluorescence quantum yield, and rapid decay time.
3. *Halide nanoscintillators*: Halide nanoscintillators, especially rare-earth fluoride-based scintillators such as NaLnF4 (where Ln stands for rare-earth elements such as Y, Gd, Lu, etc.), have excellent fluorescence emission and stability [18]. Rare-earth fluoride-based scintillators feature low phonon vibrational energy, good crystallinity, and easily tunable particle size. These properties have led to their widespread application in high-energy physics, radiomedicine, and other fields [19]. Additionally, halide nanoscintillators have good X-ray absorption properties, making them valuable materials for X-ray–excited imaging [20].
4. *Chalcogenide nanoscintillators*: Chalcogenide nanoscintillators include two types: lead-based chalcogenides and non-lead chalcogenides [21]. Both types have unique crystal structures and excellent optical properties. Lead-based chalcogenides have high X-ray absorption coefficients, wavelength tunability, and good solution processing properties, but they are toxic to the environment [20]. Therefore, researchers are actively developing nontoxic and highly stable non-lead chalcogenide scintillators [22]. Non-lead chalcogenide scintillators achieve the dual goals of being environmentally friendly and efficient by replacing elemental lead, providing a new option for X-ray-excited imaging [23].

With advancements in synthetic technology, more organic nanoscintillators have been developed [24]. Consequently, nanoscintillators can be classified into organic, inorganic, and organic-inorganic hybrid categories. Organic nanoscintillators are more attractive for in vivo medical applications than inorganic nanoscintillators because they exhibit good biocompatibility and degrade effectively in vivo.

3 Mechanism of X-Ray Excited Luminescence of Nanoscintillators

When a nanoscintillator is irradiated by X-rays, it absorbs the energy of the X-rays. This energy excites electrons in the nanoscintillator material from their ground state to higher energy levels. These excited electrons are unstable at high energy and will eventually relax back to lower energy states or the ground state, releasing energy in the process. During this relaxation, energy is redistributed within the nanoscintillator and may generate secondary electrons, establishing a nonequilibrium steady state. The transition of electrons back to the lower or ground state results in the emission of energy primarily as fluorescence [25]. The fluorescence emission process of nanoscintillators can be modulated by size and surface effects, resulting in unique fluorescence properties [26]. It should be noted that these released energies are also transferred to the surrounding molecular oxygen ($^3O_2$) to generate singlet oxygen ($^1O_2$) through energy transfer (type II PDT). Additionally, the excited electrons can be captured by the surrounding solvent and biomolecules, leading to the production of free radicals such as superoxide (O_2^-) and hydroxyl radicals (·OH) via electron transfer (type I PDT). These reactive oxygen species (ROS) are the main agents used to combat cancer cells (i.e., X-ray-induced photodynamic therapy, X-PDT).

To enhance ROS generation of X-PDT, photosensitizers are often conjugated onto the surfaces of nanoscintillators. When the photosensitizers-conjugated nonequilibria are excited by X-rays, the luminescent energy is transferred from the nanoscintillator to the photosensitizer via fluorescence resonance energy transfer (FRET). This energy triggers a photochemical reaction, in which surrounding oxygen molecules are excited to $^1O_2$, and additional free radicals such as O_2^- and OH are generated through electron transfer (see Fig. 9.1).

4 X-Ray-Excited Nanoscintillators for Cancer Therapy

1. *Nanoscintillators Alone Upon X-Ray Excitation for Cancer X-PDT*

According to the X-PDT mechanisms, nonequilibrium states alone, upon X-ray excitation, can be used for cancer therapy. However, most nonequilibrium states alone produce low levels of reactive oxygen species (ROS) after X-ray irradiation.

To effectively improve energy utilization, Huang and coworkers recently developed an organic phosphorescent nanoscintillator that functions both as a scintillator

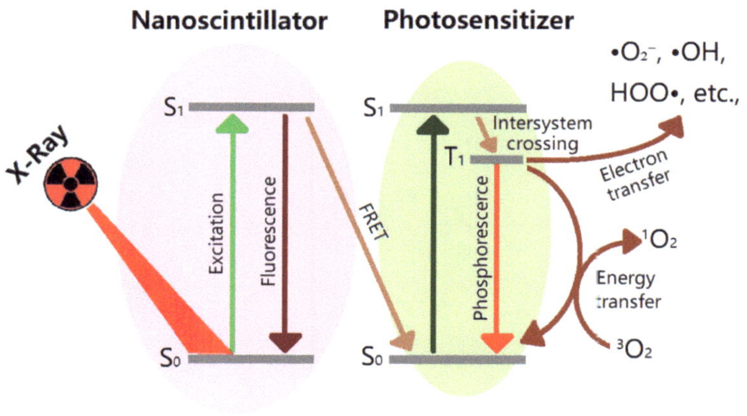

Fig. 9.1 Reactive oxygen species (ROS) produced by photosensitizers-conjugated under non-equilibrium conditions during X-ray excitation. S_0 is the ground electronic singlet state, S_1 is the first excited singlet state, and T_1 is the triplet excited state

and a photosensitizer. It directly absorbs X-ray energy and efficiently generates singlet oxygen ($^1O_2$).

The main material of this nanoscintillator is the synthesized organic molecule 9,9′-(6-iodophenoxy-1,3,5-triazine-2,4-diyl)bis(9H-carbazole), referred to as ITC. The iodine atom enhances X-ray absorption, while the triazine ring and carbazole moiety provide a rigid structure that promotes intermolecular stacking and extends the phosphorescence lifetime. Transmission electron microscopy (TEM) revealed that the nanoparticles are monodisperse and spherical, with an average diameter of 47.2 ± 0.8 nm, suitable for tumor targeting via the enhanced permeation and retention (EPR) effect. X-ray doses as low as 0.4 Gy were sufficient to inhibit tumor growth in vivo, demonstrating significantly improved efficacy compared to conventional radiotherapy. Unlike conventional inorganic scintillators (e.g., LaF_3:Tb), which rely on a two-step energy transfer (X-rays→visible light→photosensitizer), ITC acts directly as a photosensitizer, reducing energy loss and increasing singlet oxygen ($^1O_2$) production.

The results showed that this organic phosphorescent nanoscintillator could efficiently inhibit tumor growth in deep tissues at a low X-ray dose of 0.4 Gy. This discovery highlights the potential of organic phosphorescent nanoscintillator-mediated X-ray-induced photodynamic therapy (X-PDT) for applications in bioelectronics (see Fig. 9.2) [27].

For another example, Yu et al. reported on purely organic aggregation-induced emission (AIE) nanoscintillators (TBDCR NPs) that can generate both type I and type II ROS in large quantities under X-ray irradiation for tumor therapy [28]. The introduction of heteroatoms enhances X-ray absorption and ROS generation (see Fig. 9.3). AIE-activated TBDCRs exhibit aggregation-enhanced ROS production and have a reduced oxygen dependency for hydroxyl radical (·OH, type I) generation. TBDCR NPs possess a unique PEG crystalline shell that provides a rigid

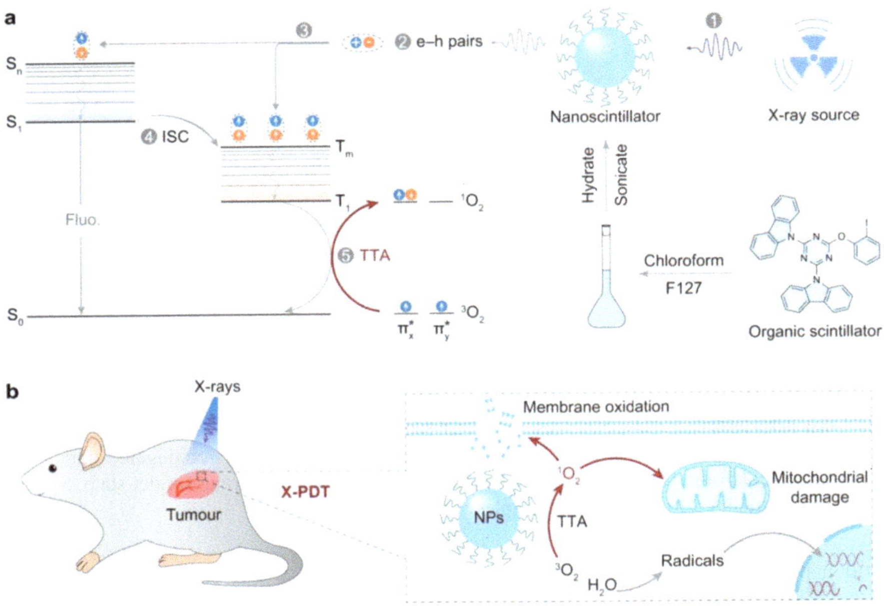

Fig. 9.2 Organic phosphorescent nanoscintillators for tumor X-ray photodynamic therapy (X-PDT). (**a**) Process for preparing organic nanoparticles in aqueous solution and their subsequent generation of singlet oxygen($^1O_2$) under X-ray irradiation. (**b**) Schematic representation of the mechanism for treating deep-seated tumors in vivo. Localized triplet oxygen($^3O_2$) molecules absorb energy to produce large quantities of singlet oxygen ($^1O_2$). The generated $^1O_2$ induces membrane oxidation and mitochondrial damage, effectively killing cancer cells [27]

intraparticle microenvironment, further amplifying ROS generation. Upon X-ray irradiation, TBDCR NPs not only produce a significant amount of singlet oxygen species but also emit bright near-infrared fluorescence, which can be used for both X-ray photodynamic therapy (X-PDT) and in vivo imaging.

2. *Nanoscintillators as X-ray Energy Transmitters for Cancer X-PDT*

Photosensitizers-conjugated nonequilibrium-mediated X-PDT is more effective than nonequilibrium therapy alone because it can generate more reactive oxygen species (ROS) to combat cancer cells. For example, Tang et al. synthesized $NaLuF_4$:Tb/Gd (15%/5%)@$NaYF_4$, a scintillator nanoparticle (ScNP) with efficient X-ray energy transduction and strong green light emission (see Fig. 9.4). This nanoparticle was combined with the photosensitizer merocyanine 540 (MC540) and loaded with the nitric oxide (NO) donor α-nitrilotriacetic acid (NEAA). The resulting nanocomposite, $NaLuF_4$:Tb/Gd (15%/5%)@$NaYF_4$@$mSiO_2$@MC540@NEAA (ScNP-MS@MC540@NEAA), produced ROS and NO under low-dose X-ray irradiation. The therapeutic process involves depletion of glutathione (GSH), self-supply of NO, mitigation of mitochondrial damage, and alleviation of hypoxia, leading to the production of ROS and reactive nitrogen species (RNS). RNS forms

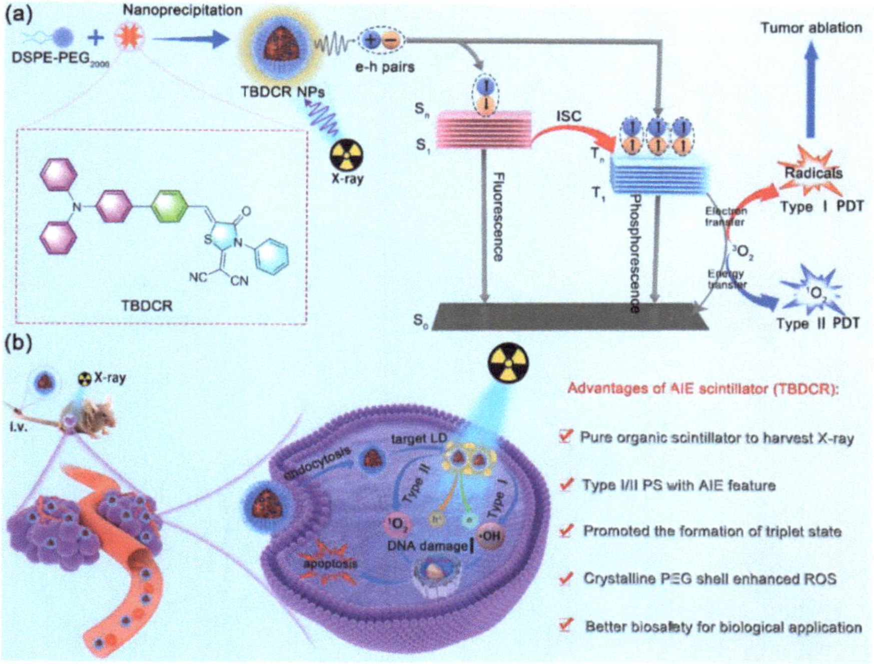

Fig. 9.3 Schematic representation of the structure and preparation of TBDCR nanoparticles (NPs) and their application in X-ray photodynamic therapy (X-PDT). (**a**) Preparation of organic TBDCR NPs: Organic TBDCR NPs were prepared, and then singlet oxygen ($^1O_2$) and radical reactive oxygen species (ROS) were generated under direct X-ray irradiation. (**b**) Application of TBDCR NPs in fluorescence imaging-guided X-PDT cancer therapy [28]

a closed-loop, anti-hypoxia, dual-mode system that synergistically enhances the antitumor effect without significant toxicity [29].

To expand the application range of X-PDT, researchers have developed additional types of nanoscintillators that emit light in the ultraviolet region upon excitation by X-rays. These can be paired with a wider variety of photosensitizers to achieve highly efficient multimodal cancer therapy. Reports have described the development of nanoparticle-based X-ray-activated anticancer "nanomedicines" comprising yttrium oxide (Y_2O_3) nanoscintillators, HIV-1 TAT peptide fragments, and psoralens. In this formulation, X-ray radiation is absorbed by the Y_2O_3 nanoscintillators, which then emit long-wave ultraviolet (UVA) light UVA photons are absorbed by nanocovalently coupled psoralens, which can crosslink adenine and thymine residues in DNA [30]. Nanoscintillators that emit UV light under X-ray excitation can be paired with a wider range of photosensitizers, provided that the absorption spectrum of the photosensitizer partially overlaps with the emission spectrum of the nanoscintillator to facilitate energy transfer or photoactivation. The absorption spectrum of psoralen spans 320–405 nm, which significantly overlaps with the 370 nm emission spectrum of Y_2O_3. This spectral compatibility allows

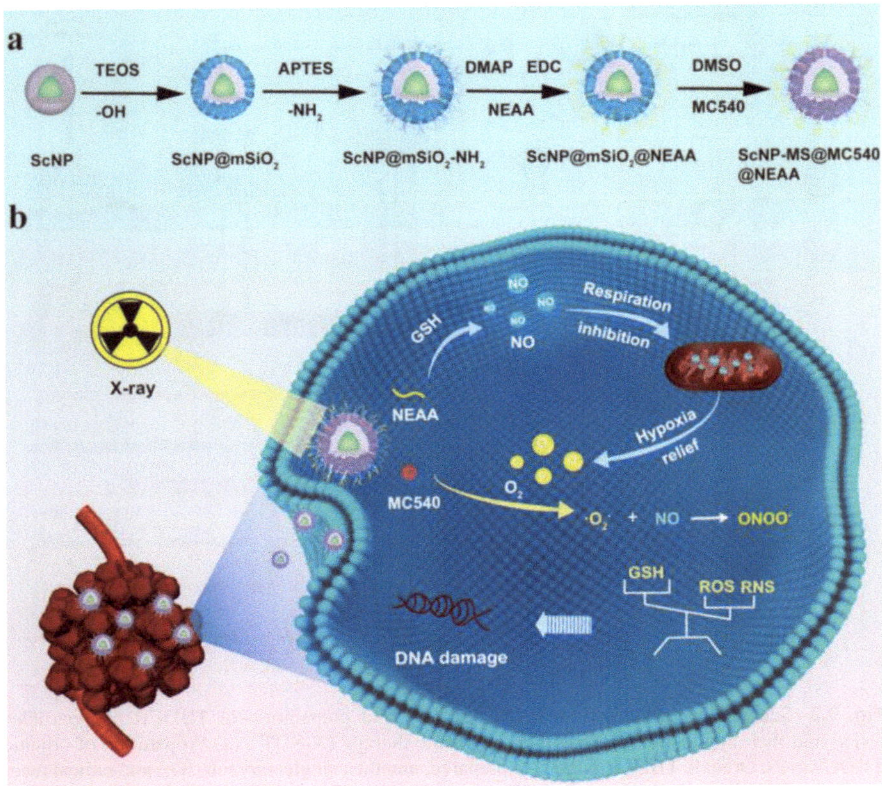

Fig. 9.4 Schematic diagram of closed-loop bimodal cancer therapy. (**a**) Synthesis process of ScNP-MS@MC540@NEAA. (**b**) Low-dose X-ray-activated nanocomposites for closed-loop anti-hypoxia bimodal inter-enhancement of NO, free radicals, and photodynamic therapy [29]

psoralens to absorb UVA photons emitted by the nanoscintillators, thereby inducing DNA cross-linking.

Kaidong Yang et al. developed a novel Gd^{3+}/Ce^{3+} co-doped $LiLuF_4$ nanoscintillator with excellent X-ray-induced UV luminescence properties and applied it to in vivo tumor therapy. The nanoscintillators were synthesized by optimizing the matrix material, dopants, and energy transfer design. Their X-ray-induced UV luminescence intensity was enhanced by approximately 18 times compared to that of conventional Ce^{3+} single-doped materials. The surface of the nanoscintillator was loaded with the drug N,N′-di-sec-butyl-N,N′-dinitroso-1,4-phenylenediamine (BNN6). BNN6 is a UV-sensitive nitric oxide precursor that decomposes in response to UV or X-ray-induced UV irradiation, releasing therapeutic nitric oxide (NO) gas. Furthermore, they combined the nanoscintillator with nitric oxide precursor to achieve a synergistic treatment involving NO release and radiotherapy under X-ray induction [31].

3. *Multimodal Therapy with X-ray Excited Nanoscintillators*

X-ray excitation of nanoscintillators can enhance radiotherapy. Under X-ray irradiation, nanoscintillators emit high-energy photons or secondary electrons, which can directly damage the DNA structure in cancer cells and thus kill them. This form of radiotherapy is precise and efficient, reducing damage to normal tissues compared to conventional radiotherapy. X-ray-excited nanoscintillators can also be combined with various treatment modalities to achieve different therapeutic effects, in addition to radiotherapy (see Fig. 9.5).

Caiju Zhang et al. designed an X-ray-activated Fe^{2+} supply platform, termed "PFCN," which contains $CaWO_4$ core nanoscintillators that emit ultraviolet (UV) light and a surface decoration of Fe_3O_4 to deliver excess Fe^{3+}. Under X-ray excitation, the UV light emitted by $CaWO_4$ catalyzes the reduction of Fe^{3+} to Fe^{2+}, which subsequently triggers the Fenton reaction to produce highly toxic hydroxyl radicals (see Fig. 9.6). Additionally, this process simultaneously induces immunogenic cell death, thereby activating immunotherapy [32]. The study achieved a triple synergy of radiotherapy, ferroptosis (iron-dependent cell death) therapy, and immunotherapy through the PFCN nanoplatform, offering a new strategy for deep tumor treatment.

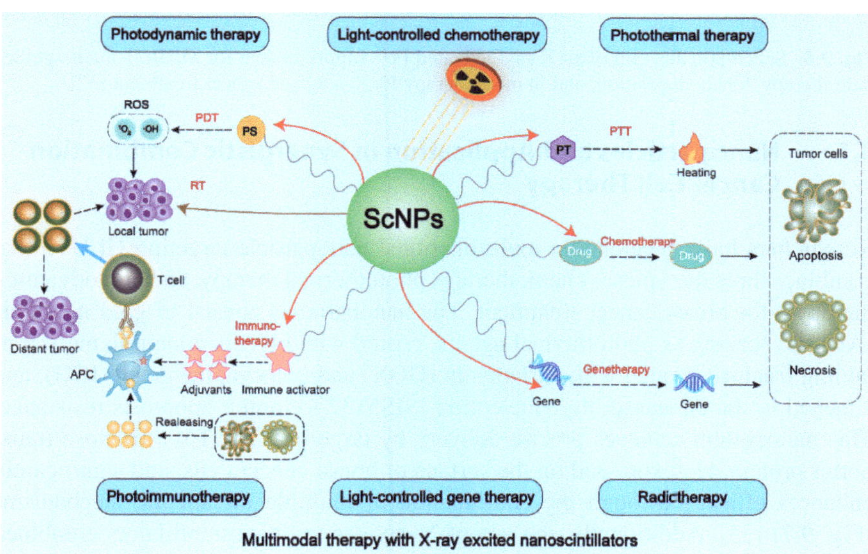

Fig. 9.5 X-ray excitation of nanoscintillator-induced multimodal therapy

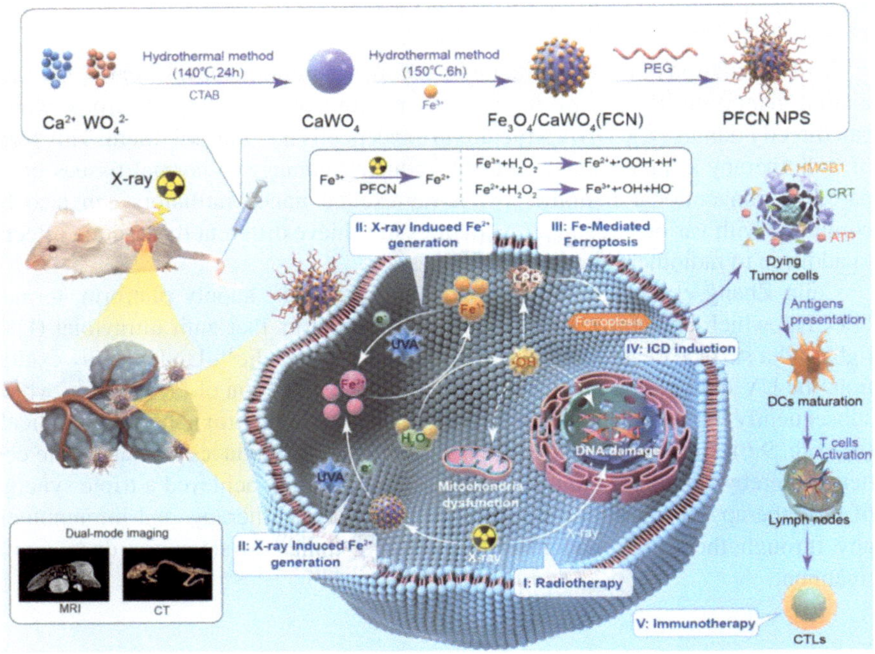

Fig. 9.6 Schematic diagram of an X-ray-activated Fe^{2+} supply system for MRI/CT image-guided radiotherapy, ferrite deposition, and immunotherapy for synergistic cancer treatment [32]

4.1 Nanoparticles and Application in Synergistic Combination Cancer Cell Therapy

Researchers have developed a multifunctional nanoparticle targeting GLUT5 that combines three therapies—chemotherapy, photothermal therapy, and photodynamic therapy—for breast cancer treatment. The nanoparticles consist of gold nanorods (AuNRs) serving as photothermal agents, coated with glycopeptide polymers containing fructose, loaded with adriamycin (DOX) and indocyanine green (ICG), and attached to the antisense oligonucleotide ISIS5132 to inhibit apoptosis resistance. This nanosystem achieves precise delivery by targeting GLUT5, a fructose transporter protein overexpressed on the surface of breast cancer cells, and significantly enhances efficacy through the combination of multiple therapeutic mechanisms (Fig. 9.7) [33]. Additionally, the use of X-ray-excited nanoscintillators combined with immunotherapy has become a prominent area of research in tumor treatment in recent years. X-ray-excited nanoscintillators can activate the body's immune system to generate specific immune responses against tumor cells by inducing immunogenic cell death (ICD). This approach has the potential to fundamentally alter the tumor growth environment and improve patient survival rates and quality of life.

A study has developed cancer cell membrane-coated mesoporous organosilicon nanoparticles (MON) containing X-ray and reactive oxygen species

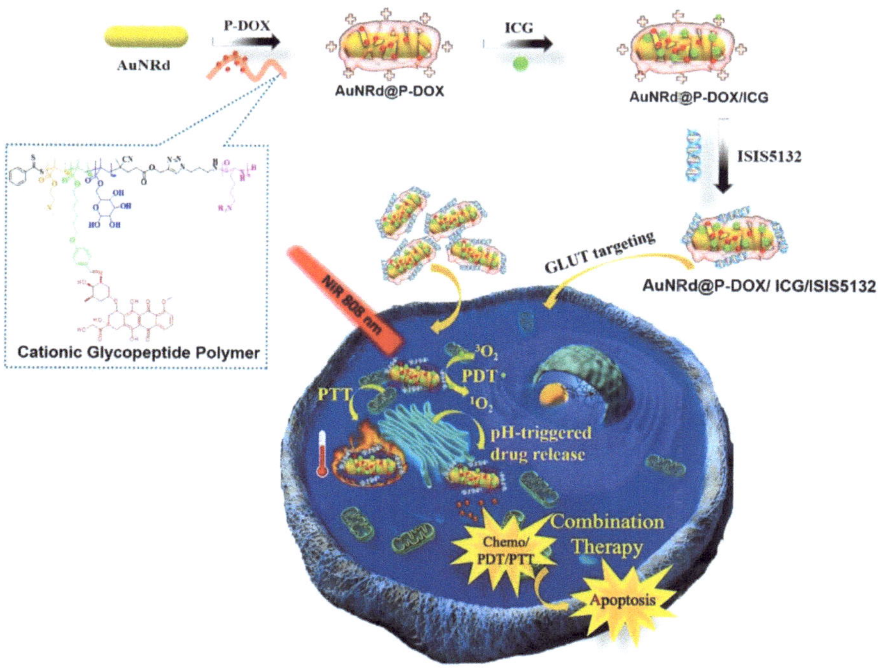

Fig. 9.7 Schematic Illustration of the preparation of multifunctional AuNRd@P-DOX/ICG/ISIS5132 phototherapeutic agents [33]

(ROS)-responsive diselenide bonds for the controlled release of doxorubicin (DOX) at the tumor site. Under low-dose X-ray irradiation, DOX-loaded MON exhibits carrier degradation-controlled release through the cleavage of the diselenide bonds, leading to DOX-mediated immunogenic cell death at the tumor site. When combined with PD-L1 checkpoint blockade, this approach further enhanced the inhibition of tumor growth and metastasis, with low systemic toxicity (see Fig. 9.8) [34]. The multimodal therapeutic strategy demonstrated in this study highlights the potential of chemoimmunotherapy by combining a chemotherapeutic drug (DOX), X-ray-responsive nanocarriers, and PD-L1 immune checkpoint blockade. This combination significantly inhibited tumor growth and metastasis in an animal model while reducing systemic toxicity.

Additionally, X-ray-excited nanoscintillators combined with gene therapy and chemotherapy can, through their carrier effect, precisely deliver therapeutic genes or chemotherapeutic drugs to tumor tissues, enabling accurate treatment. This approach can reduce the systemic toxicity of chemotherapeutic drugs and enhance therapeutic efficacy.

Platinum (Pt) nanoparticles (Pt@BSA NPs) encapsulated in bovine serum albumin (BSA) were synthesized and then combined with an anticancer drug (methotrexate, MTX) for use in radiotherapy (see Fig. 9.9). Cell viability and clone formation assays demonstrated the superior anticancer potential of Pt nanoparticles

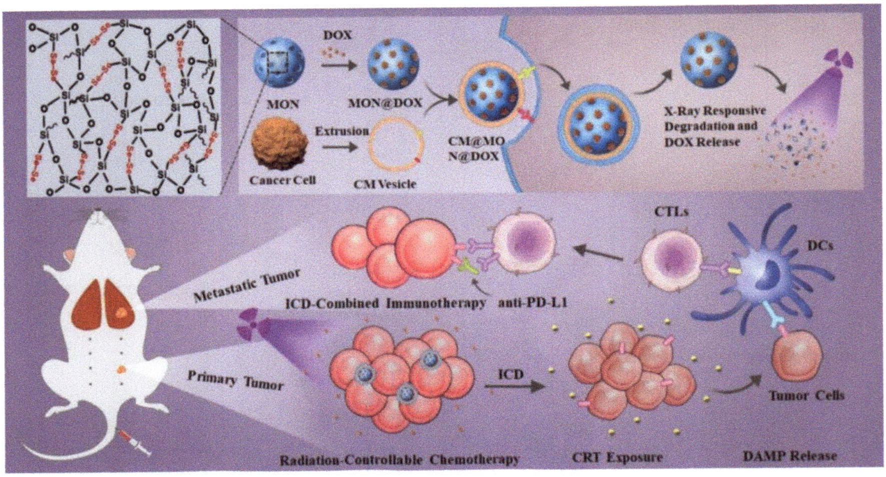

Fig. 9.8 Schematic synthesis of a diselenide-bonded bridge MON for controlled drug release using low-dose X-ray radiation [34]

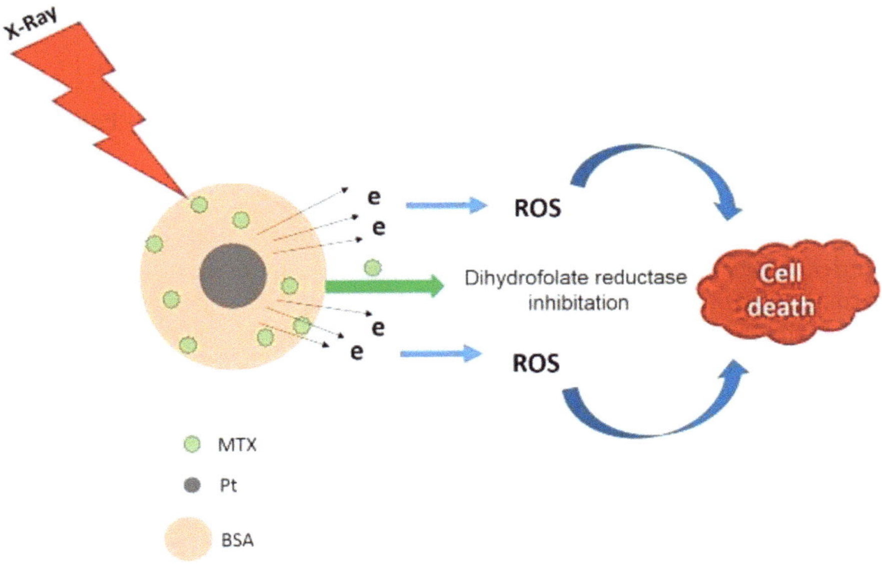

Fig. 9.9 Schematic representation of Pt@BSA-MTX [35]

when used in conjunction with MTX and X-ray irradiation, confirming the benefits of combining chemotherapy and radiotherapy for a synergistic effect [35]. This study achieved a synergistic killing effect on 4 T1 breast cancer cells through the combined application of radiotherapy and chemotherapy (MTX), utilizing the radiosensitization and targeted drug delivery functions of platinum nanoparticles to reduce the side effects associated with monotherapy.

In summary, X-ray-excited nanoscintillators can achieve maximum tumor destruction through the integration of multimodal treatments, such as inductive radiotherapy, photodynamic therapy, photothermal therapy, immunotherapy, gene therapy, and chemotherapy. This multi-modality, synergistic approach is highly precise, targeted radiosensitization, and safe, offering new directions and ideas for tumor treatment. With ongoing advancements in nanotechnology and biotechnology, the application of X-ray-excited nanoscintillators in tumor therapy is expected to expand further in the future.

5 Discussion

Advantages, disadvantages, and outlook of nanoscintillators-mediated X-PDT are as follows:

1. *Advantages:*
   (a) High Penetration: X-rays are highly penetrating and can penetrate deep into the tumor tissue [36, 37]. Therefore, X-ray-excited nanoscintillators offer significant advantages in treating deep-seated tumors.
   (b) High Energy Conversion Efficiency: Certain new nanoscintillators, such as organic phosphorescent scintillators, can directly absorb X-rays and emit phosphorescence [38], greatly improving exciton utilization rates and energy conversion efficiency. This effectively enhances the efficacy of X-ray photodynamic therapy.
   (c) Multifunctionality: By designing and synthesizing novel multifunctional nanoscintillators, such as rare-earth-doped tungstate nanomaterials, dual imaging modalities (e.g., CT and MRI) and synergistic treatments (e.g., X-PDT and RT) of tumor tissues can be achieved [39], providing more options for tumor diagnosis and treatment.
   (d) Low Toxicity and Biocompatibility: Some nanoscintillators, such as terbium-doped gadolinium tungstate nanomaterials, exhibit low biotoxicity and can be gradually metabolized in vivo, reducing the side effects in patients [40].
2. *Disadvantages:*
   (a) Performance Limitation: The current X-ray and ultraviolet (UV) luminescence performance of nanoscintillators is still weak, and achieving efficient X-ray-activated tumor therapy remains a challenge [41]. This limits the effectiveness of their application in certain tumor therapies.
   (b) Preparation Complexity: The preparation process of multifunctional nanoscintillators is relatively complicated and requires precise control of the mate-

rial's composition and structure to ensure good performance and biocompatibility [42].

(c) Clinical Application Limitation: Although X-ray-excited nanoscintillators show great potential in tumor therapy, they are still in the research stage and have not yet been applied clinically. Their safety and effectiveness need to be further verified.

3. *Future Prospects*

   (a) Performance Enhancement: By optimizing the matrix material, dopant, and energy transfer design, the X-ray ultraviolet luminescence performance of nanoscintillators can be further improved, thereby enhancing therapeutic efficacy.
   (b) Development of Multifunctional Nanoscintillators: Continued exploration and development of new nanoscintillators with additional functionalities—such as nanomaterials that combine imaging and therapeutic capabilities—are essential to meet the diverse needs of tumor treatments.
   (c) Promotion of Clinical Application: As research advances and technology matures, X-ray-excited nanoscintillators are expected to be widely used in clinical tumor treatments in the future, providing patients with more effective therapeutic options.
   (d) Safety assessment: To ensure the safety and effectiveness of nanoscintillators, their safety must be thoroughly evaluated before they are approved and promoted for clinical use. This includes long-term toxicity, biodistribution, and metabolism.

References

1. Liu J, et al. Environment factors, DNA methylation, and cancer. Environ Geochem Health. 2023;45(11):7543–68.
2. Wang X, et al. Tracking knowledge evolution, hotspots and future directions of emerging technologies in cancers research: a bibliometrics review. J Cancer. 2019;10(12):2643–53.
3. Gladović A, et al. Carcinogenic and human health risk assessment of children's and adults' exposure to toxic metal(oid)s from air PM10 in critical sites of the Republic of Serbia. Environ Sci Pollut Res. 2023;30(22):61753–65.
4. Eaves LA, et al. Identifying the transcriptional response of cancer and inflammation-related genes in lung cells in relation to ambient air chemical mixtures in Houston, Texas. Environ Sci Technol. 2020;54(21):13807–16.
5. Hendryx M, et al. Self-reported cancer rates in two rural areas of West Virginia with and without mountaintop coal mining. J Community Health. 2011;37(2):320–7.
6. Stefler D, et al. Smoking, alcohol and cancer mortality in Eastern European men: findings from the PrivMort retrospective cohort study. Int J Cancer. 2018;143(5):1128–33.
7. Nizamis K, et al. The inter-familiar issues of Greek parents facing childhood cancer. Eur J Pediatr. 2023;183(1):229–34.
8. Dabo-Trubelja A, Gottumukkala V. Review of cancer therapies for the perioperative physician. Perioper Med. 2023;12(1):25.
9. Van der Mude A. A proposed information–based modality for the treatment of cancer. Biosystems. 2022;211:104587.

10. Lu L, et al. All-inorganic perovskite nanocrystals: next-generation scintillation materials for high-resolution X-ray imaging. Nanoscale Adv. 2022;4(3):680–96.
11. Yang T, et al. Enhanced luminescence of Tb doping nanoscintillators based on NaLuF4 host matrix for X-ray imaging. J Lumin. 2024;269:120498.
12. Ahmad F, et al. Codoping enhanced radioluminescence of nanoscintillators for X-ray-activated synergistic cancer therapy and prognosis using metabolomics. ACS Nano. 2019;13(9):10419–33.
13. Dong X, et al. A heterojunction structured WO2.9-WSe2 nanoradiosensitizer increases local tumor ablation and checkpoint blockade immunotherapy upon low radiation dose. ACS Nano. 2020;14(5):5400–16.
14. Lu J, et al. X-ray-controllable release of carbon monoxide potentiates radiotherapy by ultrastable hybrid nanoreservoirs. Biomaterials. 2023;302:122313.
15. Carr Delgado H, et al. Toward "super-scintillation" with nanomaterials and nanophotonics. Nanophotonics. 2024;13(11):1953–62.
16. Crapanzano R, et al. Photo- and radio-luminescence of porphyrin functionalized ZnO/SiO2nanoparticles. Phys Chem Chem Phys. 2022;24(35):21198–209.
17. Dujardin C, et al. Inorganic Nanoscintillators: current trends and future perspectives. Adv Opt Mater. 2025;13(12):2402739.
18. Zou H, et al. A hollow NaBiF4:Tb nanoscintillator with ultra-weak afterglow for high-resolution X-ray imaging. Chem Commun. 2023;59(78):11732–5.
19. Zou H, et al. Sub-10 nm lanthanide-doped Lu6O5F8 nanoscintillators for real-time high-resolution dynamic 3D X-ray imaging. Adv Funct Mater. 2024;34(49):2409156.
20. Wang Z, et al. Mn2+-activated Cs3Cu2I5 nano-scintillators for ultra-high resolution flexible X-ray imaging. Laser Photon Rev. 2023;17(6):2200851.
21. He C, et al. Room temperature circularly polarized emission in perovskite nanocrystals through bichiral-molecule-induced lattice reconstruction. Matter. 2024;7(2):475–84.
22. Isabella Antony KJ, Jana D. Stable Mn-doped CsPbCl3 nanocrystals inside mesoporous alumina films for display and catalytic applications. ACS Appl Nano Mater. 2020;3(3):2941–51.
23. Sun S-R, et al. To increase more than 10% in efficiency of perovskite solar cells by using nanoholes SnO2. Rare Metals. 2021;40(10):2778–84.
24. Li K, et al. ROS-neutralizing surface engineering protects immunotoxicity of organic nanoscintillator-directed radiodynamic therapy. Chin Chem Lett. 2025;36:111261.
25. Huang W, et al. Photoluminescence and X-ray excited scintillating properties of Tb3+-doped borosilicate aluminate glass scintillators. Ceram Int. 2022;48(12):17178–84.
26. Zhou J, et al. The luminescence mechanism of ligand-induced interface states in silicon quantum dots. Nanoscale Adv. 2023;5(15):3896–904.
27. Wang X, et al. Organic phosphorescent nanoscintillator for low-dose X-ray-induced photodynamic therapy. Nat Commun. 2022;13(1):5091.
28. Yu Y, et al. Pure organic AIE nanoscintillator for X-ray mediated type I and type II photodynamic therapy. Adv Sci. 2023;10(26):2302395.
29. Tang X, et al. Low-dose X-ray stimulated NO-releasing nanocomposites for closed-loop dual-mode cancer therapy. Biomater Sci. 2024;12(16):4211–25.
30. <Activity of Psoralen-Functionalized Nanoscintillators against Cancer Cells upon X-ray Excitation.pdf>.
31. Yang K, et al. Designing highly UV-emitting lanthanide nanoscintillators for in vivo X-ray-activated tumor therapy. Sci China Mater. 2023;66(10):4090–9.
32. Zhang C, et al. Nanoscintillator-mediated X-ray-triggered boosting transformation of Fe3+ into Fe2+ for enhancing tumor ferroptosis/immunotherapy. Adv Funct Mater. 2023;33(45):2301462.
33. Cetin Ersen B, et al. GLUT-targeting phototherapeutic nanoparticles for synergistic triple combination cancer therapy. ACS Appl Mater Interfaces. 2023;15(7):9080–98.
34. Shao D, et al. Biomimetic diselenide-bridged mesoporous organosilica nanoparticles as an X-ray-responsive biodegradable carrier for chemo-immunotherapy. Adv Mater. 2020;32(50):e2004385.

35. Yaray K, et al. Chemoradiation therapy of 4T1 cancer cells with methotrexate conjugated platinum nanoparticles under X-ray irradiation. Inorg Chem Commun. 2023;150:110457.
36. Kamkaew A, et al. Scintillating nanoparticles as energy mediators for enhanced photodynamic therapy. ACS Nano. 2016;10(4):3918–35.
37. Lan G, et al. Nanoscale metal–organic layers for deeply penetrating X-ray-induced photodynamic therapy. Angew Chem Int Ed. 2017;56(40):12102–6.
38. Li H, et al. Constructing organic phosphorescent scintillators with enhanced triplet exciton utilization through multi-mode radioluminescence for efficient X-ray imaging. Adv Mater. 2024;36(46):e2409338.
39. Zhou R, et al. Fractionated regimen-suitable immunoradiotherapy sensitizer based on ultrasmall Fe4Se2W18 nanoclusters enable tumor-specific radiosensitization augment and antitumor immunity boost. Nano Today. 2021;36:101003.
40. Yu X, et al. CT/MRI-guided synergistic radiotherapy and X-ray inducible photodynamic therapy using Tb-doped Gd-W-Nanoscintillators. Angew Chem Int Ed. 2019;58(7):2017–22.
41. Yao B, et al. X-ray excited luminescent nanoparticles for deep photodynamic therapy. RSC Adv. 2023;13(43):30133–50.
42. Lei L, et al. Next generation lanthanide doped nanoscintillators and photon converters. eLight. 2022;2(1):17.

Magnetic Nanomaterials for Cancer Hyperthermia Therapy

10

Jun Zhang and Xiaoyu Gao

Abstract

Magnetic hyperthermia therapy (MHT) is a noninvasive treatment strategy that uses alternating magnetic fields (AMFs) to excite magnetic nanomaterials for localized heat generation and selective tumor ablation. Compared to traditional hyperthermia techniques (i.e., photothermal or radiofrequency heating), AMFs exhibit unique advantages, including deep-tissue penetration and negligible energy attenuation, demonstrating significant potential for treating visceral tumors such as hepatocellular carcinoma and glioblastoma. As representative magnetothermal materials, iron oxide nanoparticles (IONPs) can overcome the limitations of stand-alone hyperthermia through functionalization strategies such as heteroelement doping (i.e., Mn^{2+}, Co^{2+}), nucleic acid/drug conjugation, polymer hybridization, or other types of magnetic hyperthermia materials. These modifications can enable synergistic magneto-chemotherapy/PTT/PDT, gene therapy, and multimodal imaging-guided applications. This review systematically summarizes MHT nanomaterial classification, constituents, and applications in disease treatment. Additionally, it analyzes the challenges and prospects of MHT nanomaterials in tumor targeting, biosafety, and clinical translation, offering theoretical insights for developing next-generation intelligent MHT agents.

Jun Zhang and Xiaoyu Gao have equally contributed to this work.

J. Zhang (✉)
State Key Laboratory of Eye Health; Center for Basic Medical Research and Innovation in Visual System Diseases of Ministry of Education, Shanghai Key Laboratory of Orbital Diseases and Ocular Oncology, Department of Ophthalmology, Shanghai Ninth People's Hospital, Shanghai Jiao Tong University School of Medicine, Shanghai, China
e-mail: junzhang007@sjtu.edu.cn

X. Gao
School of Pharmacy, Mudanjiang Medical University, Mudanjiang, China

M. Chu (ed.), *Modern Nanomaterials and Nanotechnology for Cancer Therapy*,
https://doi.org/10.1007/978-981-95-6849-9_10

Keywords

Magnetic Hyperthermia Therapy (MHT) · Magnetic nanomaterials · Alternating magnetic field · Cancer treatment · Heat-induced cell death · Targeted therapy

1 Introduction

Despite significant advances in oncology, the treatment of deep-seated and metastatic tumors remains a formidable significant clinical challenge due to the limitations of conventional therapies, including off-target effects and restricted tissue penetrability. Within this context, magnetic hyperthermia therapy (MHT) has recently attracted attention as a spatially controlled treatment modality that uniquely combines deep-tissue accessibility with localized energy delivery. Unlike optical or ultrasound-based thermal approaches that have shortcomings, such as rapid attenuation in biological tissues, alternating magnetic fields (AMFs) can enable nearly lossless energy transmission through anatomical barriers, permitting the precise heating of tumors regardless of their depth. This serves as a distinct advantage that has strengthened research focusing on magnetic nanomaterials.

Among various magnetic nanomaterials, iron oxide nanoparticles (IONPs) have been extensively investigated due to their biocompatibility, tunable magnetic properties, and clinical translation potential. However, the therapeutic efficacy of conventional IONPs remains constrained by suboptimal heating efficiency, limited tumor targeting, and a lack of multifunctionality. Recent advances in nanomaterial engineering have addressed these challenges by using innovative strategies such as heteroelement doping (i.e., Mn^{2+}, Co^{2+}, Zn^{2+}), magnetic iron oxide/drug or nucleic acid complexes, surface modification with polymers or targeting ligands, and other types of MHT nanomaterials (Fig. 10.1). These modifications can enhance magnetothermal conversion efficiency and enable synergistic combination therapies (i.e., chemo-/PDT or PTT-therapy) with multimodal imaging capabilities, paving the way for theranostic applications.

Despite these advancements, critical challenges remain in optimizing the Curie temperature for self-regulating hyperthermia, improving tumor-specific accumulation, and ensuring long-term biosafety for clinical adoption. This mini-review comprehensively outlines the latest developments of MHT-focused nanomaterials, with an emphasis on structure-property relationships, material design strategies, and their implications for translational medicine. We further discuss emerging trends and future perspectives in the rational design of next-generation smart magnetic nanoparticles (MNPs) for precision cancer therapy.

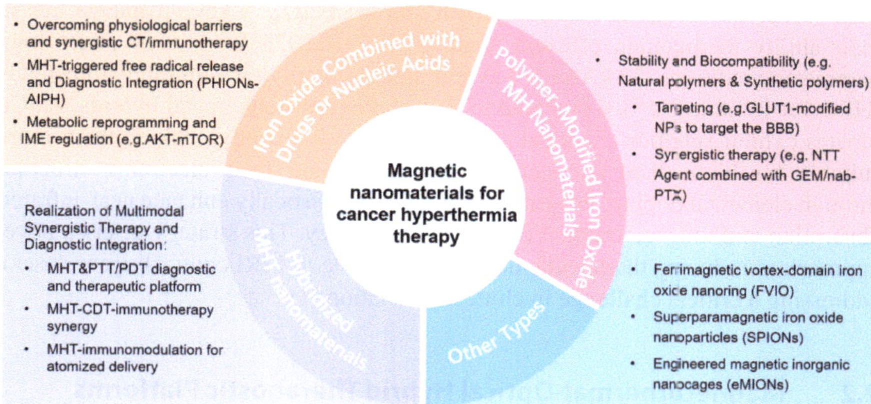

Fig. 10.1 Overview of magnetic nanomaterials for cancer hyperthermia therapy

2 Hybrid Magnetic Hyperthermia Nanomaterials: Multimodal Synergistic Therapy and Theranostic Integration

In recent years, hybrid magnetic hyperthermia nanomaterials have demonstrated significant potential for precision tumor therapy due to their unique magnetothermal effect and integratable functionalities, including photothermal conversion, chemodynamic therapy, and immunomodulation. Through rational material design and structural engineering, these nanomaterials can achieve efficient energy conversion and enable theranostic capabilities, advancing the development of multimodal synergistic therapy.

2.1 Magnetothermal-Photothermal Hybrid Nanomaterials

Conventional magneto-optical hybrid structures (i.e., Fe_3O_4/Au or magnetic nanoparticle/quantum dot composites) rely on distinct components to independently respond to magnetic and optical fields. These hybrid systems typically consist of separate magnetic and optical units, each solely contributing to either magnetothermal or photothermal heating. Because different components absorb energy from different sources, the overall heating power consists of the average of the two modes, limiting their specific heating efficiency. Therefore, minimizing the dosage of nanomaterials is crucial for avoiding potential long-term health risks. However, dual-mode hyperthermia alone often fails to significantly reduce the required dose, resulting in constrained heating efficacy.

By contrast, Zn/Co co-doped ferrite nanoparticles ($Zn_{0.3}Co_xFe_{2.7-x}O_4$, ZCFO NPs) simultaneously optimize both magnetothermal and photothermal performance. Under dual-mode excitation (27 kA/m, 430 kHz magnetic field and 2.5 W/cm^2, 808 nm laser), these materials can achieve ultrahigh specific loss power

(SLP = 14,724 W/g), far exceeding conventional Fe_3O_4. A key advantage lies in their ability to maintain efficient heating (SLP = 2023 W/g) even under low-parameter conditions (23 kA/m, 0.1 W/cm^2 laser), complying with safety thresholds ($Hf < 5 \times 10^9$ $A{\cdot}m^{-1}{\cdot}s^{-1}$). Compared to Fe/Co alloys or noble metal hybrids, Zn/Co ferrites exhibit superior chemical stability and lower toxicity, making them more suitable for biomedical applications. By modulating their electronic structure through elemental doping, these materials can synergistically enhance near-infrared absorption and magnetothermal conversion efficiency. This strategy offers a novel approach to reducing therapeutic doses (comparable to MRI contrast agent doses), addressing a critical challenge in clinical translation [1].

2.2 Magnetothermal-Optical Hybrid Theranostic Platforms

Integrating MHT with optical thermometry is essential for achieving precise thermal treatment. Opto-magnetic nanoclusters (OMNCs), designed with a core-shell architecture ($NaYF_4$@$NaYF_4$:60%Nd/20%Yb@$NaYF_4$ + Fe_3O_4), can enable multi-functional synergy. In this scenario, Fe_3O_4 nanoparticles provide magnetic targeting and MHT functionality, with the lanthanide-doped shell facilitating real-time temperature monitoring via NIR-II lifetime-based thermometry (1000 nm) and Yb^{3+} emission intensity ratio analysis, enabling depth-resolved temperature mapping in tumors. During AMF-triggered heating, these optical signals offer real-time thermal feedback, preventing overheating-induced tissue damage. This hybrid design can address the critical limitations of conventional magnetothermal materials, i.e., the lack of in situ temperature monitoring, providing a reliable tool for clinical translation [2].

In addition to the real-time temperature monitoring of tumor regions, the combination of magnetothermal and optical modalities can be used for magnetically triggered luminescence imaging, enabling simultaneous imaging during magnetic hyperthermia therapy. Composite materials consisting of magnetic ferrites and rare-earth luminophores (i.e., $Zn_{0.3}Fe_{2.7}O_4$@SiO_2@Gd_2O_3:Eu^{3+}, Eu-ZFO NCs) integrate both therapeutic and imaging functions. These composites exhibit long-lifetime luminescence (0.181 ms), which, when combined with time-gated detection technology, can effectively eliminate biological autofluorescence interference. Under a 12 kA/m magnetic field, these materials demonstrate an intrinsic loss power of 11 nHm^2/kg, achieving efficient magnetothermal conversion while maintaining excellent biocompatibility ($IC_{50} \approx 800$ μg/mL), thereby realizing theranostic synergy. At a dosage of 200 μg/mL, these materials can also induce >75% tumor cell apoptosis under AMF exposure while simultaneously enabling intracellular tracking through Eu^{3+} red emission (quantum yield = 9.3%). The SiO_2 buffer layer has been shown to isolate the magnetic core from the luminescent layer, a design that both protects the luminescent centers and maintains high magnetic responsiveness, serving as an exemplary model for the visualized therapy of tumors [3].

2.3 Magnetothermal-Chemodynamic-Immunological Hybrid Systems

Thermoelectric therapy, based on thermoelectric effects (i.e., Seebeck effect), converts thermal energy into electrical energy, generating excessive reactive oxygen species (ROS) locally in tumors for selective cancer cell killing. Conventional photothermal therapy faces limitations in treating deep-seated tumors due to insufficient laser penetration depth. However, hybrid materials can achieve multimodal synergy through energy level design, triggering cascade reactions under mild AMF conditions. Yuan et al. developed an efficient magnetothermal nanoagent ($CuFe_2O_4$, CFO), a magnetic switch, via hydrothermal synthesis. Subsequently, thermoelectric nanomaterials ($SrTiO_3$, STO) were grown in situ on the CFO surface to form magnetically triggered thermoelectric heterojunctions ($CuFe_2O_4/SrTiO_3$, CFO/STO). To enhance oral administration efficiency, the heterojunctions were subsequently coated with a fusion membrane (M) combining *Staphylococcus aureus* and macrophage cell membranes (Fig. 10.2). This design leveraged the membrane ability of *S. aureus* to target intestinal inflammatory environments and the macrophage membrane's affinity for colorectal cancer tissues, thus enabling the effective targeting of orthotopic colorectal tumors. Under magnetothermal-thermoelectric conversion mode, mild heating was found to activate electron-hole separation in STO, catalyzing the generation of O_2^-/·OH radicals. Simultaneously, the Fe^{3+}/Cu^{2+} redox cycle in CFO depletes glutathione (GSH) and promotes ·OH burst, enhancing Fenton reactions. Synergy between hyperthermia and chemodynamic therapy (CDT) was shown to induce immunogenic cell death (ICD) in tumors, characterized by calreticulin (CRT) exposure, promoting dendritic cell (DC) maturation and repolarizing tumor-associated macrophages (TAMs) toward the M1 phenotype while inhibiting the VEGF pathway. Through stepwise energy conversion, the CFO/STO system achieved a three-in-one combination of MHT, CDT, and immunotherapy, demonstrating a novel approach for comprehensive cancer treatment [4].

2.4 Aerosol-Delivered Magnetothermal-Immunomodulatory Nanomaterials

In recent years, clinical immunotherapy for lung cancer has mainly focused on adaptive immunotherapy. However, cancer cells can evade T cell immune surveillance and immune responses by releasing anti-inflammatory cytokines, reducing major histocompatibility complex expression, and overexpressing immune checkpoint molecules, leading to poor immunotherapy outcomes. The innate immune system, especially macrophages, has attracted significant attention for tumor therapy due to its diverse cell types, broad action range, and rapid response speed. Unlike other organs, the lungs possess a unique immune environment rich in alveolar macrophages (AMs), with AMs located on alveolar surfaces and accounting for 55% of all immune cells in the lungs. Activating these AMs to harness the natural power of the pulmonary immune environment offers significant potential for lung

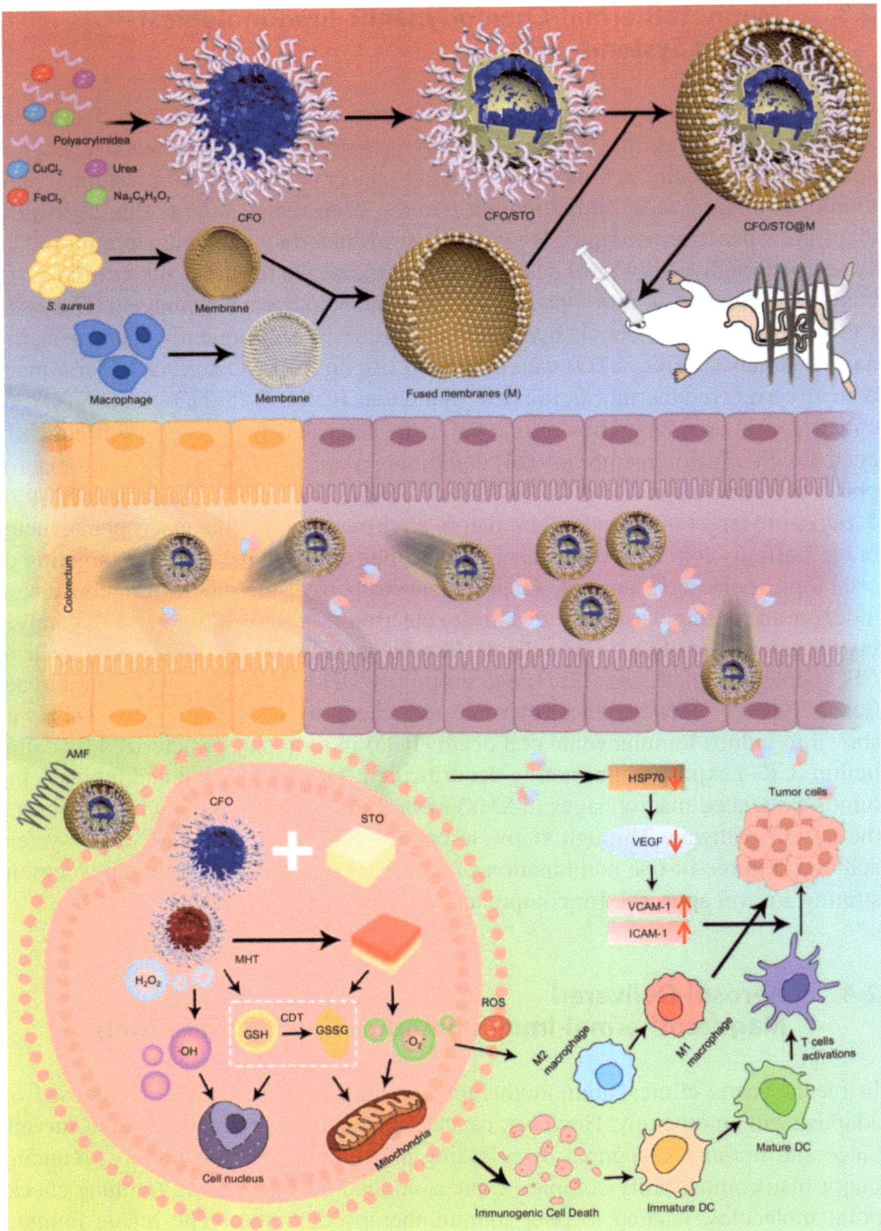

Fig. 10.2 Demonstration of targeted colorectal cancer therapy using a fusion membrane coated magnetically triggered thermoelectric heterojunction (CFO/STO@M) magnetic-thermal-electro-catalytic approach (Copyright © 2025, Xue Yuan et al. Reprinted with permission from Ref. [4])

cancer treatment. Therefore, targeting the unique immune environment of lung cancer to activate the cascade responses of macrophages and NK cells for anticancer effects holds considerable promise.

Chen et al. designed and synthesized Zn-$CoFe_2O_4$@Zn-$MnFe_2O_4$-PEG (ZCMP) nanoparticles, which could be taken up by AMs through nebulized inhalation (Fig. 10.3). Under lysosomal acidity and AMF stimulation, iron ions were released to catalyze H_2O_2 generation into ROS, activating the p-NF-κB pathway and driving AM polarization toward the M1 phenotype. M1-type AMs secreted cytokines such as IFN-β/IL-15, which upregulated STAT5/GZMB expression in the NK cells, enhancing tumor-killing capacity. This strategy leveraged the inherent immune advantages of the lungs and transformed the immunosuppressive microenvironment into an antitumor state through magnetothermal-acid-responsive hybrid materials, offering a new approach for orthotopic lung cancer treatment [5].

Hybrid magnetothermal nanomaterials have achieved multimodal synergistic therapy through component regulation (i.e., Zn/Co doping), structural design (core-shell structures, heterojunctions), and functional integration (theranostics). Combining magnetothermal therapy with photothermal therapy/chemodynamic therapy/immunotherapy has been shown to significantly improve therapeutic efficacy while reducing dosage requirements. Precise regulation through optical thermometry, luminescence imaging combined with magnetic targeting, can enhance spatiotemporal control capabilities. These materials were found to demonstrate high

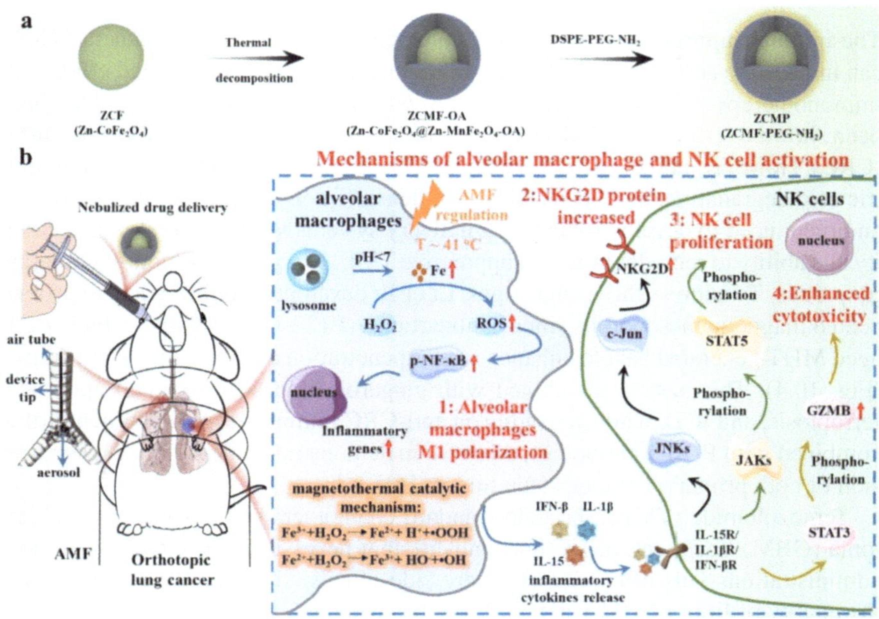

Fig. 10.3 Zn-$CoFe_2O_4$@Zn-$MnFe_2O_4$-PEG nanoparticles (ZCMP) synthesis and nebulized magnetothermal immunotherapy strategy for lung cancer, targeting the innate immune pathways (Copyright © 2024, Lizhu Chen et al. Reprinted with permission from Ref [5])

clinical translation potential due to their low toxicity, high stability, and biocompatible design that can promote practical applications. Therefore, future research should focus on the further optimization of energy conversion efficiency (i.e., through multielement co-doping or defect engineering), the investigation of cross-scale mechanisms (i.e., molecular-level analysis of magnetothermal-immune signaling pathways), and scale-up production and standardized evaluation to accelerate clinical translation.

3 Magnetic Iron Oxide/Drug or Nucleic Acid Complexes: Multifunctional Nanomaterials for Magnetothermal-assisted Tumor Therapy

Magnetic IONPs have been widely explored for tumor therapy due to their unique magnetothermal effects, biocompatibility, and multifunctionality. Through conjugation with drugs or nucleic acids, IONPs can form multifunctional composites capable of MHT, drug delivery, gene regulation, and immune activation. Below, we summarize recent advances in this field from three application perspectives.

3.1 Overcoming Physiological Barriers and Synergistic Chemo/Immunotherapy

The immunosuppressive nature of the colorectal tumor microenvironment (TME) can impede the activation of antitumor immunity, resulting in the limited efficacy of immunotherapy. In addition, gut dysbiosis and aberrant microbial metabolism have been shown to critically influence the initiation and progression of colorectal cancer (CRC). However, oral drug delivery for CRC faces multiple barriers, including gastric acid degradation, the mucus layer, and dense tumor stroma. As a result, there is an urgent need to develop robust drug delivery systems with high oral bioavailability, the ability to remodel immunosuppressive TME, and gut microbiota-modulating properties. To address these challenges, Li et al. developed mulberry leaf exosome-coated magnetic mesoporous silica nanoparticles (P127-MLL@Gins), which leveraged MHT-generated heat to enhance mucus penetration and deep tumor infiltration (Fig. 10.4). This system, combined with gingerol, induced tumor cell apoptosis, ferroptosis, and ICD, while activating an anti-CRC immune response. When further combined with PD-L1 antibodies, this system demonstrated the significant suppression of both primary and metastatic tumors [6].

Temozolomide (TMZ), the gold-standard chemotherapeutic agent for glioblastoma (GBM) treatment, has a short half-life that necessitates prolonged high-dose administration. This often causes severe side effects, significantly compromising the quality of life in GBM patients. MHT has emerged as a promising strategy for developing minimally invasive, localized, and remotely controlled antitumor therapies. Unlike conventional cancer treatments or other systemic/local hyperthermia approaches, MHT enables highly localized heating. When stimulated by an AMF, magnetic nanoparticles were found to enhance local accumulation, reduce systemic toxicity, and improve the recovery of adjacent healthy tissues.

To address these challenges, Beola et al. developed angiopep-2-modified lipid-based magnetic nanovehicles loaded with TMZ (Ang-TMZ-LMNVs) for targeted GBM therapy. The Ang-TMZ-LMNV system specifically targeted and accumulated in GBM cells, exerting combined magnetothermal and chemotherapeutic effects that effectively induced apoptosis. The results demonstrated that the Ang-TMZ-LMNVs synergistically combined MHT with TMZ chemotherapy, significantly suppressing tumor growth and prolonging survival in an orthotopic glioblastoma model [7].

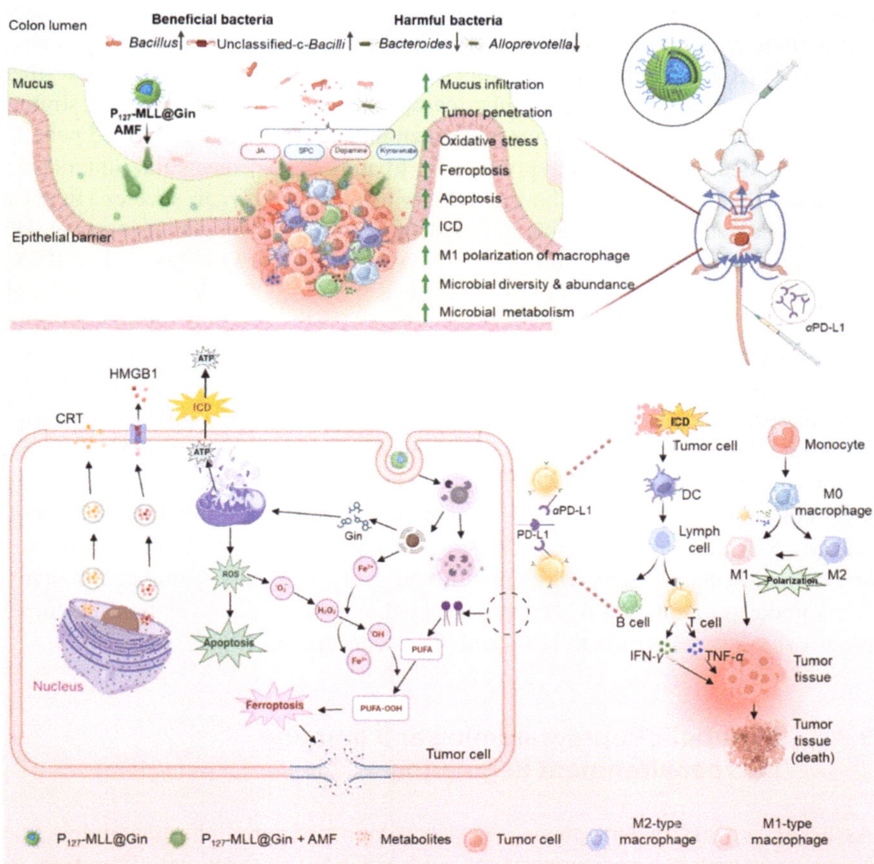

Fig. 10.4 Schematic illustration showing how oral mulberry leaf exosome-coated magnetic mesoporous silica nanoparticles (P127-MLL@Gins) promote colorectal mucus penetration, tumor targeting, and systemic antitumor immunity activation for CRC treatment (Copyright © 2024, Baoyi Li et al. Reprinted with permission from Ref. [6])

3.2 Magnetothermal-Triggered Radical Release for Theranostic Applications

The widespread clinical adoption of photodynamic therapy (PDT) remains hindered by tumor hypoxia and the limited tissue penetration depth of excitation light. As a result, there is an urgent need to develop novel therapeutic strategies capable of generating oxygen-independent free radicals without penetration depth constraints. To address this, Huang et al. developed porous hollow iron oxide nanoparticles (PHIONs) loaded with a radical initiator (AIPH) to form PHION-AIPH nanoplatforms. Under AMF stimulation, the PHIONs generated localized heat that triggered both the release and decomposition of AIPH, yielding oxygen-independent alkyl radicals. These alkyl radicals effectively killed cancer cells even under hypoxic conditions. Notably, this magnetothermally activated radical generator demonstrated remarkable therapeutic efficacy against orthotopic liver tumors in rat models. Furthermore, the PHION-AIPH system exhibited T2-weighted magnetic resonance imaging (MRI) capability, enabling a combined diagnostic and therapeutic (theranostic) approach for hepatocellular carcinoma. This magnetothermal strategy offers distinct advantages in terms of oxygen independence and unlimited penetration depth, demonstrating great promise for treating deep-seated solid tumors [8].

Another study leveraged the high levels of reactive oxygen species (ROS) in cancer cells and their vulnerable antioxidant defense systems. By using magnetothermia to elevate intratumoral H_2O_2 levels combined with chloroperoxidase (CPO) catalysis to produce O^{2-}, this approach induced apoptosis. The design formed stable supramolecular gels through enzyme-nanoparticle self-assembly, mimicking artificial neutrophil functions. Through the enzymatic interfacial self-assembly of MNPs with the peptide FmocTyr(H_2PO_3)$^-$OH and the electrostatic loading of CPO, the supramolecular nanogel served as an artificial neutrophil. The MNPs in the nanogel were shown to increase H_2O_2 levels in cancer cells under programmed AMF, similar to neutrophil activators. Meanwhile, the CPO loaded in the protective peptide nanolayer continuously converted H_2O_2 to singlet oxygen ($^1O_2$), mimicking neutrophil-released ROS for tumor therapy. The study also proposed a new concept consisting of active enzyme kinetics, which manifested as magnetically stimulated oxidative stress under programmed AMF and sustained enzymatic bio-oxidation, leading to progressive increases of both H_2O_2 and $^1O_2$ in cancer cells [9].

3.3 Metabolic Reprogramming and Immune Microenvironment Regulation

As a key process in ferroptosis, PC peroxidation requires phosphatidylcholines (PCs) containing polyunsaturated fatty acids (PUFAs) as substrates. Therefore, the regulation of fatty acid metabolism serves as a critical determinant of ferroptosis sensitivity. Fatty acid synthase (FASN), the rate-limiting enzyme in de novo lipogenesis, is frequently overexpressed in invasive tumors, including advanced prostate cancer, breast cancer, and non-small cell lung cancer, and it has been associated with poor patient prognosis and resistance to various cancer therapies. In addition, activating antitumor immune responses to suppress tumors and metastases represents a practical strategy for precision therapy. As potent effectors of innate immunity, NK cells could serve as the first line of defense against cancer, including metastatic and circulating tumor cells. Several studies have focused on activating immune responses through MHT and ferroptosis, with magnetic nanomaterials enhancing ferroptosis by modulating lipid metabolism and subsequently triggering immune responses for tumor treatment.

Building upon this mechanistic foundation, Cheng et al. designed and fabricated FiFe@RBM nanovesicles encapsulating IONPs and a FASN inhibitor (IPI-9119). These nanovesicles synergistically promoted ferroptosis through MHT by releasing Fe^{2+} and inhibiting the AKT-mTOR pathway while activating NK cells to suppress metastasis. The biomimetic nanovesicles (FiFe@RBM) consisted of red blood cell membrane (RBM)-coated IONPs and IPI-9119, which enhanced ferroptosis through lipid metabolic regulation and synergized with MHT to inhibit castration-resistant prostate cancer (CRPC) while also suppressing metastasis through activated innate immunity (Fig. 10.5). The RBM coating endowed FiFe@RBM nanovesicles with a stealth effect, prolonging blood circulation time, enabling immune evasion, and preventing macrophage phagocytosis. FiFe@RBM was shown to efficiently deliver bioactive compounds to human PC-3 CRPC cells, where it released Fe^{2+} to effectively induce ROS generation and mitochondrial dysfunction while inhibiting the AKT-mTOR pathway. By increasing PUFAs and reprogramming lipid metabolism, the system synergistically induced apoptosis and enhanced ferroptosis. Furthermore, the IONPs enabled T1/T2-weighted MRI, establishing a novel theranostic platform for combined diagnosis and treatment [10].

MHT-mediated thermal ablation treatment has shown promising clinical potential for destroying primary tumors. However, conventional MHT still faces challenges, including damage to normal tissues adjacent to the treatment site and the destruction of tumor-associated antigens due to its high initiation temperature (>50 °C). In addition, the local thermal ablation of tumors has typically demonstrated limited therapeutic inhibition of tumor metastasis. Qin et al. prepared magnetothermal phase-transition nanodroplets (RPPs) by encapsulating the immune adjuvant resiquimod (R848) and the phase-change agent perfluoropentane (PFP) within a PLGA shell; then, supermagnetic iron oxide nanoparticles (SPIOs) absorbed on the core-shell structure to form SPIO+RPPs (Fig. 10.6). Due to the cavitation effect of microbubbles generated by RPPs, the temperature threshold of MHT was reduced from 50 °C to approximately 44 °C. This cavitation effect enhanced ICD, releasing damage-associated molecular patterns such as high mobility group box 1 and ATP while exposing CRT. When combined with the immune adjuvant R848, it could reverse the immunosuppressive microenvironment and promote dendritic cell maturation and cytotoxic T lymphocyte activation, thereby simultaneously achieving tumor cell killing and metastasis inhibition [11].

Iron oxide-based magnetic materials can form composite nanomaterials by combining with apoptosis/ferroptosis-inducing drugs or nucleic acids. These composites can enhance therapeutic efficacy through multiple physical effects, such as (1)

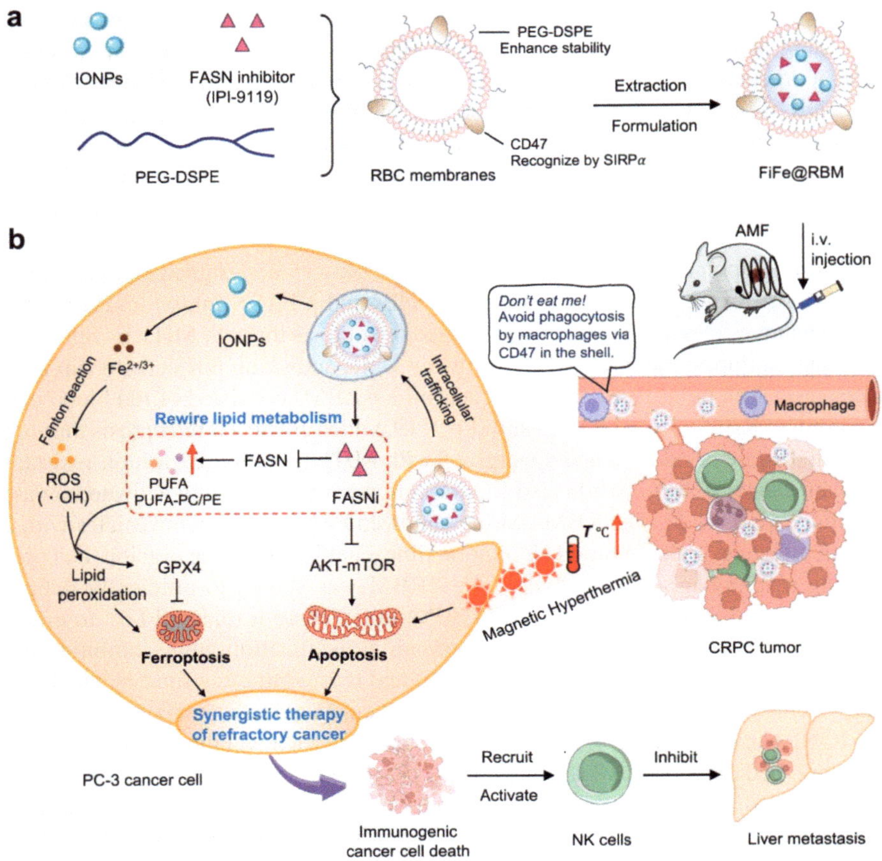

Fig. 10.5 Illustration of (**a**) red blood cell membrane (RBM)-coated IONPs and IPI-9119 (FiFe@RBM) construction and (**b**) its therapeutic action in CRPC via lipid metabolic rewiring-enhanced ferroptosis, magnetic hyperthermia synergy, and the NK cell-activated prevention of hepatic metastasis (Copyright © 2025, American Chemical Society. Reprinted with permission from Ref. [10])

improving drug penetration, (2) triggering phase transitions or radical release, (3) inducing apoptosis/ferroptosis, and (4) reprogramming metabolism (including lipid metabolism and hypoxia response). This system can promote ICD, and when combined with immune checkpoint inhibitors or adjuvants (i.e., R848), it can significantly enhance antitumor therapeutic outcomes.

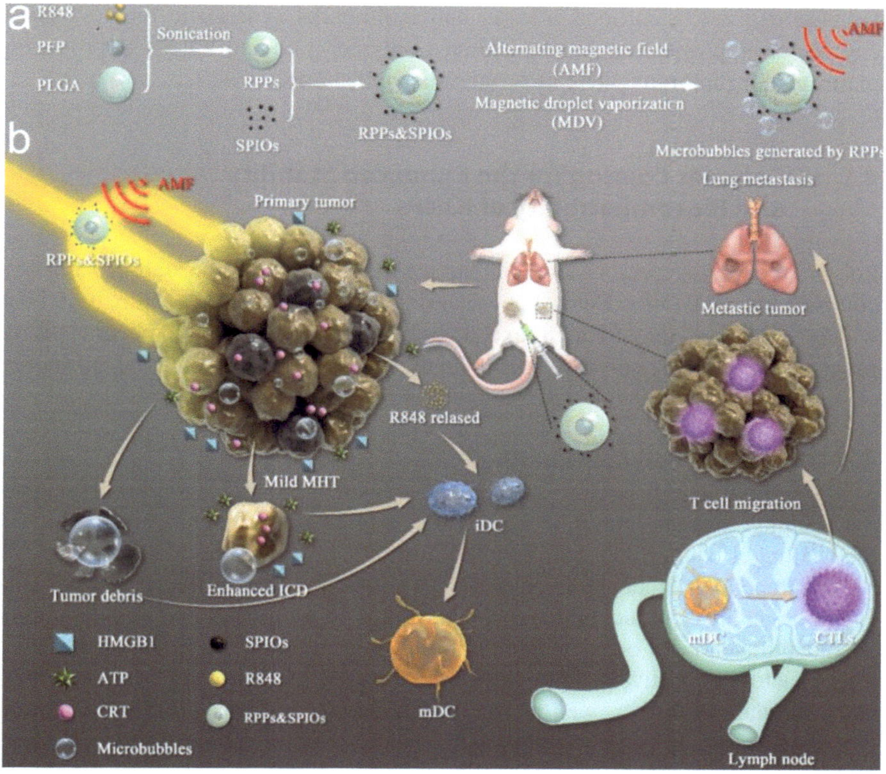

Fig. 10.6 Schematic representation of supermagnetic iron oxide nanoparticles (SPIOs) absorbed on the core-shell structure (SPIO+RPP) fabrication and their synergistic mechanism in enhancing mild MHT for tumor suppression. (**a**) Synthesis process of the SPIO+RPPs and magnetic droplet vaporization (MDV) formation. (**b**) Proposed therapeutic mechanism of RPP-potentiated mild MHT against tumor growth and metastasis (Copyright © 2023, Qiaoxi Qin et al. Reprinted with permission from Ref. [11])

4 Polymer-Modified Magnetic Hyperthermia Nanomaterials: Enhanced Stability and Targeting Through Polymer Engineering

Significant advancements have occurred in recent years in the biomedical applications of magnetic nanomaterials, especially IONPs, attracting considerable attention due to their unique magnetic properties, excellent biocompatibility, and controllable surface modifications. However, bare IONPs still face critical challenges in physiological environments, including aggregation propensity, poor stability, and the lack of targeting capability, severely limiting their clinical translation. The incorporation of polymeric materials serves as an effective solution to these limitations. Polymer modification can not only significantly enhance the colloidal stability and biocompatibility of IONPs but also enable functionalization for targeted delivery, controlled release, and multimodal synergistic therapy. These engineered improvements have positioned polymer-modified IONPs as promising platforms for tumor magnetic hyperthermia applications.

4.1 Polymer Coating for the Enhanced Stability and Biocompatibility of IONPs

Polymer coatings represent a crucial strategy for improving the stability and biocompatibility of IONPs. Natural polymers, such as chitosan and gelatin, have been widely employed due to their excellent biocompatibility and biodegradability. Meanwhile, synthetic polymers like polyethylene glycol (PEG) have gained significant attention due to their controllable physicochemical properties.

4.1.1 Natural Polymer Coatings

MNPs exhibit poor dispersion characteristics, with variations in uniformity, colloidal stability, and concentration before/after administration significantly affecting the heating efficiency and distribution uniformity within tumors. Furthermore, the rapid intracellular biodegradation of MNPs could lead to insufficient thermal dosage in target areas, reducing therapeutic efficacy. In addition, nanoparticles administered intravenously or even intratumorally may accumulate in healthy tissues. These factors collectively highlight the importance of improving MNP chemical and colloidal stability, as well as particle retention, for enhanced therapeutic outcomes. As a natural polysaccharide with excellent biocompatibility and biodegradability, chitosan serves as an ideal coating material for IONPs. Barra et al. developed chitosan-based magnetic bionanocomposite films using a simple solvent-casting method. These sub-100-μm-thick films demonstrated significant thermal effects under AMF exposure, enabling the contactless heating of cell culture media, temperature-controlled gelatin-based hydrogel regulation, and the effective killing of melanoma cells. The nanocomposite films exhibited tunable magnetic properties that facilitated localized, on-demand hyperthermia, offering novel perspectives for tumor therapy [12].

Local interventional therapy remains the primary treatment approach for early-stage tumors. In hepatocellular carcinoma treatment, the direct injection of liquid magnetic media into tumor tissues during minimally invasive procedures often encounters issues such as leakage and abnormal accumulation, leading to unintended tissue damage. Clinically approved drug-loaded hydrogels have been used in therapeutic applications, such as drug delivery, cardiac repair, and medical implants. As a commonly used natural polymer material, gelatin offers unique advantages in this field. Notably, Wu et al. developed a binary magnetic colloidal hydrogel injection system composed of superparamagnetic Fe_3O_4 nanoparticles and gelatin nanoparticles. In this system, the nanoparticles exhibited mutual repulsion after premixing, while their opposite charges in a weakly acidic environment promoted interparticle attraction, facilitating the formation of a three-dimensional gel network. Through a bottom-up synthesis strategy, this injectable hydrogel construct possessed physically cross-linked properties. Its interconnected structure could be immediately disrupted under shear stress and could subsequently self-recover without external stimuli. The system was percutaneously administered to the tumor site under ultrasound guidance, significantly reducing tumor cell viability via AMF-induced hyperthermia effects. The incorporation of ultrasonic visualization technology could not only enhance tumor imaging but could also ensure treatment precision, effectively preventing damage to normal tissues. This percutaneous thermotherapy strategy offers a novel technical solution for the minimally invasive treatment of deep-seated tumors [13].

4.1.2 Synthetic Polymer Coatings

Stand-alone MHT has several drawbacks, such as difficulty achieving complete tumor shrinkage, with high doses of magnetic nanomaterials required to induce therapeutic hyperthermia effects in tumors, compromising tumor imaging in MRI modalities used to monitor brain tumor progression. Therefore, MHT could be combined with other therapies such as chemotherapy, radiotherapy, and phototherapy to leverage the performance of MHT in cancer while using lower doses of magnetic materials. Compared to MHT, photothermal therapy (PTT) can achieve hyperthermia conditions at significantly lower doses of light-absorbing agents. However, PTT can only be applied to superficial tissues. By combining MHT with PTT, the MHT dosage can be reduced, simultaneously overcoming the limitation of insufficient light penetration in PTT-induced heating.

Mai et al. designed a self-assembled amphiphilic diblock copolymer, PEG-b-P(DiolMAm-co-BenzylMAm), which encapsulated iron oxide nanocubes (IONCs) as the magnetothermal medium and incorporated IR780 dye as the photothermal agent. Using photoinduced atom transfer radical polymerization and Diels-Alder chemistry, they designed and synthesized an amphiphilic block copolymer, PEG-b-P(DiolMAm-co-BenzylMAm), where the PEG segment provided aqueous stability, while the other segment contained diol and benzyl side groups, conferring anchoring and hydrophobic properties to the polymer. Specifically, the diol side groups provided strong affinity between the IONCs and polymer chains, enhancing compatibility with the P(DiolMAm-co-BenzylMAm) domains anchored on the IONCs. Conversely, the benzyl side groups provided strong affinity between the polymer and IR780 dye or IONC surface through π-π stacking interactions, enabling self-assembly and encapsulation. Under the synergistic effects of MHT and PTT, this system achieved the efficient killing of U87 glioma cells within the temperature range of 40–47 °C. This system not only reduced the required dosage of IONCs but also compensated for the shortcomings of PTT treatment [14].

4.2 Polymer Modification Enhances Targeting and Synergistic Therapy

The modification ability of polymeric materials can enable the targeted delivery and multifunctional integration of IONPs. Through the surface conjugation of targeting molecules, such as folic acid and RGD peptides or therapeutic drugs, polymer-modified IONPs can achieve precise tumor treatment and multimodal synergistic therapy.

4.2.1 Targeted Delivery Systems

MRI is the gold standard technique for central nervous system tumor evaluation, providing excellent noninvasive morphological and functional images while also benefiting from the use of contrast agents (CAs), such as IONPs. Current clinical CAs exhibit certain toxicity and limited specificity during use, resulting in significant off-target effects and systemic toxicity.

For glioma treatment, the presence of the blood-brain barrier (BBB) greatly limits the delivery efficiency of nanomedicines to brain tumors. Therefore, improving the targeting ability of CAs and reducing peripheral effects have become major issues that need to be solved. In glioblastoma patients, glucose transporter type 1 and 3 (GLUT1 and GLUT3) are overexpressed, especially in glioblastoma stem cells. This overexpression provides an opportunity for functionalized NPs to actively target glioblastoma. Caro et al. developed glucuronic acid-functionalized IONPs (NPs@GlcA), which utilized hypoglycemia-induced GLUT1 overexpression to promote the transport of nanoparticles across the BBB. NPs@GlcA were efficiently taken up by GLUT1 and GLUT3 on glioblastoma microvascular endothelial cells, significantly increasing the drug concentration at tumor sites while serving as an MRI contrast agent for tumor diagnosis. Glucuronic acid has been linked to PEGs through amide bonds, and they are modified on IONPs. The core size of NPs@GlcA was 20 nm, with a hydrodynamic diameter of 50 nm, exhibiting a T2 relaxivity higher than 100 $mM^{-1}\ s^{-1}$ under a high magnetic field (9.4 T). Therefore, it could be used for MRI tracking, and it could act as a nano-heater when exposed to AMF [15].

4.2.2 Synergistic Therapeutic Strategy

In addition to the aforementioned antitumor effects, MHT can disrupt stromal barriers. Under AMF exposure, MNPs can sensitize tumor cells to chemotherapy, with MHT modulating the tumor extracellular matrix by altering the organization of collagen fibers and enabling better drug penetration. Polymer-modified IONPs have been shown to integrate multiple therapeutic modalities to achieve synergistic effects. Díaz-Riascos et al. employed dextran-coated magnetic fluids (NTT agent) in combination with gemcitabine/albumin-bound paclitaxel (GEM/nab-PTX) for the treatment of pancreatic ductal adenocarcinoma (PDAC). MHT can directly kill tumor cells as well as disrupt the dense tumor stroma, significantly enhancing the penetration and antitumor efficacy of chemotherapeutic drugs. In vivo experiments demonstrated that a combination of low-dose chemotherapy with MHT exhibited highly synergistic effects, offering new insights for PDAC treatment [16].

Polymer-modified magnetic hyperthermia nanomaterials offer innovative solutions for tumor treatment by integrating functions, such as enhanced stability, targeted delivery, controlled release, and theranostics integration. Although challenges remain in clinical translation, with the interdisciplinary convergence of materials science, nanotechnology, and biomedicine, this field is likely to achieve broad development prospects.

5 Other Types of Magnetic Hyperthermia Nanomaterials

Sorafenib (SRF) serves as one of the most effective monotherapies for hepatocellular carcinoma (HCC); however, drug resistance constitutes a major obstacle in SRF treatment of HCC. The primary mechanism of SRF in liver cancer treatment involves inducing ferroptosis, a novel form of cell death closely associated with tumor therapy. Tang et al. utilized ferrimagnetic vortex-domain iron oxide (FVIO) nanoring as an effective magnetothermal therapeutic agent while simultaneously supplying iron required for ferroptosis in tumor cells (Fig. 10.7). The iron delivered by FVIO significantly enhanced the susceptibility of HCC cells to SRF-induced ferroptosis. Furthermore, the nanoscale heat generated by FVIO during MH increased ROS levels, stimulated lipid peroxidation (LPO), and inactivated GPX4, further exacerbating the ferroptosis process. The combined application of FVIO-mediated MH and SRF demonstrated a more significant reduction in tumor cell viability compared to either treatment alone [17].

The application of magnetic hyperthermia nanomaterials in tumor therapy continues to achieve new breakthroughs. In addition to traditional drug/nucleic acid composite strategies, studies have found that rotating magnetic fields (RMFs) can effectively improve the tumor penetration of superparamagnetic iron oxide nanoparticles (SPIONs). RMFs can break the chain-like arrangement of SPIONs to form smaller aggregates, thereby enhancing deep penetration and improving magnetic heating efficiency, achieving treatment with lower intensity and shorter duration. This represents a simple method to improve tumor tissue penetration while ensuring

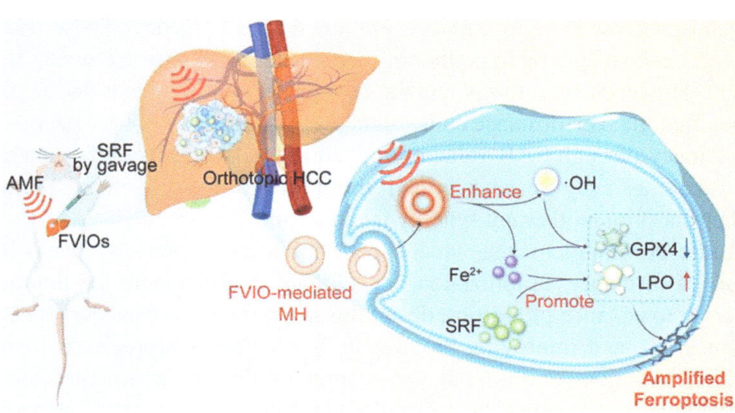

Fig. 10.7 Mechanistic schematic of ferrimagnetic vortex-domain iron oxide (FVIO)-based magnetic hyperthermia combined with SRF for enhanced ferroptosis in HCC treatment (Copyright © 2024, American Chemical Society. Reprinted with permission from Ref. [17])

the efficiency of MHT and enhancing tumor-killing effects [18]. To address the insufficient efficiency of traditional magnetic thermal materials, Zhang et al. developed genetically engineered magnetic inorganic nanocages (eMIONs). The encapsulin protein consists of an iron-storage protein nanocage found in *Myxococcus xanthus*, featuring excellent biocompatibility, appropriate cavity size, and an iron oxidase center. Through biomineralization technology, a highly crystalline Fe_3O_4/FeO structure was obtained by encapsulating the encapsulin protein, achieving a specific absorption rate (SAR) value as high as 2390 W/g (300 kHz, 15 kA/m), representing nearly a fivefold increase over traditional materials. The eMIONs combined peroxidase-like activity with magnetocatalytic properties, doubling the survival period in liver cancer models and demonstrating significant clinical translation potential. These innovations offer new approaches to overcoming the penetration depth and energy efficiency bottlenecks in MHT [19].

6 Summary

Magnetic hyperthermia nanomaterials offer promising potential in tumor therapy, enabling multimodal synergistic treatments. Through component regulation (i.e., Zn/Co doping) and structural design (core-shell/heterojunction), these materials can achieve combined magnetic hyperthermia-photothermal/chemodynamic/immunotherapy. In addition, they can facilitate theranostic integration, significantly improving therapeutic efficacy while reducing treatment dosage and enabling precise spatiotemporal control when combined with optical thermometry and luminescence imaging technologies.

Simultaneously, magnetic hyperthermia nanomaterials have achieved innovations in therapeutic mechanisms. For example, magnetic iron oxide-based materials can synergize with apoptosis/ferroptosis drugs to enhance drug penetration through physical effects, induce tumor cell apoptosis/ferroptosis to reprogram tumor metabolism, and promote ICD to enhance immunotherapy efficacy. These therapeutic advantages make magnetic hyperthermia nanomaterials highly promising for tumor treatment. However, limitations remain during their development and application, including material performance constraints (i.e., the need to improve energy conversion efficiency through SAR value optimization) and unresolved therapeutic challenges, such as drug resistance and the need to further elucidate magnetic hyperthermia-immunotherapy synergy mechanisms.

In future development, magnetic hyperthermia nanomaterials should focus on material performance optimization, developing multielement co-doping/defect engineering to enhance properties, designing smart responsive carriers (pH/temperature-sensitive), and improving biomimetic modification strategies (exosome/cell membrane coating). An in-depth exploration of disease treatment mechanisms should include investigating the molecular mechanisms of magnetic hyperthermia-immunity pathways, exploring novel synergistic strategies such as magnetic hyperthermia-gene editing, and establishing standardized efficacy evaluation systems.

Advancing clinical translation should also focus on overcoming large-scale production bottlenecks and improving long-term safety assessments.

Through continuous material innovation and mechanism exploration, magnetic hyperthermia nanomaterials continue to evolve into an important platform for comprehensive tumor therapy. Future progress requires coordinated efforts across material design, mechanism research, and clinical application to overcome current limitations and achieve safer, more effective tumor treatments.

References

1. Yu X, Yang T, Liu R, Wu Da, Tian D, Zhou T, Yan H, He S, Zeng H. Simultaneous enhancement of magnetothermal and photothermal responses by Zn, Co co-doped ferrite nanoparticles. Small. 2022;18(52):2205037.
2. Li D, Jia M, Du J, Li Y, Jia T, Chen G. Optomagnetic heater-thermometer nanoplatform for tumor magnetothermal therapy with operando temperature feedback. ACS Nano. 2025;19(8):8328–37.
3. Das A, Meena R, Kumar R, Kuanr BK. Sensitizer-passivated magnetic ferrite/Eu3+-doped oxide nanocomposite: a novel luminescent probe for facile bioimaging with superior magnetic hyperthermia performance. ACS Appl Mater Interfaces. 2025;17(14):20872–84.
4. Yuan X, Kang Y, Li R, Niu G, Shi J, Yang Y, Fan Y, Ye J, Han J, Pei Z, Zhang Z, Ji X. Magnetically triggered thermoelectric heterojunctions with an efficient magnetic-thermo-electric energy cascade conversion for synergistic cancer therapy. Nat Commun. 2025;16(1):2369.
5. Chen L, Hu P, Fang W, Wu T, Shi J. Nebulized immunotherapy of orthotopic lung cancer by mild magnetothermal–based innate immunity activations. Angew Chem Int Ed. 2025;64(1):e202413127.
6. Li B, Zu M, Jiang A, Cao Y, Wu J, Shahbazi M-A, Shi X, Reis RL, Kundu SC, Xiao B. Magnetic natural lipid nanoparticles for oral treatment of colorectal cancer through potentiated antitumor immunity and microbiota metabolite regulation. Biomaterials. 2024;307:122530.
7. Beola L, Iturrioz-Rodríguez N, Pucci C, Bertorelli R, Ciofani G. Drug-loaded lipid magnetic nanoparticles for combined local hyperthermia and chemotherapy against glioblastoma Multiforme. ACS Nano. 2023;17(18):18441–55.
8. Huang G, Qiu Y, Yang F, Xie J, Chen X, Wang L, Yang H. Magnetothermally triggered free-radical generation for deep-seated tumor treatment. Nano Lett. 2021;21(7):2926–31.
9. Zhang Q, Wu J, Wang J, Wang X, Wu C, Chen M, Wu Q, Lesniak MS, Mi Y, Cheng Y, Wang Q. A neutrophil-inspired supramolecular nanogel for magnetocaloric–enzymatic tandem therapy. Angew Chem Int Ed. 2020;59(9):3732–8.
10. Cheng X, Xu J, Cui Y, Liu J, Chen Y, He C, Cui L, Liu Y, Song B, Gong C, Mi P. Nanovesicles for lipid metabolism reprogram-enhanced ferroptosis and magnetotherapy of refractory tumors and inhibiting metastasis with activated innate immunity. ACS Nano. 2025;19(7):7213–30.
11. Qin Q, Zhou Y, Li P, Liu Y, Deng R, Tang R, Wu N, Wan L, Ye M, Zhou H, Wang Z. Phase-transition nanodroplets with immunomodulatory capabilities for potentiating mild magnetic hyperthermia to inhibit tumour proliferation and metastasis. J Nanobiotechnol. 2023;21(1):131.
12. Barra A, Wychowaniec JK, Winning D, Cruz MM, Ferreira LP, Rodriguez BJ, Oliveira H, Ruiz-Hitzky E, Nunes C, Brougham DF, Ferreira P. Magnetic chitosan bionanocomposite films as a versatile platform for biomedical hyperthermia. Adv Healthc Mater. 2024;13(11):2303861.
13. Wu W-S, Yan X, Chen S, Du Y, Hu J, Song Y, Zha Z, Xu Y-J, Cao B, Xuan S-H, Liu X, Chen B, Dong L, Lu Y, Yu S-H. Minimally invasive delivery of percutaneous ablation agent via magnetic colloidal hydrogel injection for treatment of hepatocellular carcinoma. Adv Mater. 2024;36(26):2309770.

14. Mai BT, Fernandez-Cabada T, Conteh JS, Nucci GEP, Fiorito S, Gavilán H, Debellis D, Gjurgjaj L, Pellegrino T. Nanoplatforms for magnetic-photo-heating of thermo-resistant tumor cells: singular synergic therapeutic effects at mild temperature. Small. 2024;20(51):2310522.
15. Caro C, Paez-Muñoz JM, Pernía Leal M, Carayol M, Feijoo-Cuaresma M, García-Martín ML. Metabolically-driven active targeting of magnetic nanoparticles functionalized with glucuronic acid to glioblastoma: application to MRI-tracked magnetic hyperthermia therapy. Adv Healthc Mater. 2025;14(3):2404391.
16. Díaz-Riascos ZV, Llaguno-Munive M, Lafuente-Gómez N, Luengo Y, Holmes S, Volatron J, Ibarrola O, Mancilla S, Sarno F, Aguirre JJ, Razafindrakoto S, Southern P, Terán FJ, Keogh A, Salas G, Prina-Mello A, Lacal JC, del Pozo A, Pankhurst QA, Hidalgo M, Gazeau F, Somoza Á, Schwartz S Jr, Abasolo I. Preclinical development of magnetic nanoparticles for hyperthermia treatment of pancreatic cancer. ACS Appl Mater Interfaces. 2025;17(2):2924–39.
17. Tang Q, Wang Y, Yan B, Zhang J, Wang T, Fang Y, Ye Z, Zhang N, Zhang N, Wu Z, Fan H, Lyu Y, Liu X, Wu R. Intracellular magnetic hyperthermia sensitizes sorafenib to orthotopic hepatocellular carcinoma via amplified ferroptosis. ACS Nano. 2024;18(43):29804–19.
18. Choi J, Kim D-i, Kim J-y, Pané S, Nelson BJ, Chang Y-T, Choi H. Magnetically enhanced intracellular uptake of superparamagnetic iron oxide nanoparticles for antitumor therapy. ACS Nano. 2023;17(16):15857–70.
19. Zhang Y, Wang X, Chu C, Zhou Z, Chen B, Pang X, Lin G, Lin H, Guo Y, Ren E, Lv P, Shi Y, Zheng Q, Yan X, Chen X, Liu G. Genetically engineered magnetic nanocages for cancer magneto-catalytic theranostics. Nat Commun. 2020;11(1):5421.

Application of Microneedles in Antitumor Therapy

11

Jiaqi Li, Nianping Feng, and Yongtai Zhang

Abstract

As a new drug delivery platform, microneedles offer advantages such as painless, minimally invasive administration and rapid high-dose drug delivery. They can efficiently and precisely deliver drugs to target sites, playing a crucial role in enhancing therapeutic efficacy and reducing toxicity. Microneedle technology has been applied to the delivery of anticancer drugs, and when combined with photothermal and photodynamic therapies, immunotherapy, and other approaches, it provides new solutions and perspectives for anticancer treatment. Additionally, microneedle technology is utilized for drug monitoring and diagnostics. By detecting tumor-specific signaling factors in bodily fluids like interstitial fluid, it enables predictive analysis, aiding in early treatment and prevention. Consequently, microneedle technology holds promise for delivering efficient, precise, and safe personalized treatment plans for cancer patients in the future, injecting new momentum into the fight against cancer.

Keywords

Microneedles · Transdermal drug delivery · Anti-tumor therapy · Localized cancer treatment · Minimally invasive delivery

1 Introduction

The application of microneedle technology in the field of tumor therapy is a rapidly developing research direction (Fig. 11.1). Microneedles (MNs) deliver drugs directly to the tumor tissues in a minimally invasive way, thereby improving

J. Li · N. Feng (✉) · Y. Zhang (✉)
Shanghai University of Traditional Chinese Medicine, Shanghai, China
e-mail: npfeng@shutcm.edu.cn; analysisdrug@126.com

M. Chu (ed.), *Modern Nanomaterials and Nanotechnology for Cancer Therapy*,
https://doi.org/10.1007/978-981-95-6849-9_11

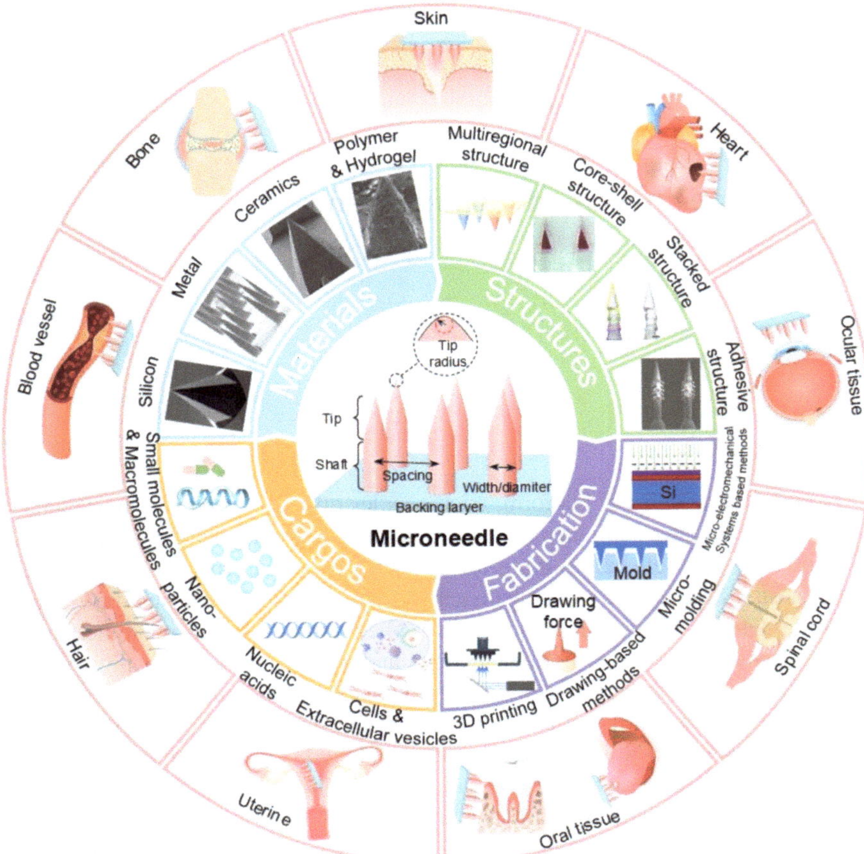

Fig. 11.1 Construction and application of microneedles. (Copyright © 2023 Science Direct Ltd. Reprinted with permission from ref. [2])

treatment effectiveness and reducing side effects. Microneedle technology is an emerging transdermal drug delivery method. It allows microneedles to pierce in the skin to make macromolecules penetrate the barrier of the cuticle with minimal invasion, increase the efficiency of transdermal penetration, and reach a specific depth to play local or systemic effects. The length of microneedles is at the micron level. Compared with traditional injection methods, microneedle technology has the advantages of being minimally invasive, painless, and highly patient-compliant. In addition, microneedles have the advantages of macromolecule delivery, easy management, good tolerance, and easy realization of intelligent detection. Microneedles have been developed for chemotherapy, light therapy, and immunotherapy to treat solid tumors. Intra-microneedle drug delivery can reduce systemic toxicity by delivering drugs efficiently and precisely to the lesion area [1, 2].

2 Application of Microneedles in Antitumor Therapy

2.1 Chemotherapy

A straightforward application involves delivering antitumor drugs via microneedles. Topical administration of chemotherapeutic agents is an effective method for treating skin tumors. Microneedle technology enables high-dose drug penetration into tumor tissues (Fig. 11.2). One study employed roller-type microneedles combined with doxorubicin hydrochloride (DOX) and celecoxib co-loaded liposomes to treat melanoma. Microneedles pretreated with liposome gel enhanced DOX penetration by approximately twofold compared to passive administration, demonstrating excellent targeting efficacy [3]. Resveratrol, a potent anticancer phytochemical, shows significant promise in breast cancer treatment. However, its short half-life and low bioavailability pose major obstacles to effective application. Nanostructured lipid carriers delivering resveratrol to breast tissue via microneedle arrays have been developed, exhibiting enhanced skin permeability, anticancer activity, and reduced skin retention—representing an effective strategy for topical anti-breast cancer therapy [4]. Separate reports have encapsulated cisplatin within pH-responsive lipid nanoparticles, delivered to tumor sites via dissolving microneedles. Upon encountering acidic environments, these nanoparticles release the drug, inducing DNA damage in tumor cells, blocking the cell cycle, and promoting apoptosis [5].

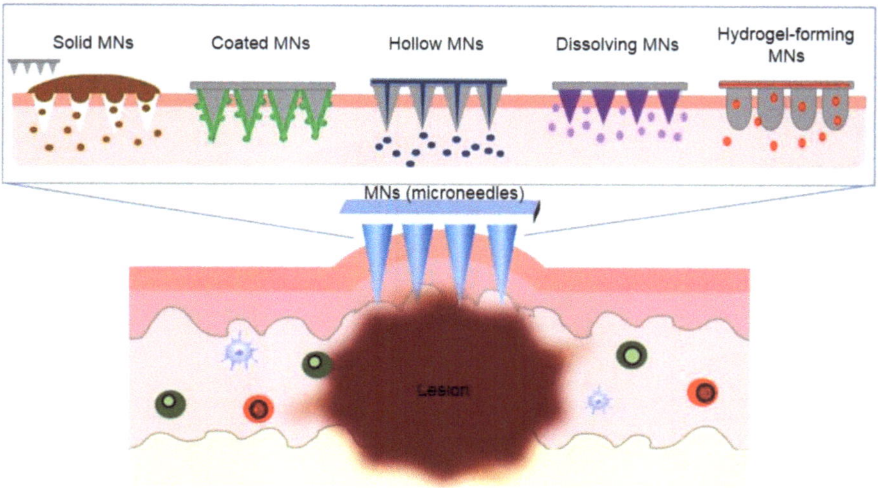

Fig. 11.2 Microneedles-assisted antitumor drug delivery to the tumor site. (Copyright © 2021 Elsevier Ltd. Reprinted with permission from ref. [3])

2.2 Photothermal Therapy and Photodynamic Therapy

Phototherapy encompasses photodynamic therapy (PDT) and photothermal therapy (PTT). PTT is an emerging tumor treatment modality that utilizes photothermal agents (PTAs) to convert light energy into heat energy under near-infrared (NIR) irradiation, thereby eliminating tumor cells. The photothermal conversion efficiency and biocompatibility of PTAs significantly influence therapeutic outcomes. The integration of microneedle technology with PTT offers novel therapeutic strategies. Microneedles can deliver photothermal agents, which are activated by NIR irradiation to generate localized high temperatures, thereby eliminating tumor cells. This approach offers the advantage of minimizing damage to surrounding healthy tissues while enhancing treatment precision and safety.

In a recent report, a research team developed a new multifunctional microneedle patch for synergistic photothermal-gas therapy of head and neck malignant melanoma and associated skin defects. This microneedle platform generates photothermal effects under NIR irradiation while releasing nitric oxide to promote tumor cell apoptosis and accelerate tissue regeneration through sustained angiogenesis. This strategy not only eliminates tumor cells through the synergistic effects of photothermal and gas therapies but also promotes the repair of postoperative facial skin defects [6] (Fig. 11.3a). Furthermore, the study enhances photothermal efficacy by regulating tumor-associated fibroblasts (CAFs) to prevent tumor recurrence and metastasis (Fig. 11.3b. CAFs play a pivotal role in the tumor microenvironment, and modulating their activity amplifies the effectiveness of PTT [7].

PDT is primarily used to treat tumors, such as skin cancer, esophageal cancer, lung cancer, and bladder cancer. The fundamental principle of this therapy involves

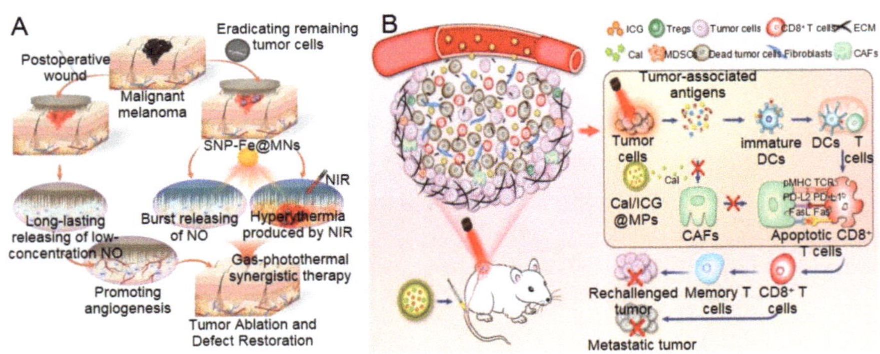

Fig. 11.3 Microneedle-assisted drug delivery system combined with photothermal therapy. (**a**) Microneedle patch for synergistic photothermal-gas therapy of melanoma and associated skin defects. *NO* nitric oxide, *NIR* near-infrared laser. SNP-Fe@MNs: a novel multifunctional sodium nitroprusside and Fe^{2+} ion-loaded microneedles platform. (**b**) Enhancing photothermal efficacy by regulating cancer-associated fibroblasts (CAFs). Cal/ICG@MPs: tumor cell-derived microparticles co-delivering calcipotriol and Indocyanine green. *ECM* extracellular matrix, *MDSCs* myeloid-derived suppressor cells, *DCs* dendritic cells. (Copyright ©2024 Springer Nature Ltd. Reprinted with permission from ref. [6]; Copyright © 2022 Springer Nature Ltd. Reprinted with permission from ref. [7])

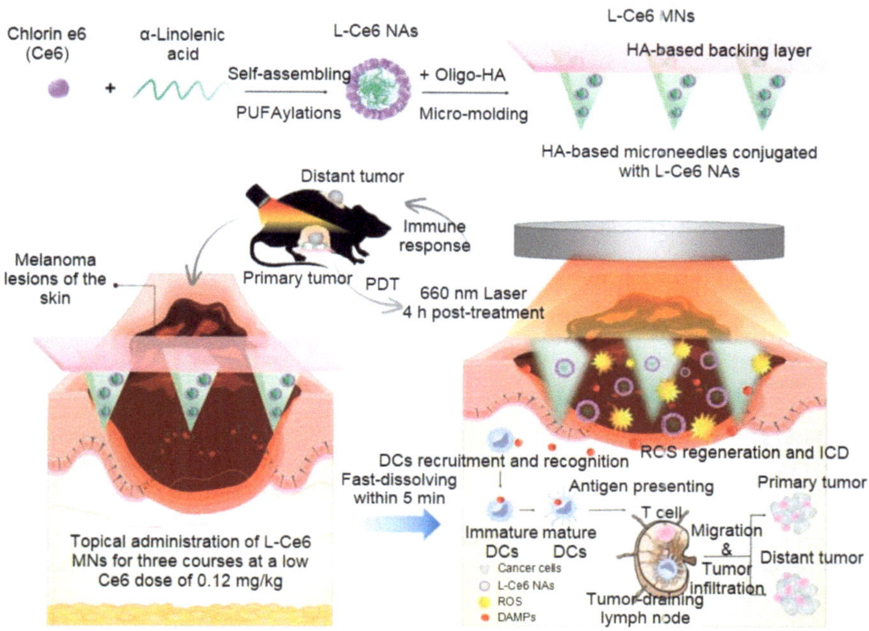

Fig. 11.4 Microneedle-assisted drug delivery system combined with photodynamic therapy. *L-Ce6 NAs* nanoassemblies of chlorin e6. *PUFA* polyunsaturated fatty acid, *HA* hyaluronic acid, *DCs* dentric cells, *ROS* reactive oxygen species. (Copyright ©2021 ACS Publication Ltd. Reprinted with permission from ref. [8])

irradiating the target area with light of a specific wavelength. This causes photosensitizers that have been ingested and accumulated in the diseased tissue to transition from their ground state to an excited state. This generates highly reactive singlet oxygen, which undergoes oxidative reactions with nearby biomolecules, triggering cytotoxic responses that ultimately promote cell death. Combined with PDT, microneedles carrying photosensitizers generate localized high temperatures upon NIR activation, thereby killing tumor cells. This approach offers advantages in minimizing damage to surrounding healthy tissues while enhancing treatment precision and safety. To reduce photosensitizer dosage while maintaining efficacy, Ce6-integrated rapid-dissolution microneedle patches have been developed [8]. In a mouse melanoma model, microneedles containing low-dose Ce6 (0.12 mg/kg) promoted reactive oxygen species generation via PDT, effectively ablating the primary tumor. In PDT, the hypoxic environment within tumor tissues may limit therapeutic efficacy. To address this, researchers developed a microneedle system capable of generating gaseous oxygen to improve photosensitizer penetration into tumor tissues and alleviate hypoxia. This system utilizes solid hydrogen peroxide within the microneedles to react with surrounding fluids, forming gaseous oxygen bubbles. The resulting fluid flow promotes photosensitizer migration to deeper tissue layers, thereby enhancing PDT therapeutic outcomes [9] (Fig. 11.4).

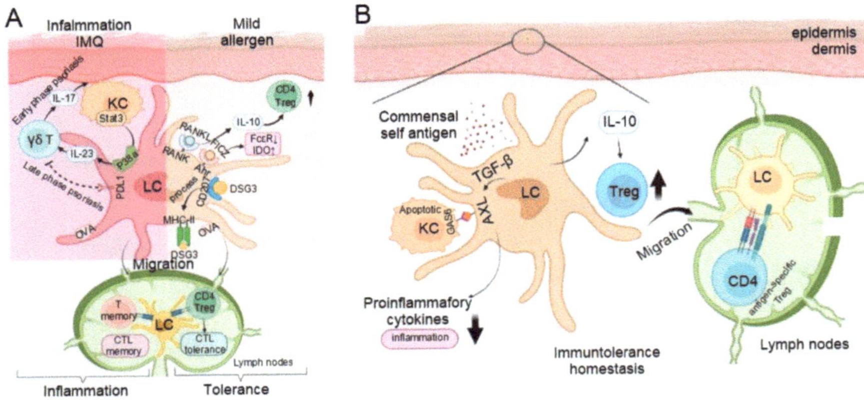

Fig. 11.5 Different statements of LC. (**a**) Langerhans cells in skin diseases. *LC* Langerhans cells, *KC* keratinocytes, *IMQ* imiquimod-induced (IMQ) psoriasis-like dermatitis model, *RANK* nuclear factor κB, *RANKL* nuclear factor κB ligand, *AhR* aryl hydrocarbon receptor, *FICZ* a potent AhR agonist, *OVA* ovalbumin, *IDO* indoleamine 2,3-dioxygenase, *CTL* cytotoxic T lymphocyte. (**b**) Langerhans cells in steady state skin immune homeostasis. *TGF-β* transforming growth factor β. *GAS6* growth arrest specific 6. (Copyright ©2022 MDPI Ltd. Reprinted with permission from ref. [14])

2.3 Immunotherapy

2.3.1 Skin and Immunity

The skin typically functions as a "wall" or "fence," shielding internal organs from external threats, such as physical, mechanical, chemical, and pathogenic microorganisms, while maintaining a stable internal environment. Structurally, the skin consists of the epidermis, dermis, and subcutaneous tissue from superficial to deep layers. Beyond its physical barrier role, the immune barrier represents a crucial line of defense (Fig. 11.5).

Within the epidermis, keratinocytes play a pivotal role in transmitting signals of external invasion while releasing multiple cytokines to facilitate immune cell entry into the skin. The skin is rich in antigen-presenting cells (APCs), including dendritic cells (DCs), monocytes/macrophages, and B cells. These cells can capture and migrate polymeric nanoparticles delivered by microneedles to lymph nodes, thereby inducing immune cell proliferation and differentiation.

DCs, as specialized antigen-presenting cells, play a central role in initiating immune responses. While quiescent DCs are inefficient at presenting antigens to T cells, activated DCs migrate to lymph nodes and effectively activate naive T cells. DCs are primarily divided into two major subsets: myeloid DCs (mDCs) and lymphoid DCs (LDCs) or plasmacytoid DCs (pDCs). Langerhans cells belong to the mDC category. These cells possess potent capabilities for recognizing, phagocytosing, and presenting foreign antigens. As the sole APCs in the epidermis, they mediate immune responses to construct the skin's foremost immune barrier. Their localized presence in the epidermis makes them an ideal target for transdermal vaccination strategies. Langerhans cells function both as APCs—directly phagocytosing pathogens and allergens within the epidermis—and as APCs migrating to draining lymph nodes. There, they activate T lymphocytes through antigen presentation, thereby initiating adaptive immune responses or inducing immune tolerance to maintain skin homeostasis [9–14].

In microneedle-mediated antitumor therapy, immune activation is often combined with phototherapy. Researchers have developed a microneedle delivery system containing gold nanorods and the immune checkpoint inhibitor anti-PD-1 peptide. This system enhances immune responses, prolongs survival in mice, and causes minimal side effects in normal tissues. It has been reported that core-shell microneedles designed for melanoma immunotherapy have been developed. Within the microneedle structure, the core component—an inhibitor of indoleamine 2,3-dioxygenase (IDO)—is immobilized within a cross-linked polymer gel, while the microneedle shell encapsulates nanoparticles containing the photosensitizer indocyanine green. Under NIR irradiation, the photosensitizer converts light energy into heat, causing tumor cell ablation and inducing antigen release. This promotes the chemotaxis and maturation of DCs along with cytokine secretion. The IDO inhibitor further alleviates immune suppression in the tumor microenvironment, activating and enhancing immune responses to achieve immunotherapy for both primary and metastatic tumors [15].

2.3.2 Vaccines

The application of microneedle technology in the field of antitumor vaccines is a currently active research direction. Numerous reports have documented the use of microneedles to deliver biomacromolecules such as monoclonal antibodies, proteins, and nucleic acids for tumor immunotherapy. Some studies have combined nanoparticle vaccines with microneedles to enhance antitumor efficacy [16]. Residual tumor cells following malignant melanoma resection remain a pressing challenge. Biomaterial-mediated immunotherapy has emerged as a priority strategy for antitumor applications, making effective drug delivery a critical issue. For malignant melanoma, a unique bilayer microneedle patch was designed: Hydroxyapatite was self-assembled with short-chain peptides from tumor cells to create a biomimetic nanovaccine, which was loaded onto the tips of microneedles in a gelatin methacryloyl-based microneedle patch. This technology effectively suppressed malignant melanoma proliferation and induced apoptosis, promoting regeneration in a full-thickness skin defect model in Sprague-Dawley rats within 15 days.

Microneedle technology has demonstrated unique advantages in antitumor vaccine delivery, notably for its ability to effectively activate the immune system [17] (Fig. 11.6a). Microneedle systems enable the combined delivery of tumor antigens and immune adjuvants, thereby inducing more potent immune responses [18] (Fig. 11.6b). Microneedles effectively activate skin-resident T cells, enhancing their cytotoxic effects against tumor cells. Researchers developed whole-cell encapsulated antitumor vaccine microneedle patches, combined with transdermal co-delivery of granulocyte-macrophage colony-stimulating factor and autophagy-promoting agents (Tat-beclin 1). This system delivers the vaccine

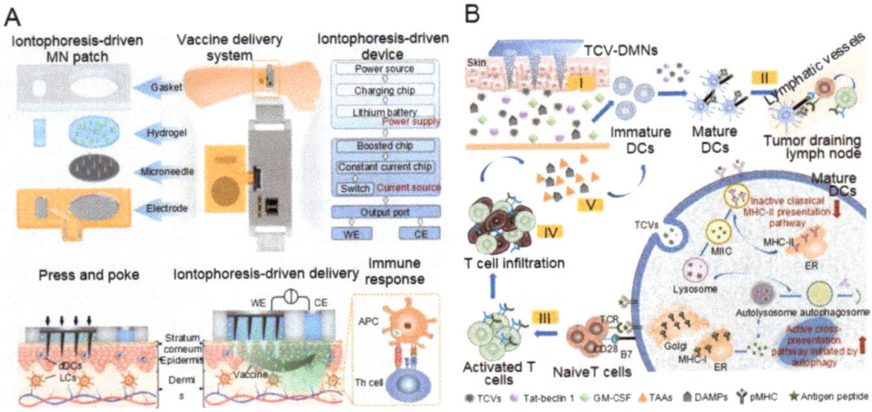

Fig. 11.6 Microneedle drug delivery systems function as vaccines. (**a**) Microneedle patch for the active transdermal delivery of vaccine macromolecules. *WE* working electrode, *CE* counter electrode. (**b**) Microneedle-assisted vaccination combined with autophagy regulation for antitumor immunotherapy. Cancer-immunity cycle: I. Tumor antigen release; II. Tumor antigen is presented by DCs; III. Activate effector T cells; IV. T cell infiltration in tumor tissues; V. T cells kill tumor cells and release tumor antigen, *TCV-DMNs*: a whole cell-encapsulated antitumor vaccine microneedle patch, *TCVs* tumor cell vaccines, *GM-CSF* granulocyte-macrophage colony-stimulating factor, *TAAs* tumor-associated antigens, *DAMPs* damage-associated molecular patterns, *MHC* major histocompatibility complex, *pMHC* peptide-MHC complexes. Tat-beclin 1: autophagy promoter. ER: red. Golgi: white. (Copyright ©2023 Nature Ltd. Reprinted with permission from ref. [17]; Copyright ©2023 Elsevier Ltd. Reprinted with permission from ref. [18])

directly into the skin via microneedles, activating APCs within the skin to subsequently activate T cells and generate an antitumor immune response. The study demonstrated that this microneedle vaccine effectively suppressed melanoma challenges, induced regression of established malignant tumors, and achieved recurrence-free survival exceeding 40 days. This indicates that whole-cell encapsulated microneedle-assisted transdermal vaccination combined with autophagy regulation induces potent antitumor immune responses by enhancing transdermal delivery efficiency, promoting antigen internalization and cross-presentation, and boosting T cell activity. Microneedle-delivered vaccines demonstrated significant antitumor effects and markedly improved survival rates in mouse models. To stimulate target cells for immunogenicity induction, soluble microneedles have been engineered for delivering induction vaccines and adjuvants [19]. This microneedle comprises a bioabsorbable polypeptide matrix with a nanopolyplex, which includes a cationic amphiphilic conjugate with ovalbumin-expressing plasmid OVA (pOVA) and immunostimulant polyinosine: polycytidylic acid (poly (I: C)), in which cationic nanomultichains effectively combine pOVA and poly (I: C), facilitating their transport into intracellular compartments of DCs and macrophages. In mice, subcutaneous administration of the microneedle-free nanocomplex induced a stronger antigen-specific antibody response than subcutaneous implantation of the microneedle mixture.

2.3.3 Immunosuppressants

Blocking immune checkpoints is a crucial immunotherapy strategy, with its core principle centered on disrupting the mechanisms by which tumor cells evade immune surveillance. Within the tumor microenvironment, T cell immune function is frequently suppressed, preventing them from effectively recognizing and attacking tumor cells. Tumor cells employ multiple mechanisms to inhibit T cell function and evade immunity, one of which involves secreting immunosuppressive molecules such as programmed death protein-1 (PD-1) and cytotoxic T lymphocyte-associated antigen 4 (CTLA-4). Taking the PD-1/PD-L1 pathway as an example, PD-1 is a receptor primarily expressed on T cell surfaces, while PD-L1 is its ligand, mainly expressed on specific tumor cells and immune cells. When PD-L1 binds to PD-1, it transmits an inhibitory signal to T cells, hindering their activation and proliferation. Many tumor cells evade immune surveillance and clearance by overexpressing PD-L1, which binds to PD-1 on T cells and suppresses T cell-mediated immune responses. Consequently, PD-1/PD-L1 inhibitors block this pathway, lifting the suppression on T cells and enhancing the immune system's ability to attack tumors [20] (Fig. 11.7a).

By blocking immune checkpoint receptors on activated T cells to enhance T cell function, immune checkpoint T cell PD-1 blocking antibodies have been approved by the US Food and Drug Administration for treating advanced melanoma. Therefore, a microneedle-based transdermal delivery method was designed to achieve localized targeted therapy against the immunosuppressive receptors PD-1 and immunosuppressive enzyme indoleamine 2,3-dioxygenase (IDO) in melanoma treatment. Immunotherapy nanovesicles were constructed by encapsulating anti-PD-1 antibody (aPD1) within hyaluronic acid nanocarriers modified with the IDO inhibitor 1-methyl-DL-tryptophan. This combination strategy, based on "drug A embedded within a carrier formed by drug B," significantly enhances the microneedle's drug-carrying capacity. In vivo efficacy studies demonstrate that these microneedles effectively combat tumors, markedly enhance T cell immunity, and alleviate local immune suppression [21]. The combined delivery of anti-PD-L1 and IDO inhibitors via microneedles modulates immune suppression in the tumor microenvironment, yielding superior antitumor effects compared to intratumoral injection [22] (Fig. 11.7b). Concurrently, the microneedle matrix of polyvinyl alcohol inhibits IDO inhibitor crystallization, facilitating drug loading.

2.4 Monitoring and Diagnosis

Microneedles are employed in biomonitoring and diagnostics by implanting them into the skin to absorb interstitial fluid (ISF), which contains various metabolites serving as biomarkers. Microneedle-based biosensors provide a robust platform for high-throughput, rapid disease state diagnosis and monitoring. Microneedles are primarily categorized into solid microneedles, coated microneedles, hollow microneedles, soluble microneedles, and hydrogel microneedles. The selection of different microneedle types depends on specific application requirements and drug

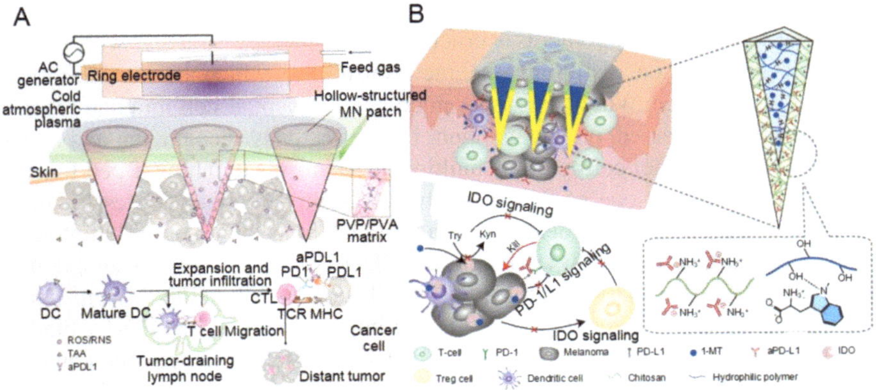

Fig. 11.7 Microneedle-assisted drug delivery systems used for immunosuppressant delivery. (**a**) Microneedle-mediated cold atmospheric plasma (CAP) and ICB therapy. *TAA* tumor-associated antigen, *PVP* poly(vinyl pyrrolidone), *PVA* poly(vinyl alcohol), *RNS* reactive nitrogen species. (**b**) Microneedle-assisted co-delivery of aPD-L1 and 1-MT for melanoma immunotherapy. *PD-*1 programmed cell death receptor-1, *PD-L*1 programmed cell death receptor ligand-1, *IDO* indoleamine 2,3-dioxygenase, *Try* tryptophan, *Kyn* kynurenine. (Copyright ©2020 ACS Publications Ltd. Reprinted with permission from ref. [20]; Copyright ©2020 Elsevier Ltd. Reprinted with permission from ref. [22])

delivery mechanisms [23]. For instance, solid microneedles are primarily used for drug delivery, while hollow microneedles are suitable for extracting biological fluids for subsequent in vitro testing. Furthermore, microneedle technology has been integrated with sensor technology to enable real-time, continuous monitoring of multiple physiological indicators, providing crucial data support for precision medicine. In tumor detection, microneedle technology has seen widespread application in melanoma, breast cancer, and other fields [24, 25] (Fig. 11.8a). However, challenges remain in advancing this technology toward clinical implementation and industrial production, with no microneedle products based on biomonitoring and diagnostics currently entering clinical trials. Despite this, microneedle products integrated with smart devices have been emerging, indicating promising future prospects for smart microneedles [26].

Abnormal expression of microRNAs (miRNAs) is highly correlated with tumorigenesis and metastasis, making them potential biomarkers for early tumor diagnosis and clinical treatment. Detecting miRNAs in ISF poses significant challenges due to

Fig. 11.8 Microneedle system for tumor monitoring and diagnosis. (**a**): Microneedle-assisted breast cancer early detection. MBL: a kind of thin film material. (**b**) Microneedles with programmed DNA hydrogels for sampling and sensing of certain microRNA in dermal ISF. TOE: recombinant protein. (**c**) Spatial tumor biopsy with fluorescence PCR microneedle array. PEGDA: polyethylene glycol diacrylate. (Copyright ©2020 Elsevier Ltd. Reprinted with permission from ref. [24]; Copyright ©2022 ACS Publication Ltd. Reprinted with permission from ref. [27]; Copyright ©2023 Elsevier Ltd. Reprinted with permission from ref. [28])

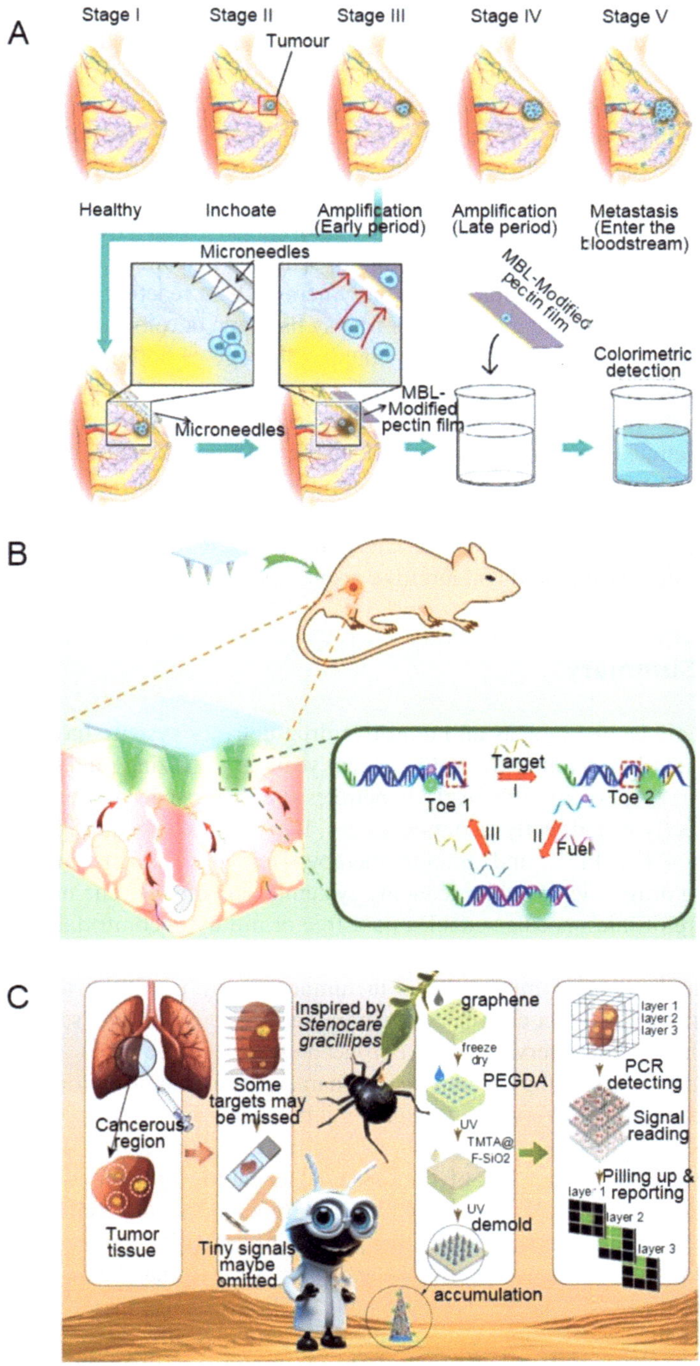
A
Stage I
Stage II
Stage III
Stage IV
Stage V
Tumour
Healthy
Inchoate
Amplification (Early period)
Amplification (Late period)
Metastasis (Enter the bloodstream)
Microneedles
MBL-Modified pectin film
Colorimetric detection
Microneedles
MBL-Modified pectin film
B
Target
I
Toe 1
Toe 2
III
II
Fuel
C
Inspired by Stenocare gracilipes
graphene
freeze dry
PEGDA
UV
TMTA@ F-SiO2
UV
demold
accumulation
Cancerous region
Tumor tissue
Some targets may be missed
Tiny signals maybe omitted
layer 1
layer 2
layer 3
PCR detecting
Signal reading
Pilling up & reporting
layer 1
layer 2
layer 3

their low expression levels and limited sample volume. To address this issue, researchers developed a microneedle patch system encapsulating methacrylate hyaluronic acid (MeHA) within a smart DNA hydrogel matrix. The MeHA/DNA matrix exhibits high hydrophilicity, enabling sufficient ISF extraction (0.97 ± 0.2 mg) within 5 min. Furthermore, through cascade toehold-mediated DNA strand displacement signal amplification technology, low-abundance miRNAs (as low as 241.56 pM) were sensitively detected. This microneedle design offers potential for personalized diagnostics and real-time monitoring [27] (Fig. 11.8b).

To achieve visual, selective, and highly sensitive detection of target DNA and identify tumor markers, a fluorescent polymerase chain reaction (F-PCR) microarray (MNA) platform for spatial tumor biopsy has been developed [28] (Fig. 11.8c). This microarray platform employs a coupled strategy of bottom-up self-assembly and top-down lithography. It consists of a substrate assembled from hydrophobic silica nanoparticles and graphene aerogel-hydrogel hybrid microneedles that effortlessly penetrate tissue samples. This enables stereotactic collection of tumor nucleic acid biomarkers for three-dimensional sampling. Furthermore, each microneedle tip functions as an independent compartment, enabling biochemical reactions to proceed without interference. This facilitates direct F-PCR reactions for tumor marker detection, delivering rapid and intuitive result reporting.

3 Summary

Microneedle technology, as an innovative drug delivery system, has garnered significant attention in tumor therapy in recent years. By utilizing miniaturized needle arrays, this technology effectively penetrates the skin barrier to deliver antitumor drugs directly or indirectly to tumor tissues. It combines with multiple therapies—including PTT, PDT, and immunotherapy—for antitumor treatment, thereby enhancing drug efficacy while reducing systemic side effects. The minimally invasive nature of microneedles enables pain-free or mildly uncomfortable drug delivery, significantly improving patient compliance and quality of life.

The application of microneedles in tumor therapy manifests in two primary aspects: first, as a direct delivery vehicle for anticancer drugs and, second, as a carrier for therapeutic interventions. Microneedles are engineered to customize drug release profiles according to tumor tissue characteristics, enabling controlled release and targeted drug delivery. Additionally, microneedles are employed for minimally invasive extraction of tumor markers, facilitating early diagnosis [24, 25].

However, the application of microneedle technology in tumor treatment still faces several challenges, including material biocompatibility and degradability, mechanical strength and stability, mass production, and cost control. To meet clinical demands and enhance the efficacy and safety of microneedles in drug delivery, further research and optimization are required. For instance, first, as microneedles constitute an invasive drug delivery method, potential adverse reactions such as allergies and inflammation may occur despite rapid wound healing. This can be addressed by employing antimicrobial microneedle substrates [29]. Second, the residual presence of microneedles and their substrates in the body poses potential safety risks, necessitating the use of biodegradable and biocompatible materials. Furthermore, technical bottlenecks persist in the industrial-scale batch production of microneedles; future advancements in 3D printing technology hold promise for achieving large-scale microneedle fabrication.

Despite numerous challenges, microneedle technology holds promising prospects for research and application in cancer treatment. With ongoing advancements in materials science, bioengineering, pharmaceutical technology, and related fields, microneedles are poised to deliver more efficient, precise, and safe personalized treatment options for cancer patients in the future, injecting new momentum into the fight against cancer.

References

1. Vora LK, Sabri AH, McKenna PE, Achmad H, Aaron RJH, Usanee D, Alejandro JP, Eneko L, Ryan FD. Microneedle-based biosensing. Nat Rev Bioeng. 2024;2:64–81.
2. Lyu S, Dong A, Xu X, Bei HP, Yuen HY, Cheung CWJ, Wong MS, He Y, Zhao X. Going below and beyond the surface: microneedle structure, materials, drugs, fabrication, and applications for wound healing and tissue regeneration. Bioact Mater. 2023;27:303–26.
3. Ahmed KS, Shan X, Mao J, Qiu L, Chen J. Derma roller® microneedles – mediated transdermal delivery of doxorubicin and celecoxib co-loaded liposomes for enhancing the anticancer effect. Mater Sci Eng C Mater Biol Appl. 2019;99:1448–58.
4. Gadag S, Narayan R, Nayak AS, Catalina AD, Sant S, Nayak Y, Garg S, Nayak UY. Development and preclinical evaluation of microneedle-assisted resveratrol loaded nanostructured lipid carriers for localized delivery to breast cancer therapy. Int J Pharm. 2021;606:120877.
5. Yang D, Chen M, Sun Y, Jin Y, Lu C, P X, Quan G, Wu C. Microneedle-mediated transdermal drug delivery for treating diverse skin diseases. Acta Biomater. 2021;121:119–33.
6. Dong S, Zhang Y, Zhang Y, Mei Y, Sina A, Zou R, Niu L. A novel multifunctional microneedle patch for synergistic photothermal- gas therapy against maxillofacial malignant melanoma and associated skin defects. J Nanobiotechnol. 2024;22:199.
7. Li X, Yong T, Wei Z, Bie N, Zhang X, Zhan G, Li J, Qin J, Yu J, Zhang B, Gan L, Yang X. Reversing insufficient photothermal therapy-induced tumor relapse and metastasis by regulating cancer-associated fibroblasts. Nat Commun. 2022;13:2794.
8. Bian Q, Huang L, Xu Y, Wang R, Gu Y, Yuan A, Ma X, Hu J, Rao Y, Xu D, Wang H, Gao J. A facile low-dose photosensitizer-incorporated dissolving microneedles-based composite system for eliciting antitumor immunity and the abscopal effect. ACS Nano. 2021;15:19468–79.
9. Liu P, Fu Y, Wei F, Ma T, Ren J, Xie Z, Wang S, Zhu J, Zhang L, Tao J, Zhu J. Microneedle patches with O2 propellant for deeply and fast delivering photosensitizers: towards improved photodynamic therapy. Adv Sci (Weinh). 2022;9(25):e2202591.
10. Naik S, Bouladoux N, Linehan JL, Han SJ, Harrison OJ, Wilhelm C, Conlan S, Himmelfarb S, Byrd AL, Deming C, Quinones M, Brenchley JM, Kong HH, Tussiwand R, Murphy KM, Merad M, Segre JA, Belkaid Y. Commensal-dendritic-cell interaction specifies a unique protective skin immune signature. Nature. 2015;520(7545):104–8.
11. Jakob T, Ring J, Udey MC. Multistep navigation of Langerhans/dendritic cells in and out of the skin. J Allergy Clin Immunol. 2001;108(5):688–96.
12. Ronchese F, Hilligan KL, Mayer JU. Dendritic cells and the skin environment. Curr Opin Immunol. 2020;64:56–62.
13. Wilcox NC, Taheri G, Halievski K, Talbot S, Silva JR, Ghasemlou N. Interactions between skin-resident dendritic and Langerhans cells and pain-sensing neurons. J Allergy Clin Immunol. 2024;154(1):11–9.

14. Zhou L, Jiang A, Veenstra J, Ozog DM, Mi QS. The roles of skin Langerhans cells in immune tolerance and cancer immunity. Vaccine. 2022;10(9):1380.
15. Weng J, Zheng G, Wen J, Yang J, Yang Q, Zheng X, Yan Q. Construction and application of microneedle-mediated photothermal therapy and immunotherapy combined anti-tumor drug delivery system. Drug Deliv. 2023;30(1):2232950.
16. Chen Z, Guo Z, Hu T, Huang B, Zheng Q, Du X, Huang L, Hu W. Double-layered microneedle patch loaded with bioinspired nano-vaccine for melanoma treatment and wound healing. Int J Biol Macromol. 2024;262:129961.
17. Zheng Y, Ye R, Gong X, Yang J, Liu B, Xu Y, Nie G, Xie X, Jiang L. Iontophoresis-driven microneedle patch for the active transdermal delivery of vaccine macromolecules. Microsyst Nanoeng. 2023;9:35.
18. Yang D, Chen M, Sun Y, Shi C, Wang W, Zhao W, Wen T, Liu T, Fu J, Lu C, Wu C, Quan G, Pan X. Microneedle-assisted vaccination combined with autophagy regulation for antitumor immunotherapy. J Control Release. 2023;357:641–54.
19. Duong HTT, Yin Y, Thambi T, Kim BS, Jeong JH, Lee DS. Highly potent intradermal vaccination by an array of dissolving microneedle polypeptide cocktails for cancer immunotherapy. J Mater Chem B. 2020;8(6):1171–81.
20. Han X, Li H, Zhou D, Chen Z, Gu Z. Local and targeted delivery of immune checkpoint blockade therapeutics. Acc Chem Res. 2020;53(11):2521–33.
21. Ye Y, Wang J, Hu Q, Gabrielle MH, Xin H, Wang C, Gu Z. Synergistic transcutaneous immunotherapy enhances antitumor immune responses through delivery of checkpoint inhibitors. ACS Nano. 2016;10(9):8956–63.
22. Yang P, Lu C, Qin W, Chen M, Quan G, Liu H, Wang L, Bai X, Pan X, Wu C. Construction of a core-shell microneedle system to achieve targeted co-delivery of checkpoint inhibitors for melanoma immunotherapy. Acta Biomater. 2020;104:147–57.
23. Ma S, Li J, Pei L, Feng N, Zhang Y. Microneedle-based interstitial fluid extraction for drug analysis: advances, challenges, and prospects. J Pharm Anal. 2023;13(2):111–26.
24. Chen L, Zhang C, Xiao J, You J, Zhang W, Liu Y, Xu L, Liu A, Xin H, Wang X. Local extraction and detection of early stage breast cancers through a microneedle and nano-Ag/MBL film based painless and blood-free strategy. Mater Sci Eng C Mater Biol Appl. 2020;109:110402.
25. Sharifi M, Avadi MR, Attar F, Dashtestani F, Ghorchian H, Rezayat SM, Saboury AA, Falahati M. Cancer diagnosis using nanomaterials based electrochemical nanobiosensors. Biosens Bioelectron. 2019;126:773–84.
26. Zheng M, Sheng T, Yu J, Gu Z. Microneedle biomedical devices. Nat Rev Bioeng. 2024;2:324–42.
27. Yang Q, Wang Y, Liu T, Wu C, Li J, Cheng J, Wei W, Yang F, Zhou L, Zhang Y, Yang S, Dong H. Microneedle array encapsulated with programmed DNA hydrogels for rapidly sampling and sensitively sensing of specific microRNA in dermal interstitial fluid. ACS Nano. 2022;16(11):18366–75.
28. Zhang X, Chen G, Wang Y, Zhao Y. Spatial tumor biopsy with fluorescence PCR microneedle array. Innovation (Camb). 2023;5(1):100538.
29. Shi Y, Zhao J, Li H, Yu M, Zhang W, Qin D, Qiu K, Chen X, Kong M. A drug-free hair follicle cycling regulatable, separable, antibacterial microneedle patch for hair regeneration therapy. Adv Healthc Mater. 2022;11(19):e2200908.

Nanomedicine-Based Strategies for Cancer Stem Cells Therapy

12

Xin Xie

Abstract

Cancer has long been one of the major diseases threatening human health. Despite significant achievements in cancer treatment, cancer recurrence and metastasis remain the primary causes of high mortality rates among clinical patients. While chemotherapy and radiation therapy continue to be the mainstay of clinical cancer treatment, drug resistance persists, leading to cancer relapse and poor prognosis. A wealth of literature indicates a strong correlation between drug resistance and cancer stem cells (CSCs). Cancer stem cells, also known as tumor-initiating cells, constitute a small subset of malignant tumor cells. They exhibit characteristics of both stem cells and cancer cells, including self-renewal and differentiation properties. Despite their small proportion within tumor tissue, tumor stem cells are considered culprits in tumorigenesis, treatment resistance, metastasis, and relapse. Advances in cell biology have revealed several mechanisms through which tumor stem cells acquire resistance to chemotherapy and radiotherapy. These mechanisms include elevated levels of ATP-binding cassette (ABC) transporters, exceptional DNA repair capacity, overexpression of anti-apoptotic proteins, and enhanced protection against reactive oxygen species (ROS). Furthermore, tumor stem cells maintain their stem cell properties by overactivating various transcription factors and signaling pathways, such as Wnt/β-catenin, Notch, Hedgehog, and Nanog. This self-renewal and proliferation of tumor stem cells contribute to tumor progression. Typically, CSCs are implicated in treatment failure, relapse, and metastasis due to their stem-like characteristics. Consequently, targeting CSCs represents an effective anticancer therapeutic strategy. Currently, several strategies have been developed to eradicate cancer stem cells (CSCs), including targeting CSC surface markers, inhibiting ATP-

X. Xie (✉)
School of Pharmacy, Guangdong Medical University, Dongguan, China
e-mail: xiexin@gdmu.edu.cn

M. Chu (ed.), *Modern Nanomaterials and Nanotechnology for Cancer Therapy*,
https://doi.org/10.1007/978-981-95-6849-9_12

binding cassette transporters, intervening in critical signaling pathways, and inducing differentiation of tumor stem cells. Additionally, phototherapy (both photodynamic therapy and photothermal therapy), as a noninvasive and spatiotemporally controlled approach, achieves indiscriminate killing of tumor cells and tumor stem cells by converting light into heat or reactive oxygen species. Consequently, it serves as an alternative method for eliminating tumor stem cells. Furthermore, recent literature suggests that tumor stem cells exhibit significant sensitivity to ferroptosis, making ferroptosis another viable approach for eradicating tumor stem cells.

Keywords

Cancer Stem Cells (CSCs) · Nanomedicine · Targeted cancer therapy · Drug resistance · Tumor recurrence

1 Introduction

Cancer has long been one of the major diseases threatening human health. Despite significant achievements in cancer treatment, cancer recurrence and metastasis remain the primary causes of high mortality rates among clinical patients. While chemotherapy and radiation therapy continue to be the mainstay of clinical cancer treatment, drug resistance persists, leading to cancer relapse and poor prognosis [1–3]. A wealth of literature indicates a strong correlation between drug resistance and cancer stem cells (CSCs). Cancer stem cells, also known as tumor-initiating cells, constitute a small subset of malignant tumor cells. They exhibit characteristics of both stem cells and cancer cells, including self-renewal and differentiation properties. Despite their small proportion within tumor tissue, tumor stem cells are considered culprits in tumorigenesis, treatment resistance, metastasis, and relapse [4–8]. Advances in cell biology have revealed several mechanisms through which tumor stem cells acquire resistance to chemotherapy and radiotherapy. These mechanisms include elevated levels of ATP-binding cassette (ABC) transporters, exceptional DNA repair capacity, overexpression of antiapoptotic proteins, and enhanced protection against reactive oxygen species (ROS) [9–16]. Furthermore, tumor stem cells maintain their stem cell properties by overactivating various transcription factors and signaling pathways, such as Wnt/β-catenin, Notch, Hedgehog, and Nanog. This self-renewal and proliferation of tumor stem cells contribute to tumor progression [7, 17–22]. Typically, CSCs are implicated in treatment failure, relapse, and metastasis due to their stem-like characteristics. Consequently, targeting CSCs represents an effective anticancer therapeutic strategy. Currently, several strategies have been developed to eradicate cancer stem cells (CSCs), including targeting CSC surface markers, inhibiting ATP-binding cassette transporters, intervening in critical signaling pathways, and inducing differentiation of tumor stem cells [23]. Additionally, phototherapy (both photodynamic therapy and photothermal therapy), as a noninvasive and spatiotemporally controlled approach, achieves indiscriminate killing of tumor cells and tumor stem cells by converting light into heat or reactive oxygen species. Consequently, it serves as an alternative method for eliminating tumor stem cells [24, 25]. Furthermore, recent literature suggests that tumor stem cells exhibit significant sensitivity to ferroptosis, making ferroptosis another viable approach for eradicating tumor stem cells [26].

Recently, there has been significant interest in eradicating cancer stem cells (CSCs) using nanotherapeutic strategies. Conventional chemotherapy drugs and anti-CSC agents often suffer from poor physicochemical properties, low bioavailability, limited specificity, and substantial toxicity [27–30]. In contrast, nanomedicines exhibit superior pharmacokinetic characteristics, enhancing therapeutic efficacy while minimizing adverse effects. Notably, nanoparticles with specific sizes (100–200 nm) and surface modifications tend to accumulate actively or passively within tumor tissues, significantly improving treatment outcomes [28, 31, 32]. In light of this, researchers have developed a range of nanocarrier systems, including polymer-based nanoparticles, inorganic nanoparticles, and liposomal nanoparticles [33, 34]. One strategy involves coupling nanomedicines with ligands specific to biomarkers expressed on CSCs, achieving targeted delivery to these stem cells [35, 36]. Another approach utilizes nanosystems to co-deliver anti-CSC agents and chemotherapy drugs, simultaneously targeting CSCs and a large population of cancer cells [10]. This dual action not only eliminates primary tumors but also mitigates issues related to drug resistance, relapse, and metastasis. In this article, we review the application of nanomedicines in antitumor stem cells by introducing some representative studies (Fig. 12.1).

2 Nanomedicine Strategy Against CSC

2.1 Nanomedicines Targeting Surface Biomarkers of CSCs

The concept of ligand-receptor interactions has been applied to the design of targeted nanomedicines [37]. Driven by advances in cellular and molecular biology, researchers have identified various biomarkers on the cell membrane surface of cancer stem cells (CSCs). While these CSC biomarkers vary depending on the tissue of origin, CD44, CD133, EpCAM, CD20, and aldehyde dehydrogenase (ALDH) are well-known markers that play crucial roles in CSC maintenance [36, 38–40]. These biomarkers provide an opportunity for nanomedicines to actively target tumor stem cells.

CD44, a transmembrane glycoprotein receptor specific to CSCs, is associated with high proliferation and antiapoptotic properties [41]. Hyaluronic acid (HA), a naturally occurring polysaccharide with excellent biocompatibility and biodegradability, exhibits particularly strong binding affinity for CD44 [42, 43]. Consequently, HA serves as an attractive ligand for actively targeting tumor stem cells. Li et al. developed a targeted nanoparticle system based on functionalized cationic albumin with hyaluronic acid (HA), specifically designed to target tumor stem cells overexpressing CD44 [44]. All-trans retinoic acid (ATRA), an effective differentiation

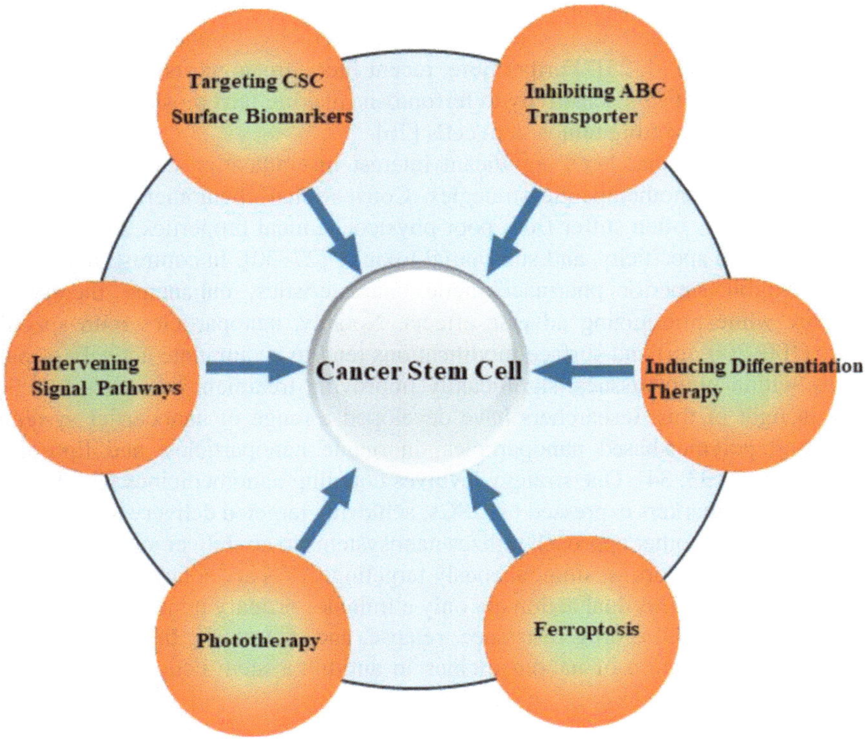

Fig. 12.1 Possible nanomedicine-based strategies that can eliminate CSCs

agent, was encapsulated within the nanoparticles (referred to as HA-eNPs) at an impressive encapsulation rate of 93% and a drug loading efficiency of 8.37%. Morphological analysis revealed that the prepared nanoparticles exhibited a spherical shape with uniform size, surrounded by a coating composed of HA. Due to the surface conjugation of HA, the resulting HA-eNPs demonstrated high affinity and specific binding to CD44-rich B16F10 cells. In vivo imaging confirmed the accumulation of HA-eNPs in tumor-bearing mouse lungs. Furthermore, cell toxicity assays demonstrated that ATRA-loaded HA-eNPs exhibited superior cytotoxicity against B16F10 cells compared to free drug or HA-eNPs, highlighting their excellent targeting performance. Additionally, in vitro experiments showed that HA-eNPs/ATRA treatment significantly reduced the side population of B16F10 cells. In an in vivo experimental model, HA-eNPs/ATRA significantly inhibited tumor growth in lung metastasis mice. To specifically target prostate cancer stem cells (CSCs) and cancer cells overexpressing the CD44 receptor, Turkekul et al. engineered quercetin-loaded hyaluronic acid-modified lipid nanoparticles (referred to as LP-Quer-HA) [45]. In cell experiments, treatment with free quercetin or an equivalent concentration of LP-Quer-HA for 72 h resulted in distinct outcomes. While free quercetin only modestly reduced the viability of PC3 cells by 16%, LP-Quer-HA significantly increased cell death rates to 60%. Compared to free quercetin, LP-Quer-HA upregulated E-cadherin expression while concurrently suppressing cell migration and reducing fibronectin, N-cadherin, and MMP9 expression. Tumor spheroids experiment further demonstrated that LP-Quer-HA treatment significantly decreased the number of CD44-positive cells in PC3 cells, along with reduced expression of CD44, Oct3/4, and Wnt. Additionally, LP-Quer-HA inhibited p-ERK expression while enhancing p38/MAPK and NF-κB protein expression. Similar significant effects were observed in LNCaP cells. Utilizing HA-modified lipid nanoparticles as a quercetin delivery system enhanced its efficacy at lower concentrations, effectively impeding prostate cancer cell proliferation and migration.

To develop a nanocarrier capable of specific targeting of cancer stem cells (CSCs) and efficient drug delivery, Jia et al. first employed an integrated microfluidic device to screen for the natural compound isoliquiritigenin (ISL), which inhibits the phenotypes of lung cancer stem cells (Fig. 12.2) [46]. Subsequently, they designed a hollow structure composed of manganese dioxide (H-MnO_2) to deliver the CSC-targeting compound ISL and the photothermal agent indocyanine green (ICG). Finally, by modifying this structure with hyaluronic acid, they created H-MnO_2@(ICG + ISL)@HA, achieving highly efficient targeting of CD44-overexpressing tumor stem cells. The prepared nanocomposite material exhibited excellent chemotherapeutic and photothermal capabilities, along with near-infrared-triggered drug release, demonstrating outstanding CSC-killing effects and tumor inhibition. Furthermore, upon tail vein injection of the nanocomposite material, near-infrared-triggered phototherapy, combined with potent chemotherapy, effectively suppressed tumor growth without significant side effects.

CD133, another transmembrane glycoprotein, is associated with tumor invasiveness and has been utilized as a recognition factor for cancer stem cells (CSCs) in various cancers, including prostate cancer, neuroblastoma, renal carcinoma, hepatocellular carcinoma, and lung cancer [47]. As shown in Fig. 12.3, Crous et al. constructed a nanobiocomplex termed NBC by utilizing anti-CD133 antibodies (Abs), the photosensitizer AlPcS4Cl, and gold nanoparticles (AuNPs) [48]. The photosensitizer AlPcS4Cl was conjugated to the AuNP surface via ligand absorption and exchange methods, while the anti-CD133 antibodies were covalently linked to the AuNP surface for active tumor biomarker recognition. Subcellular localization results demonstrated that due to the antibody's affinity for the CD133 biomarker on lung tumor stem cells, NBC nanosystems significantly improved the intracellular accumulation of AlPcS4Cl within both the cytoplasm and the cell nucleus compared to free AlPcS4Cl. Cells cultured with NBC exhibited markedly enhanced cytotoxicity and apoptosis effects, along with a sharp reduction in the proliferation and activity of lung tumor stem cells compared to other groups. These findings suggest that photosensitizers, when actively targeted using AuNP-Ab conjugates, can enhance drug delivery within lung CSCs in vitro, thereby augmenting the overall effectiveness of photodynamic therapy (PDT) against lung cancer cells.

EpCAM is typically present in untransformed tissues, particularly in epithelial cells. However, it is overexpressed under pathological conditions such as tumors. As an oncogenic factor, EpCAM promotes tumor cell proliferation and migration,

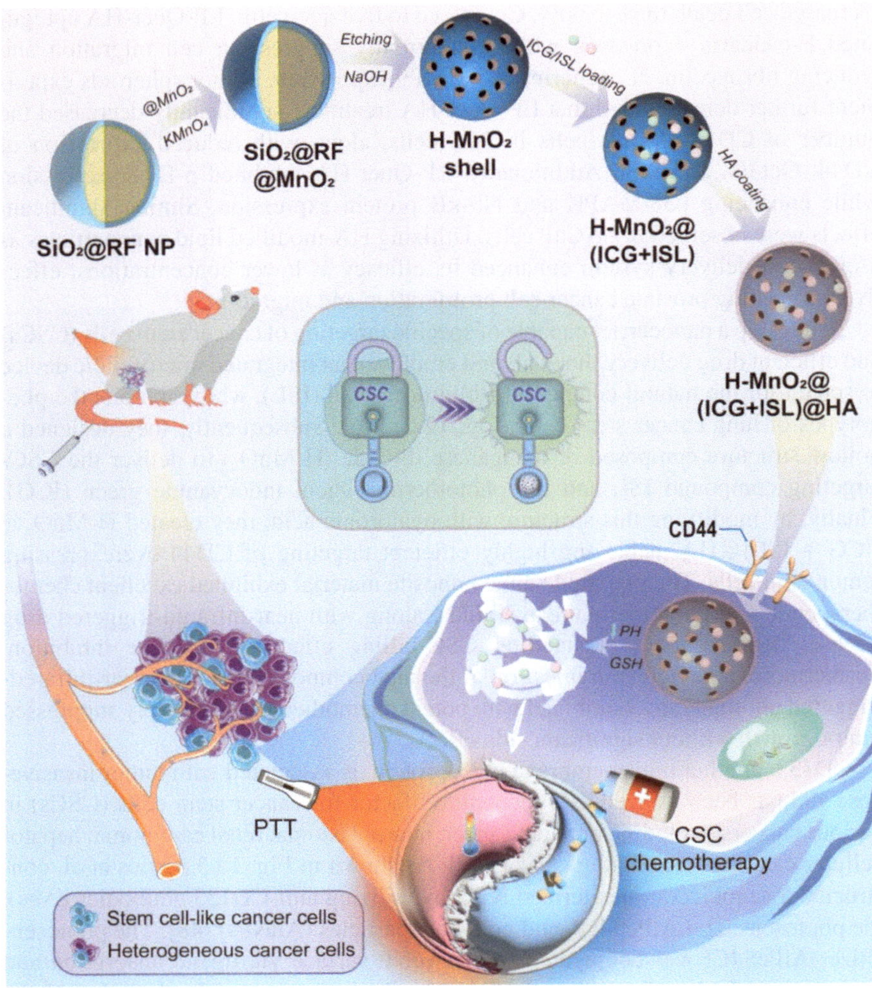

Fig. 12.2 Schematic illustration of designing CSC-targeting H-MnO_2@(ICG + ISL)@HA nanoparticles (NPs) for combined photothermal therapy and chemotherapy. (Adapted with permission from Ref. [46]. Copyright 2023 American Chemical Society)

making it a potential candidate target for cancer therapy [49]. Locatelli et al. developed a drug delivery carrier capable of simultaneously transporting two different therapeutic agents, namely, adriamycin (Adr) and gold nanorods (GNRs) [50]. By modifying the carrier's surface with EpCAM antibodies, they achieved target delivery to the lesion site. When decorated with EpCAM antibodies, Adr/GNRs@PMs-antiEpCAM specifically targeted cancer stem cells (CSCs) and accumulated at the tumor site, subsequently leading to complete CSC eradication upon laser irradiation. Additionally, photoacoustic imaging effectively detected enhanced GNR aggregation in the tumor region, enabling both diagnosis and treatment.

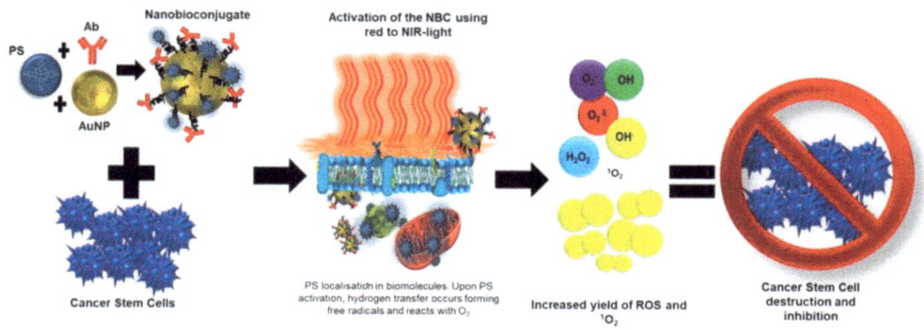

Fig. 12.3 Schematic illustration of designing gold nanoparticle-antibody-mediated drug delivery for photodynamic therapy. (Adopted from Crous et al. (2020) [48] under the Creative Commons License CY 4.0, doi.org/10.3390/ijms21113742)

Previous literature demonstrated that CD20+ melanoma CSCs could proliferate rapidly to form tumor spheres and could differentiate into different cell types [51]. To target melanoma tumor stem cells overexpressing CD20, the authors designed salt-mycin-loaded lipid-polymer nanoparticles functionalized with anti-CD20 aptamers (CD20-SA-NPs) [52]. The CD20-SA-NPs were prepared using a one-step nanoprecipitation method, where sulfur-modified anti-CD20 aptamers were conjugated to SA-NPs via a maleimide-thiol reaction. These CD20-SA-NPs exhibited a particle size of 96.3 nm, encapsulation efficiency exceeding 60%, and sustained drug release capability. Notably, CD20+ melanoma CSCs exhibited significantly higher uptake of CD20-SA-NPs compared to SA-NPs and salt-mycin, resulting in enhanced cytotoxic effects in vitro. Furthermore, CD20-SA-NPs demonstrated optimal therapeutic efficacy in inhibiting tumor sphere formation and increasing the proportion of CD20+ cells in melanoma cell lines, further validating their targeting efficiency. In a melanoma xenograft mouse model, CD20-SA-NPs outperformed SA-NPs and salt-mycin in suppressing melanoma growth. Thus, CD20 represents a superior target for drug delivery to melanoma CSCs.

2.2 Nanomedicines Based on ATP-Binding Cassette Transporter Inhibitors

Cancer stem cells (CSCs) exhibit increased expression of ATP-binding cassette (ABC) multidrug efflux pumps, including ABCB1 (P-glycoprotein or P-gp) and MDR1. These ABC transporters, membrane-bound proteins, mediate ATP-dependent efflux of various molecules, including cytotoxic anticancer drugs [53–55]. Consequently, intracellular drug concentrations decrease, leading to chemotherapy resistance. Given that CSCs possess multiple mechanisms of chemoresistance, monotherapy or combination chemotherapy often fails. Therefore, the combined use of chemotherapy drugs and ABC multidrug efflux pump inhibitors holds promise as a prospective approach to address CSC-mediated drug resistance.

P-glycoprotein (P-gp), a typical ATP-binding cassette (ABC) efflux transporter, exhibits broad specificity. High expression of P-gp has been identified as a key feature of the CSC population [53]. Notably, chemotherapy can further elevate P-gp expression, exacerbating chemotherapy resistance. Haag's group developed a nanomedicine with CSC-specific targeting, simultaneously delivering the chemotherapeutic drug DOX and the P-gp inhibitor tariquidar (Tar) to enhance chemotherapy and overcome multidrug resistance in breast CSCs (Fig. 12.4) [56]. The $mSiO_2$-dPG nanocarrier demonstrates high loading capacity, excellent pH-responsive behavior, and good biocompatibility. Under the combined action of CSC-specific targeting (HA) and P-gp inhibitor Tar, $mSiO_2$-dPG nanomedicine significantly enhances DOX delivery in a three-dimensional breast CSC-resistant model, thereby augmenting chemotherapy efficacy against breast CSCs. Additionally, the expression of stemness-related genes and tumor sphere formation capacity is markedly suppressed, highlighting the excellent ability of this nanosystem to overcome MDR in breast CSCs. This strategy holds promise for effectively eliminating CSCs and overcoming MDR in breast CSCs.

From human MDA-MB-231 breast cancer cells, Sun et al. isolated a subpopulation of cells with the characteristic (CD44+/CD24-), designated as 231-CSC [57]. The result demonstrated that 231-CSC exhibits enhanced proliferation, migration,

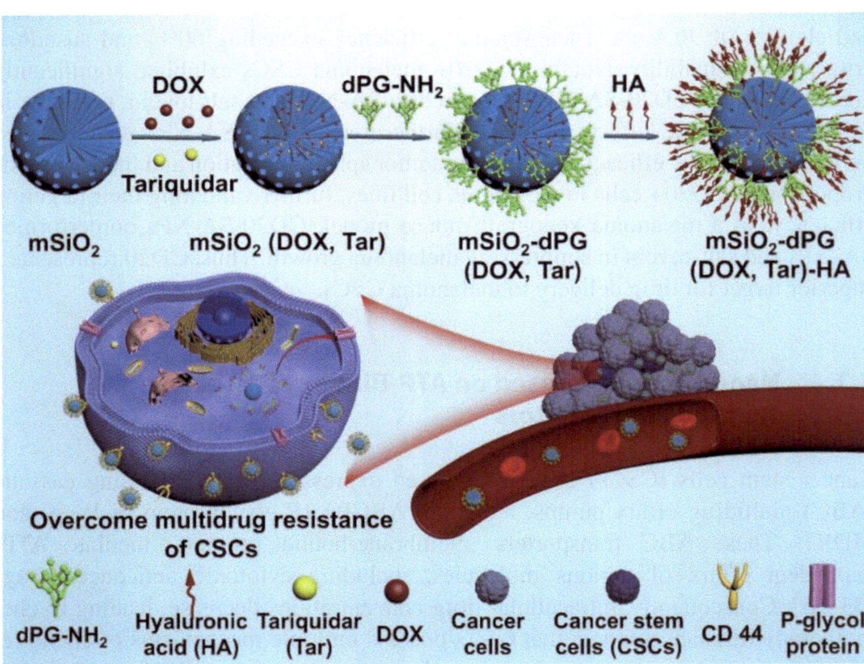

Fig. 12.4 Synthetic scheme of dendritic polyglycerol-conjugated mesoporous silica-based target nanocarriers co-delivery of doxorubicin and tariquidar to overcome multidrug resistance in breast cancer stem cells. (Adapted with permission from ref. [56] Copyright 2021, Elsevier)

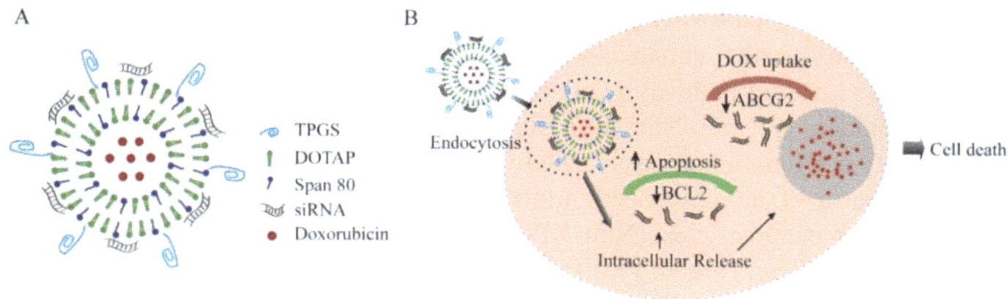

Fig. 12.5 Co-delivery in one niosomal delivery system of DOX simultaneously with siRNAs targeted to ABCG2 and BCL2 mRNA. (Adapted with permission [57]. Copyright 2015, Elsevier)

tumor formation, and chemoresistance. To address the issue of multidrug resistance in CSCs, they designed a niosome-based co-delivery system using nonionic surfactants. This system simultaneously delivers siRNAs targeting the ABC transporter protein (ABCG2) and the antiapoptotic defense gene (BCL2), along with doxorubicin (DOX) (Fig. 12.5). By downregulating the drug-resistant gene ABCG2, the nanosystem sensitizes CSCs to DOX while also inducing apoptosis by reducing BCL2 expression. Compared to the control group, the co-delivery system successfully transported siRNA and DOX to both the cytoplasm and nucleus of CSCs, resulting in a 60% reduction in ABCG2 mRNA and a 65% reduction in BCL2 mRNA. Additionally, Western blot experiments confirmed a significant decrease in the corresponding protein expression. Furthermore, this nanosystem significantly lowered the IC_{50} of DOX in CSCs.

To reverse BCSC drug resistance and enhance the anticancer efficacy of ATRA, Li et al. developed a pH-sensitive nanosystem loaded with the hydrophobic MDR reverser schisandrin B (SchB) [58]. Initially, ATRA was grafted onto pH-sensitive polymer PBAE via esterification to obtain ATRA-g-PBAE (AP), which served as the carrier for encapsulating the active molecule SchB to create SchB/AP NPs. Cell experiments demonstrated that this nanosystem achieves efficient targeted delivery of SchB, reversing multidrug resistance by inhibiting ATPase activity and affecting mitochondrial function related to P-gp energy supply and transport. The delivery of SchB provides a novel therapeutic option for overcoming ATRA resistance in BCSCs.

2.3 Nanomedicines Based on CSCs Signaling Pathways

During tumorigenesis or the formation of cancer stem cells (CSCs), numerous signaling pathways—typically involved in maintaining cancer stem cell characteristics—become aberrantly activated or suppressed. These pathways can influence the expression of specific genes within CSCs, including those related to growth factor secretion, cytokine regulation, pro-apoptotic and antiapoptotic proteins, metastasis, and proliferation. Notable pathways affecting CSCs include Wnt, Hedgehog, Notch, BMP, Bmi, PI3K/Akt, integrins, and STAT [59, 60]. Targeting and disrupting these abnormal signaling pathways offer promising therapeutic approaches specifically tailored to CSCs, further enhancing cancer treatment outcomes.

The Wnt signaling pathway regulates cell proliferation, differentiation, apoptosis, and stemness [61, 62]. Numerous studies have found that tumor cells with high Wnt activity are functionally associated with cancer stem cells (CSCs). For example, Mohapatra et al. developed nimbolide (Nim)-encapsulated poly(lactic-co-glycolic acid) nanoparticles (Nim-NPs) and investigated their anti-CSC effects in vitro (tumorspheres) and in vivo (zebrafish and xenograft mouse models) [63]. In vitro results demonstrated that Nim-NPs significantly reduced several intrinsic features of breast cancer stem cells (BCSCs) through epigenetic reprogramming via the DNMTs-SFRP1-Wnt/β-catenin signaling axis, compared to free Nim. Moreover, in two distinct preclinical models, Nim NPs enhanced antitumor and antimetastatic efficacy compared to free Nim. Additionally, tumor tissue slice experiments further validated that Nim achieves its antitumor effects through epigenetic reprogramming via the DNMTs-SFRP1-Wnt/β-catenin signaling axis.

The overactivation of the Hedgehog pathway is associated with tumor initiation and the maintenance of cancer stem cells (CSCs) [39]. To sensitize cancer cells to chemotherapy, Lu et al. developed a nanostructured metal-organic framework (EA-Cu) by coordinating the Hedgehog pathway inhibitor ellagic acid (EA) with Cu^{2+} (Fig. 12.6) [64]. They then loaded doxorubicin (DOX) onto this framework and further modified it with chondroitin sulfate (CS), which has targeted effects on tumor stem cells. Notably, EA suppresses stemness maintenance by inhibiting the Hedgehog pathway, while Cu^{2+} disrupts mitochondrial metabolism, further reducing the stem cell characteristics of tumor cells and enhancing DOX-mediated chemotherapy efficacy. Additionally, EA synergistically induces mitochondrial dysfunction and copper-mediated cell death, effectively lowering ATP levels and subsequently inhibiting the activity of P-glycoprotein (P-gp), thus reducing drug efflux and sensitizing DOX-mediated chemotherapy. Furthermore, the modified CS imparts tumor stem cell-specific targeting properties to the nano-drug, while the EA-Cu component within the nano-drug exhibits pH/glutathione (GSH)-responsive release behavior. This system demonstrates excellent antitumor effects by inducing copper-mediated cell death and significantly inhibiting cancer cell stemness, holding great potential for overcoming chemotherapy resistance in cancer treatment.

Research findings indicate that Notch is associated with invasive and metastatic growth as well as treatment resistance. Moreover, Notch signaling transduction is crucial for the maintenance and self-renewal of cancer stem cells (CSCs) [65]. Mamaeva et al. identified specific phenotypic features of breast CSCs and utilized these characteristics to design nanoparticles for efficient delivery of stem cell inhibitors [66]. Studies demonstrate that breast CSCs exhibit enhanced glucose uptake compared to normal breast cancer cells. Therefore, by functionalizing mesoporous silica nanoparticles (MSN) with a glucose-like moiety, researchers effectively delivered Notch signaling inhibitors (γ-secretase inhibitors) to CSCs. Data indicate that the prepared functionalized particles are efficiently internalized by tumor stem cells, and both in vitro and in vivo experiments confirm the reduction of the CSC

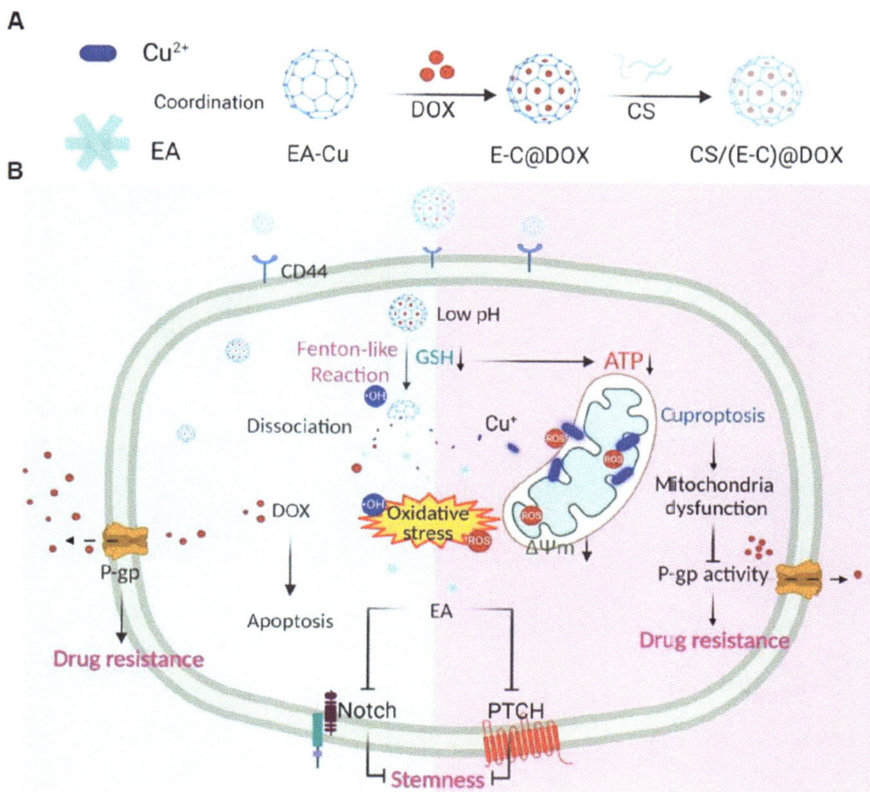

Fig. 12.6 (**a**) Preparation of CS/NPs. (**b**) The underlying mechanism of CS/NPs for therapeutic resistance. (Adapted with permission from Ref. [64] Copyright 2023, Wiley-VCH)

population and inhibition of tumor growth in the chick embryo chorioallantoic and xenograft mouse models.

Furthermore, previous literature has investigated other signaling pathways, such as those related to mammalian target of rapamycin (mTOR), Janus kinases/signal transducers and activators of transcription (JAK/STAT), and focal adhesion kinase (FAK), to target the signaling pathways of tumor stem cells [22]. As displayed in Fig. 12.7, Wang et al. reported a strategy for STAT3 degradation based on a nanoscale proteolysis-targeting chimera (PROTAC), which effectively chemically reprograms CSCs associated with hepatocellular carcinoma (HCC) [67]. This approach not only inhibits CSC growth but also triggers an anti-HCC immune response. The synthesis of the PROTAC prodrug involves linking the STAT3 binding domain (inS3) with the E3 ligand caged with thioketal (VL-TK) using a polyethylene glycol linker (OEG) with adjustable length and flexibility. Encapsulation of this complex with cRGD-modified cationic liposomes achieves CSC-targeted delivery while promoting lysosomal escape. Activation of CSCs by elevated ROS levels leads to effective STAT3 degradation through chemical reprogramming, impairing their stemness features

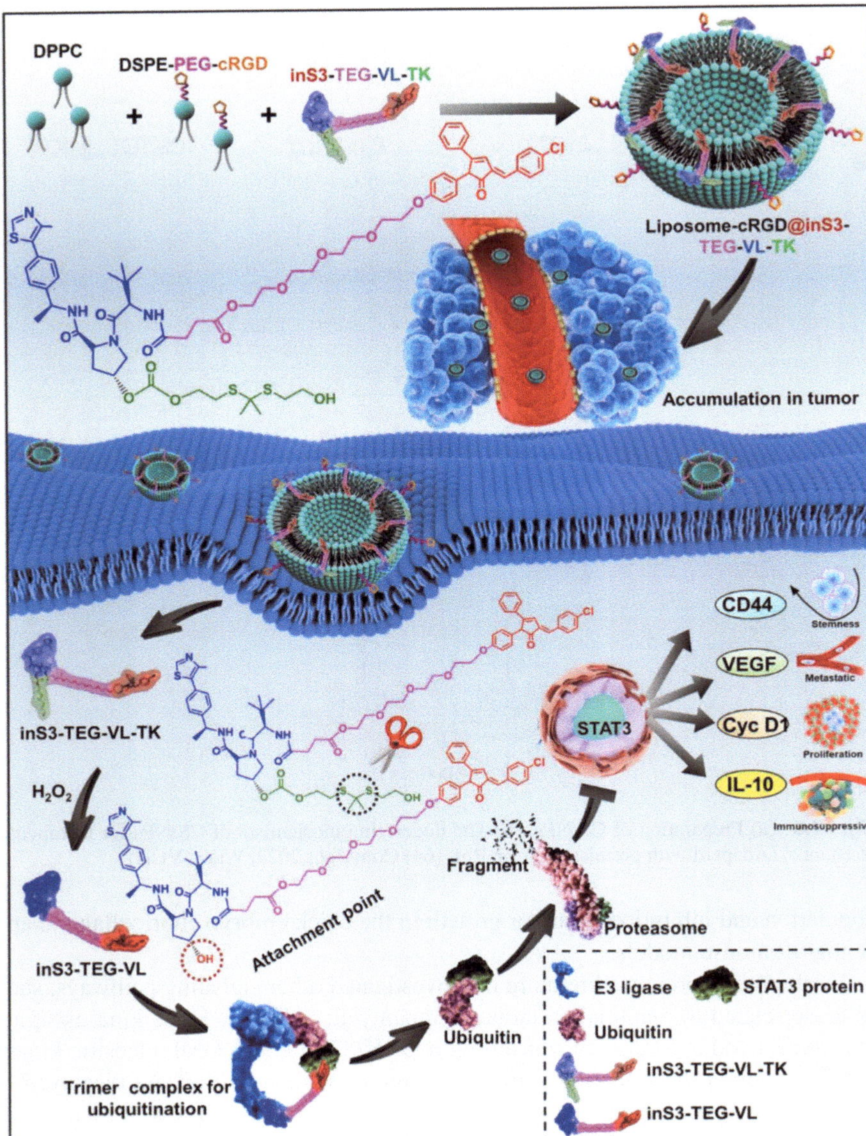

Fig. 12.7 Construction of the STAT3-degrading liposomal PROTAC prodrug and the anti-CSC mechanisms for improving HCC treatment efficacy. (Adapted with permission from Ref. [67]. Copyright 2024 American Chemical Society)

and reshaping the immunosuppressive tumor microenvironment (TME) into an immune-supportive state, thereby enhancing anti-HCC immunity.

In another effort, Wang et al. have fabricated a novel nanomedicine that enhances chemotherapy efficacy by simultaneously targeting cancer stem cells (CSCs) within

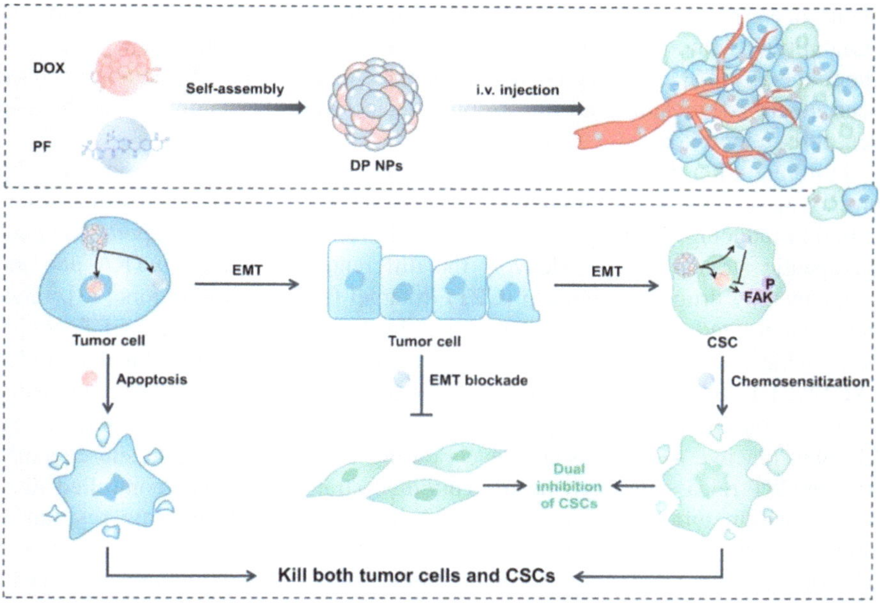

Fig. 12.8 Schematic illustration for the DP NPs assembled by DOX and PF to prevent tumor progression and metastasis via simultaneously targeting CSCs and preventing tumor cells from dedifferentiating into CSCs. (Adapted with permission [68]. Copyright 2023, Elsevier)

tumor tissue and preventing tumor cells from converting into CSCs [68]. This nanomedicine self-assembles through the combination of doxorubicin (DOX) and focal adhesion kinase (FAK) inhibitor PF-573228 (PF) (Fig. 12.8). After accumulating at the tumor site, DOX induces tumor cell death, while PF overcomes CSC resistance to DOX by inhibiting the phosphorylation of tyrosine 397 on FAK (FAK pY397) in CSCs. Additionally, PF suppresses the acquisition of CSC phenotypes in tumor cells by inhibiting the epithelial-mesenchymal transition (EMT) process. As a result, this designed nanomedicine exhibits potent capabilities in addressing tumor progression and metastasis.

2.4 Nanomedicines to Induce Differentiation of Cancer Stem Cells

The hallmark of malignant tumors is the loss of cellular differentiation and uncontrolled proliferation. To address this challenge, differentiation therapy aims to restore endogenous differentiation programs and induce maturation in tumor cells. When this occurs, tumor cells differentiate into more benign phenotypes, resulting in reduced growth and metastatic potential, while differentiation markers increase [69]. However, chemotherapy drugs primarily target rapidly dividing tumor cells and do not affect slowly proliferating cancer stem cells (CSCs), often leading to recurrence. The emergence of dedifferentiated CSCs in solid tumors contributes to variability in tumor growth, dissemination, and recurrence after radiotherapy or chemotherapy [70]. Several potential drugs inducing cancer cell differentiation have been identified, including all-trans retinoic acid (ATRA), vitamin D, and histone deacetylase inhibitors (HDACi) [71].

Differentiation therapy represents a clinically effective anticancer approach by promoting malignant cell differentiation into more mature phenotypes using differentiation-inducing agents [72]. As shown in Fig. 12.9, Mo's group reported a nanotherapeutic strategy that co-delivers the differentiation inducer ATRA and the chemotherapy drug camptothecin (CPT) to target CSCs within tumors [73]. First, they synthesized nitroimidazole-modified hyaluronic acid–oxalate–CPT conjugates, which were then self-assembled to prepare ATRA/CPT-loaded nanoparticles (ATRA/CPT-NPs). In this nanosystem, all-trans retinoic acid is released under hypoxic conditions, leading to CSC differentiation within the hypoxic niche. Elevated reactive oxygen species levels during CSC differentiation subsequently trigger CPT release and cell death. This dual strategy achieves controlled drug release within tumor stem cells and reduces stemness-related drug resistance, enhancing chemotherapy response. In a mouse model of breast tumors, treatment with the nano-drug ATRA/CPT-NPs suppressed tumor growth and prevented postoperative tumor recurrence and metastasis.

Xie et al. designed a small-molecule self-assembling nanotherapeutic composed of differentiation inducers, photothermal agents, and chemotherapy drugs for combined treatment [74]. As illustrated in Fig. 12.10, the hydrophilic chemotherapy drug irinotecan (IRI) was conjugated to the hydrophobic differentiation inducer all-trans retinoic acid (ATRA) via an ester bond, resulting in the amphiphilic IRI-ATRA conjugate. Its amphiphilic properties enabled self-assembly into nanoparticles in the aqueous phase. Additionally, the photothermal agent IR825 was encapsulated through hydrophobic/hydrogen bonding interactions, yielding the IR825@IRI-ATRA nanotherapeutic. IR825@IRI-ATRA exhibited dual-responsive behavior under esterase and/or acidic conditions. Furthermore, IR825 served as a fluorescence/photoacoustic (PA) imaging agent, enabling tracking of drug distribution. To reduce blood clearance during systemic circulation and enhance accumulation in tumor tissue, DSPE-PEG-RGD was incorporated into IR825@IRI-ATRA, creating a tumor-targeted nanotherapeutic, IR825@IRI-ATRA/RGD. Within the tumor microenvironment, IR825@IRI-ATRA/RGD rapidly disintegrated in response to low pH and overexpressed esterases, leading to the release of ATRA and IRI. These agents, respectively, induced differentiation of cancer stem cells (CSCs) and killed ordinary tumor cells. Moreover, IR825 not only facilitated photothermal therapy (PTT)-mediated cell death in both CSCs and regular tumor cells but also enabled in vivo fluorescence/PA imaging. Importantly, systemic administration of IR825@IRI-ATRA/RGD demonstrated excellent antitumor and antimetastatic efficacy in tumor-bearing mice.

Triple-negative breast cancer (TNBC) is highly invasive and originates from a small subset of TNBC stem cells (TNBCSCs). These TNBCSCs are responsible for chemotherapy resistance, tumor metastasis, and recurrence. To explore novel

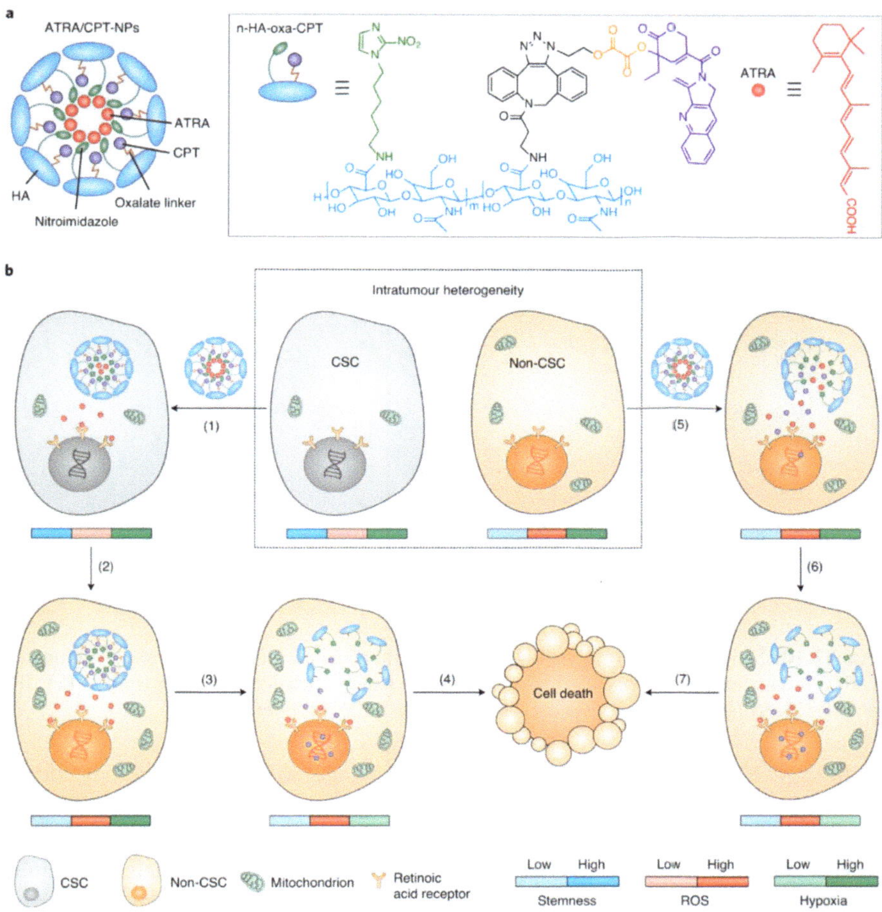

Fig. 12.9 Schematic of a convergent nanotherapeutic strategy to overcome intratumour heterogeneity-imposed therapeutic obstacle for enhanced anticancer efficacy. (Adapted with permission [73]. Copyright 2021, Springer Nature)

strategies for eradicating TNBCSCs, Wu et al. reported a self-assembling nanoprodrug based on disulfide bonds. This nanoprodrug enables co-delivery of ferroptosis-inducing agent (Fc), differentiation inducer (ATRA), and chemotherapy drug (DOX), simultaneously targeting TNBCSCs and TNBC (Fig. 12.11) [75]. Initially, a molecular prodrug (Fc-SS-ATRA) was synthesized by linking ferrocene and ATRA through disulfide bonds, which self-assembles into a nano-sized drug carrier. Furthermore, this carrier can encapsulate DOX, resulting in the nano-drug DOX@Fc-SS-ATRA. To enhance circulation time in the bloodstream and improve tumor stem cell targeting efficiency, DSPE-PEG-HA was incorporated into the nanosystem, yielding an actively targeted nano-drug, HA/DOX@Fc-SS-ATRA. In this system, the differentiation inducer transforms TNBCSCs into normal TNBC cells,

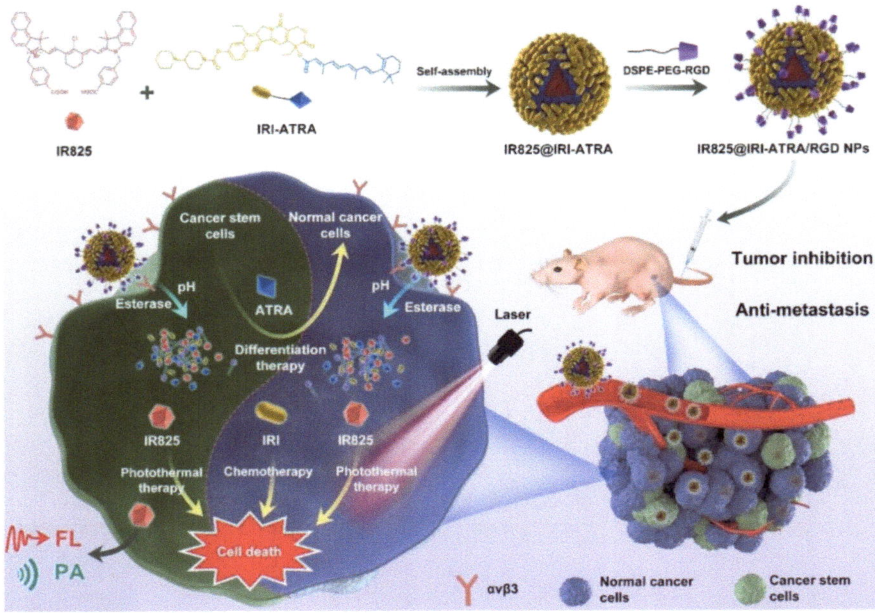

Fig. 12.10 Schematic illustration of the tumor inhibition and anti-metastasis through eliminating both CSCs and non-CSCs enabled by the co-delivery nanoparticle system comprising ATRA, IRI, and IR825. (Adapted with permission [74]. Copyright 2022, Elsevier)

indirectly eliminating TNBCSCs when combined with chemotherapy drugs. Notably, iron death therapy differs fundamentally from apoptosis-induced differentiation or chemotherapy-induced cell death, as the former leads to cell death in both TNBCSCs and normal TNBC cells. In various TNBC mouse models, this multifunctional nanoprodrug significantly enhances antitumor efficacy and effectively suppresses tumor metastasis. By controlling drug release and reducing stemness-related drug resistance, this comprehensive strategy improves the chemosensitivity of TNBC treatment.

2.5 Phototherapy-Based Nanomedicine

Phototherapy (comprising photothermal and photodynamic therapy) is currently an efficient modality for treating tumors. Compared to traditional treatment approaches, phototherapy demonstrates improved specificity, reduced likelihood of drug resistance, and lower cellular toxicity. Furthermore, phototherapy lacks selectivity in its cytotoxic effects on normal tumor cells and cancer stem cells (CSCs) [76, 77]. As a result, it can simultaneously target both normal tumor cells and CSCs, addressing the challenges posed by drug resistance and metastasis associated with CSCs.

In 2023, Yao et al. developed a novel iridium nanosheet, Ir Nts-Ab, which combines CD44 antibody modification with enzymatic activity (Fig. 12.12) [78]. Ir

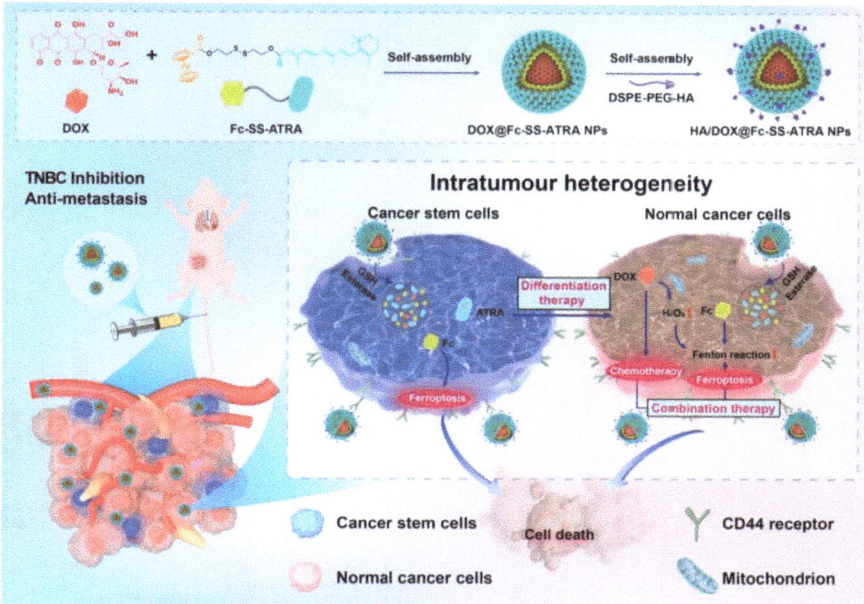

Fig. 12.11 Schematic illustration for self-assembly nano-prodrug that can co-deliver ferroptosis drug (Fc), differentiation-inducing agent (ATRA), and chemotherapeutics (DOX) to achieve the combination therapy of ferroptosis–differentiation–chemotherapy on tumor inhibition and anti-metastasis of CSC-enriched TNBC mouse model. (Adapted with permission from Ref. [75] Copyright 2023, Wiley-VCH)

Nts-Ab effectively eradicates CSCs for cancer treatment. Leveraging the active targeting properties of the antibody, Ir Nts-Ab efficiently accumulates in tumor tissue. Additionally, it exhibits superoxide dismutase (SOD)-like and catalase (CAT) enzyme activities, enabling the elimination of excess reactive oxygen species and promoting oxygen generation. This alleviates hypoxia and inflammation within the tumor microenvironment (TME), thereby reducing the proportion of CSCs and inhibiting metastasis. Moreover, Ir Nts-Ab specifically targets both CSCs and normal cancer cells through near-infrared II (NIR-II) photothermal therapy (PTT). Additionally, photoacoustic imaging facilitates monitoring of drug accumulation and TME improvement within the tumor tissue. Importantly, Ir Nts-Ab demonstrates excellent biocompatibility, making it suitable for biomedical applications.

Additionally, Ning et al. designed a type I aggregation-induced emission (AIE) photosensitizer-loaded biomimetic mesoporous organosilica nanosystem (PMT) for precise eradication of cancer stem cells (CSCs) and prevention of tumor recurrence after radiotherapy (Fig. 12.13) [79]. This nanosystem comprises a type I AIE photosensitizer (TBP-2) loaded into mesoporous organosilica nanoparticles (MON) coated with platelet membranes. The efficient targeting of CD44 protein by platelet membranes enables PMT to selectively target CSCs. Upon cellular uptake, overexpressed glutathione triggers MON degradation and TBP-2 release. Type I

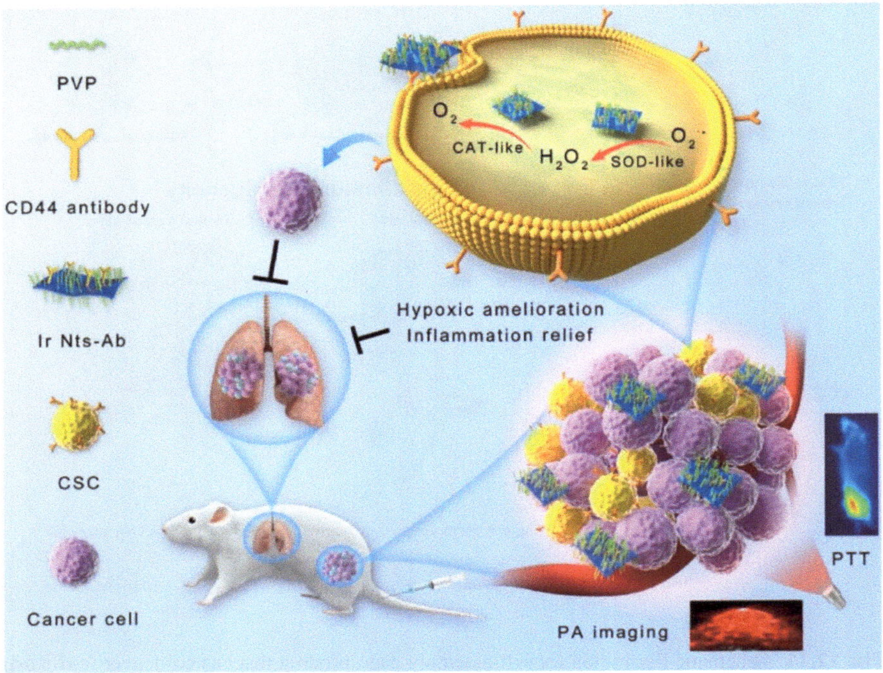

Fig. 12.12 Schematic of the multifunctional Ir Nts-Ab with CAT- and SOD-like activity for the eradication of CSCs by targeting TME regulation and NIR-II PTT. (Adapted with permission [78]. Copyright 2023, Elsevier)

photodynamic therapy (PDT), activated by exposure to white light, generates abundant hydroxyl radicals, promoting CSC death. In vivo experimental results demonstrate that PMT treatment specifically eliminates CSCs, with no observed tumor recurrence when combined with radiotherapy. However, tumor recurrence is observed in mice receiving radiotherapy alone. Furthermore, PMT treatment significantly reduces the expression of CSC markers in tumor slices. Thus, this nanosystem based on photodynamic therapy efficiently removes CSCs, preventing tumor recurrence.

2.6 Ferroptosis-Based Nanomedicines

Currently, the primary strategy for eliminating cancer stem cells (CSCs) relies on apoptosis-dependent approaches. However, CSCs often resist cell apoptosis through dormancy, enhanced DNA damage repair, and interactions with their supportive microenvironment, leading to treatment inefficacy [80, 81]. Iron-dependent non-apoptotic cell death, known as ferroptosis, has emerged as a promising avenue in anti-CSC therapeutic strategies. Notably, recent research highlights CSCs' strong

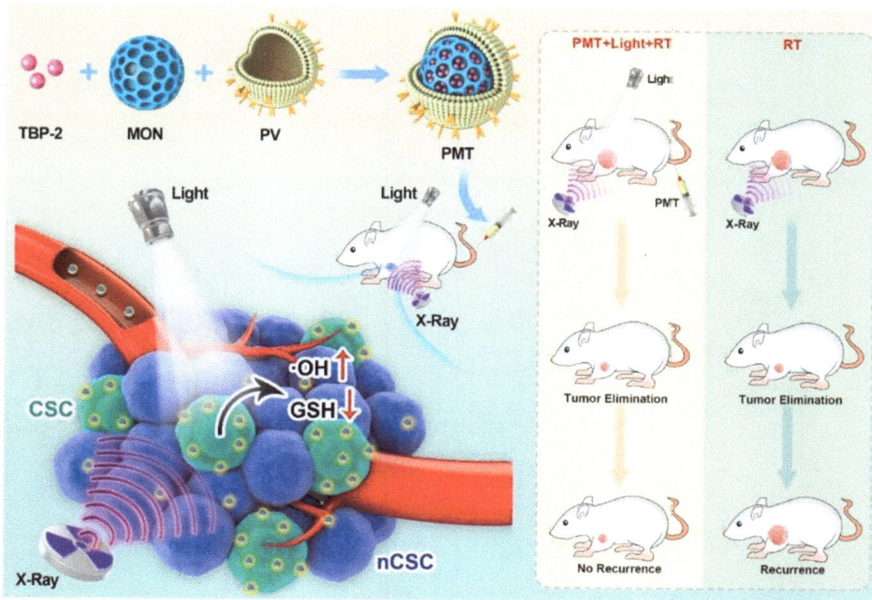

Fig. 12.13 Schematic illustration of a type I AIE photosensitizer-loaded biomimetic nanosystem for precise depletion of cancer stem cells and prevention of tumor recurrence after radiotherapy. (Adapted with permission [79]. Copyright 2023, Elsevier)

dependence on iron and their sensitivity to ferroptosis, suggesting that targeting iron metabolism within CSCs could enhance cancer treatment efficacy [82].

As shown in Fig. 12.14, Wu et al. designed a nanosystem amplifying iron death by targeting hyaluronic acid (HA)-coated siProminin2-loaded FeOOH nanoparticles (FeOOH/siPROM2@HA) [83]. This nanosystem releases Fe^{3+} and siProminin2 in the acidic tumor microenvironment. Released siProminin2 inhibits Fe^{3+} efflux, while Fe^{3+} undergoes an oxidation–reduction reaction with endogenous glutathione, generating Fe^{2+}. Elevated lipid peroxidation (LPO) subsequently triggers iron death via a Fenton reaction. Both in vitro and in vivo experiments demonstrate that this system significantly suppresses tumor growth by downregulating glutathione peroxidase 4 (GPX4). Thus, this work showcases a "three-pronged" strategy, simultaneously increasing intracellular iron levels, inhibiting iron efflux, and depleting GSH, to accelerate iron death in breast cancer stem cells (BCSCs) and provide a potential approach for targeted CSC cancer therapy.

Recently, Feng et al. have developed a nebulized nanocatalytic drug strategy in which iron-based nanoparticles effectively target the cancer stem cell (CSC) niche by modulating tumor-associated macrophages (TAMs), thereby triggering CSC ferroptosis (Fig. 12.15) [84]. Following aerosol administration, the iron-based nanoparticles can efficiently accumulate in pulmonary cancer microlesions through TAM targeting mediated by dextran. Consequently, internalized TAMs play a dominant role in elevating iron levels around the CSC niche and disrupting redox balance by

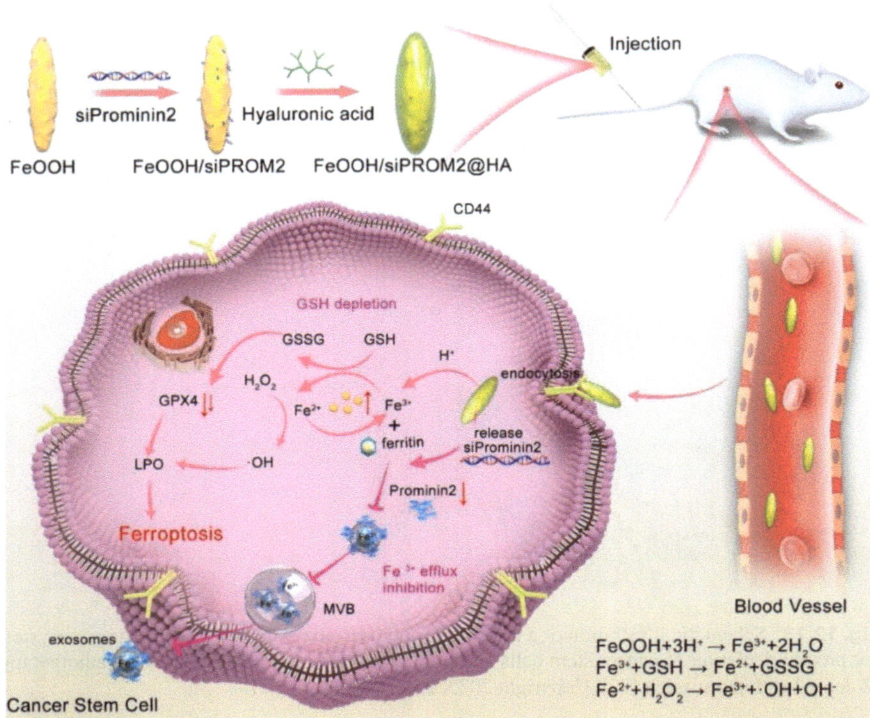

Fig. 12.14 Schematic diagram of FeOOH/siPROM2@HA-induced ferroptosis in breast cancer stem cells. (Adapted with permission [83]. Copyright 2023, Elsevier)

downregulating glucose-6-phosphate metabolism in lysosomes and iron metabolism. The altered microenvironment enhances CSC sensitivity to ferroptosis. In an orthotopic mouse model of lung cancer, aerosol-induced CSC ferroptosis significantly reduces CSC stemness and effectively suppresses early-stage lung cancer progression.

3 Summary and Outlook

Cancer stem cells (CSCs) constitute the fundamental cause of cancer recurrence and adverse prognosis, possessing heightened self-renewal and differentiation capabilities. Additionally, they exhibit resistance to conventional cancer treatments, including chemotherapy and radiotherapy. Consequently, identifying and selectively targeting CSCs has emerged as a novel therapeutic strategy for complete tumor eradication. Among the most compelling carriers, nanodrug delivery systems (NDDS) efficiently transport drugs to tumor sites. This article provides an overview of CSC-targeting strategies based on nanotechnology, including the following:

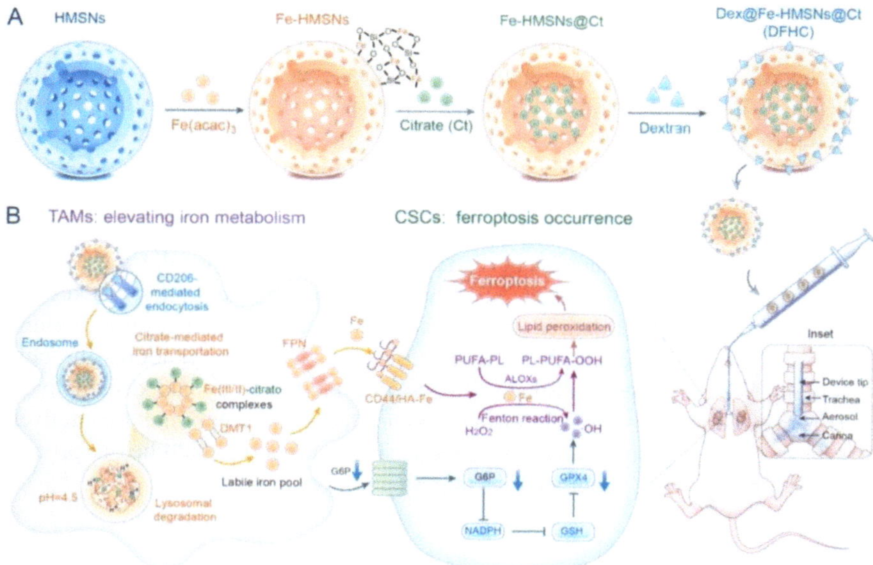

Fig. 12.15 DFHC-regulated TAMs induce ferroptosis of CSCs in the nebulized therapy of early orthotopic lung cancer. (Adapted with permission from Ref. [84]. Copyright 2023 American Chemical Society)

1. Nanodrugs Targeting CSC Surface Biomarkers: These nanodrugs specifically recognize CSCs through surface biomarkers, enhancing precision in therapy.
2. Nanodrugs Based on ATP-Binding Cassette Transporter Inhibitors: Leveraging ATP-binding cassette transporters, these nanodrugs modulate drug release in response to specific tumor microenvironments and intracellular signals.
3. Nanodrugs Targeting CSC Signaling Pathways: These nanodrugs interfere with CSC-specific signaling pathways, disrupting their survival mechanisms.
4. Nanodrugs Inducing CSC Differentiation: By promoting CSC differentiation, these nanodrugs reduce stemness and enhance vulnerability to treatment.
5. Phototherapy-Based Nanodrugs: Nanodrugs utilizing phototherapy mechanisms selectively damage CSCs upon light activation.
6. Nanodrugs Harnessing Ferroptosis: These nanodrugs exploit ferroptosis, a form of regulated cell death, to target CSCs effectively.

Notably, collaborative delivery systems using nanocarriers have been developed. These systems respond to specific tumor microenvironments and intracellular signals, allowing on-demand drug release. This approach enhances synergistic effects in eliminating drug-resistant CSCs and abundant tumor cells. Despite the achievements in overcoming resistance to stemness-related chemotherapy through nanotherapeutic strategies, challenges persist.

1. Diverse CSC Resistance Mechanisms: Due to the heterogeneity and diversity of CSCs, multiple resistance mechanisms are mediated by these cells. Additionally, pathways maintaining CSC stemness are numerous. Thus, solely inhibiting a single resistance mechanism or signaling pathway may yield insignificant results and prove insufficient against various tumor types. Combining CSC elimination strategies with conventional chemotherapy and radiation therapy is essential for achieving eradication.
2. Holistic Approach to Nanotherapy: While current research primarily focuses on whether constructed nanosystems can eliminate CSCs, the overall treatment process should not be overlooked. Effective nanotherapy necessitates five key features: prolonged circulation, tumor accumulation, deep penetration, cellular internalization, and controlled drug release [85]. In summary, addressing challenging CSCs involves multifaceted strategies, facilitated by nanodrug delivery. However, achieving clinically ideal outcomes requires exceptional drug delivery efficiency and rigorous safety evaluations. These aspects should be our primary focus for advancing nanotherapies targeting tumor stem cells.

References

1. Seoane J. Cancer: division hierarchy leads to cell heterogeneity. Nature. 2017;549(7671):164–6.
2. Shibue T, Weinberg RA. EMT, CSCs, and drug resistance: the mechanistic link and clinical implications. Nat Rev Clin Oncol. 2017;14(10):611–29.
3. Huang T, Song X, Xu D, Tiek D, Goenka A, Wu B, Sastry N, Hu B, Cheng SY. Stem cell programs in cancer initiation, progression, and therapy resistance. Theranostics. 2020;10(19):8721–43.
4. Rosen JM, Jordan CT. The increasing complexity of the cancer stem cell paradigm. Science. 2009;324(5935):1670–3.
5. Shrestha S, Banstola A, Jeong JH, Seo JH, Yook S. Targeting cancer stem cells: therapeutic and diagnostic strategies by the virtue of nanoparticles. J Control Release. 2022;348:518–36.
6. Bayik D, Lathia JD. Cancer stem cell-immune cell crosstalk in tumour progression. Nat Rev Cancer. 2021;21(8):526–36.
7. Lytle NK, Barber AG, Reya T. Stem cell fate in cancer growth, progression and therapy resistance. Nat Rev Cancer. 2018;18(11):669–80.
8. Reya T, Morrison SJ, Clarke MF, Weissman IL. Stem cells, cancer, and cancer stem cells. Nature. 2001;414(6859):105–11.
9. Fuchs E. The tortoise and the hair: slow-cycling cells in the stem cell race. Cell. 2009;137(5):811–9.
10. Liu H, Liu M, Zhao Y, Mo R. Nanomedicine strategies to counteract cancer stemness and chemoresistance. Explor Target Antitumor Ther. 2023;4(4):630–56.
11. Nathansen J, Meyer F, Muller L, Schmitz M, Borgmann K, Dubrovska A. Beyond the double-strand breaks: the role of DNA repair proteins in cancer stem-cell regulation. Cancers (Basel). 2021;13(19):4818.
12. Dzobo K, Senthebane DA, Ganz C, Thomford NE, Wonkam A, Dandara C. Advances in therapeutic targeting of cancer stem cells within the tumor microenvironment: an updated review. Cells. 2020;9(8):1896.
13. Toh TB, Lim JJ, Chow EK. Epigenetics in cancer stem cells. Mol Cancer. 2017;16(1):29.
14. Peitzsch C, Tyutyunnykova A, Pantel K, Dubrovska A. Cancer stem cells: the root of tumor recurrence and metastases. Semin Cancer Biol. 2017;44:10–24.

15. Zhou HM, Zhang JG, Zhang X, Li Q. Targeting cancer stem cells for reversing therapy resistance: mechanism, signaling, and prospective agents. Signal Transduct Target Ther. 2021;6(1):62.
16. Lah TT, Novak M, Breznik B. Brain malignancies: glioblastoma and brain metastases. Semin Cancer Biol. 2020;60:262–73.
17. Takebe N, Miele L, Harris PJ, Jeong W, Bando H, Kahn M, Yang SX, Ivy SP. Targeting notch, hedgehog, and Wnt pathways in cancer stem cells: clinical update. Nat Rev Clin Oncol. 2015;12(8):445–64.
18. Willert K, Brown JD, Danenberg E, Duncan AW, Weissman IL, Reya T, Yates JR 3rd, Nusse R. Wnt proteins are lipid-modified and can act as stem cell growth factors. Nature. 2003;423(6938):448–52.
19. Kelly KF, Ng DY, Jayakumaran G, Wood GA, Koide H, Doble BW. Beta-catenin enhances Oct-4 activity and reinforces pluripotency through a TCF-independent mechanism. Cell Stem Cell. 2011;8(2):214–27.
20. ten Berge D, Kurek D, Blauwkamp T, Koole W, Maas A, Eroglu E, Siu RK, Nusse R. Embryonic stem cells require Wnt proteins to prevent differentiation to epiblast stem cells. Nat Cell Biol. 2011;13(9):1070–5.
21. Jamieson CH, Ailles LE, Dylla SJ, Muijtjens M, Jones C, Zehnder JL, Gotlib J, Li K, Manz MG, Keating A, Sawyers CL, Weissman IL. Granulocyte-macrophage progenitors as candidate leukemic stem cells in blast-crisis CML. N Engl J Med. 2004;351(7):657–67.
22. Annett S, Robson T. Targeting cancer stem cells in the clinic: current status and perspectives. Pharmacol Ther. 2018;187:13–30.
23. Barbato L, Bocchetti M, Di Biase A, Regad T. Cancer stem cells and targeting strategies. Cells. 2019;8(8):926.
24. Suo X, Zhang J, Zhang Y, Liang XJ, Zhang J, Liu D. A nano-based thermotherapy for cancer stem cell-targeted therapy. J Mater Chem B. 2020;8(18):3985–4001.
25. Wang B, Wang W, Xu Y, Liu R, Li R, Yang P, Zhao C, Dai Z, Wang Y. Manipulating redox homeostasis of cancer stem cells overcome chemotherapeutic resistance through photoactivatable biomimetic nanodiscs. Small. 2024;20(28):e2308539.
26. Cosialls E, El Hage R, Dos Santos L, Gong C, Mehrpour M, Hamai A. Ferroptosis: cancer stem cells rely on iron until "to die for" it. Cells. 2021;10(11):2981.
27. Shen S, Xia JX, Wang J. Nanomedicine-mediated cancer stem cell therapy. Biomaterials. 2016;74:1–18.
28. Bozzato E, Bastiancich C, Preat V. Nanomedicine: a useful tool against glioma stem cells. Cancers (Basel). 2020;13(1):9.
29. Singh P, Pandit S, Mokkapati V, Garg A, Ravikumar V, Mijakovic I. Gold nanoparticles in diagnostics and therapeutics for human cancer. Int J Mol Sci. 2018;19(7):1979.
30. Oliveira BSA, de Assis ACC, Souza NM, Ferreira LFR, Soriano RN, Bilal M, Iqbal HMN. Nanotherapeutic approach to tackle chemotherapeutic resistance of cancer stem cells. Life Sci. 2021;279:119667.
31. Nunes T, Hamdan D, Leboeuf C, El Bouchtaoui M, Gapihan G, Nguyen TT, Meles S, Angeli E, Ratajczak P, Lu H, Di Benedetto M, Bousquet G, Janin A. Targeting cancer stem cells to overcome Chemoresistance. Int J Mol Sci. 2018;19(12):4036.
32. Tong R, Cheng J. Anticancer polymeric nanomedicines. Polym Rev. 2007;47(3):345–81.
33. Soltani F, Ramezani M, Amel Farzad S, Mokhtarzadeh A, Hashemi M. Comparison study of the effect of alkyl-modified and unmodified PAMAM and PPI dendrimers on solubility and antitumor activity of crocetin. Artif. Cells Nanomed. Biotechnol. 2017;45(7):1356–62.
34. Aghebati-Maleki A, Dolati S, Ahmadi M, Baghbanzhadeh A, Asadi M, Fotouhi A, Yousefi M, Aghebati-Maleki L. Nanoparticles and cancer therapy: perspectives for application of nanoparticles in the treatment of cancers. J Cell Physiol. 2020;235(3):1962–72.
35. Durymanov M, Reineke J. Non-viral delivery of nucleic acids: insight into mechanisms of overcoming intracellular barriers. Front Pharmacol. 2018;9:971.

36. Mokhtarzadeh A, Hassanpour S, Vahid ZF, Hejazi M, Hashemi M, Ranjbari J, Tabarzad M, Noorolyai S, de la Guardia M. Nano-delivery system targeting to cancer stem cell cluster of differentiation biomarkers. J Control Release. 2017;266:166–86.
37. Mi P, Cabral H, Kataoka K. Ligand-installed nanocarriers toward precision therapy. Adv Mater. 2020;32(13):e1902604.
38. D'Andrea V, Panarese A, Tonda M, Biffoni M, Monti M. Cancer stem cells as functional biomarkers. Cancer Biomark. 2017;20(3):231–4.
39. Najafi M, Farhood B, Mortezaee K. Cancer stem cells (CSCs) in cancer progression and therapy. J Cell Physiol. 2019;234(6):8381–95.
40. Prasad S, Ramachandran S, Gupta N, Kaushik I, Srivastava SK. Cancer cells stemness: a doorstep to targeted therapy. Biochimica et Biophysica Acta (BBA). 2020;1866(4):165424.
41. Marquardt S, Solanki M, Spitschak A, Vera J, Putzer BM. Emerging functional markers for cancer stem cell-based therapies: understanding signaling networks for targeting metastasis. Semin Cancer Biol. 2018;53:90–109.
42. Choi KY, Han HS, Lee ES, Shin JM, Almquist BD, Lee DS, Park JH. Hyaluronic acid-based activatable nanomaterials for stimuli-responsive imaging and therapeutics: beyond CD44-mediated drug delivery. Adv Mater. 2019;31(34):e1803549.
43. Dzobo K, Sinkala M. Cancer stem cell marker CD44 plays multiple key roles in human cancers: immune suppression/evasion, drug resistance, epithelial-mesenchymal transition, and metastasis. OMICS. 2021;25(5):313–32.
44. Li Y, Shi S, Ming Y, Wang L, Li C, Luo M, Li Z, Li B, Chen J. Specific cancer stem cell-therapy by albumin nanoparticles functionalized with CD44-mediated targeting. J Nanobiotechnol. 2018;16(1):99.
45. Turkekul K, Erdogan S. Potent suppression of prostate cancer cell growth and eradication of cancer stem cells by CD44-targeted nanoliposome-quercetin nanoparticles. J Cancer Prev. 2023;28(4):160–74.
46. Jia Y, Sun J, Yang J, Chen C, Zhang Z, Yang K, Shen P, Qu S, He B, Song Y, Han X. Tumor microenvironment-responsive Nanoherb delivery system for synergistically inhibition of cancer stem cells. ACS Appl Mater Interfaces. 2023;15(13):16329–42.
47. Brugnoli F, Grassilli S, Al-Qassab Y, Capitani S, Bertagnolo V. CD133 in breast cancer cells: more than a stem cell marker. J Oncol. 2019;2019:7512632.
48. Crous A, Abrahamse H. Effective gold nanoparticle-antibody-mediated drug delivery for photodynamic therapy of lung cancer stem cells. Int J Mol Sci. 2020;21(11):3742.
49. Walcher L, Kistenmacher AK, Suo H, Kitte R, Dluczek S, Strauss A, Blaudszun AR, Yevsa T, Fricke S, Kossatz-Boehlert U. Cancer stem cells-origins and biomarkers: perspectives for targeted personalized therapies. Front Immunol. 2020;11:1280.
50. Locatelli E, Li Y, Monaco I, Guo W, Maturi M, Menichetti L, Armanetti P, Martin RC, Comes Franchini M. A novel theranostic gold nanorods- and Adriamycin-loaded micelle for EpCAM targeting, laser ablation, and photoacoustic imaging of cancer stem cells in hepatocellular carcinoma. Int J Nanomedicine. 2019;14:1877–92.
51. Fang D, Nguyen TK, Leishear K, Finko R, Kulp AN, Hotz S, Van Belle PA, Xu X, Elder DE, Herlyn M. A tumorigenic subpopulation with stem cell properties in melanomas. Cancer Res. 2005;65(20):9328–37.
52. Zeng YB, Yu ZC, He YN, Zhang T, Du LB, Dong YM, Chen HW, Zhang YY, Wang WQ. Salinomycin-loaded lipid-polymer nanoparticles with anti-CD20 aptamers selectively suppress human CD20+ melanoma stem cells. Acta Pharmacol Sin. 2018;39(2):261–74.
53. Robey RW, Pluchino KM, Hall MD, Fojo AT, Bates SE, Gottesman MM. Revisiting the role of ABC transporters in multidrug-resistant cancer. Nat Rev Cancer. 2018;18(7):452–64.
54. Xiao Q, Zhou Y, Lauschke VM. Ethnogeographic and inter-individual variability of human ABC transporters. Hum Genet. 2020;139(5):623–46.
55. Baniebrahimi G, Mir F, Khanmohammadi R. Cancer stem cells and oral cancer: insights into molecular mechanisms and therapeutic approaches. Cancer Cell Int. 2020;20:113.

56. Pan Y, Zhou S, Li Y, Parshad B, Li W, Haag R. Novel dendritic polyglycerol-conjugated, mesoporous silica-based targeting nanocarriers for co-delivery of doxorubicin and tariquidar to overcome multidrug resistance in breast cancer stem cells. J Control Release. 2021;330:1106–17.
57. Sun M, Yang C, Zheng J, Wang M, Chen M, Le DQS, Kjems J, Bunger CE. Enhanced efficacy of chemotherapy for breast cancer stem cells by simultaneous suppression of multidrug resistance and antiapoptotic cellular defense. Acta Biomater. 2015;28:171–82.
58. Li W, Fu Y, Sun J, Gong H, Yan R, Wang Y. Construction and in vitro evaluation of pH-sensitive nanoparticles to reverse drug resistance of breast cancer stem cells. Discov Oncol. 2024;15(1):21.
59. An SM, Ding Q, Zhang J, Xie J, Li L. Targeting stem cell signaling pathways for drug discovery: advances in the Notch and Wnt pathways. Sci China Life Sci. 2014;57(6):575–80.
60. Krishnamurthy N, Kurzrock R. Targeting the Wnt/beta-catenin pathway in cancer: update on effectors and inhibitors. Cancer Treat Rev. 2018;62:50–60.
61. Zhan T, Rindtorff N, Boutros M. Wnt signaling in cancer. Oncogene. 2017;36(11):1461–73.
62. Ruan Y, Ogana H, Gang E, Kim HN, Kim YM. Wnt signaling in the tumor microenvironment. Adv Exp Med Biol. 2021;1270:107–21.
63. Mohapatra P, Madhulika S, Behera S, Singh P, Sa P, Prasad P, Swain RK, Sahoo SK. Nimbolide-based nanomedicine inhibits breast cancer stem-like cells by epigenetic reprogramming of DNMTs-SFRP1-Wnt/beta-catenin signaling axis. Mol Ther Nucleic Acids. 2023;34:102031.
64. Lu S, Tian H, Li B, Li L, Jiang H, Gao Y, Zheng L, Huang C, Zhou Y, Du Z, Xu J. An ellagic acid coordinated copper-based Nanoplatform for efficiently overcoming cancer chemoresistance by cuproptosis and synergistic inhibition of cancer cell stemness. Small. 2024;20(17):e2309215.
65. Krishna BM, Jana S, Singhal J, Horne D, Awasthi S, Salgia R, Singhal SS. Notch signaling in breast cancer: from pathway analysis to therapy. Cancer Lett. 2019;461:123–31.
66. Mamaeva V, Niemi R, Beck M, Ozliseli E, Desai D, Landor S, Gronroos T, Kronqvist P, Pettersen IK, McCormack E, Rosenholm JM, Linden M, Sahlgren C. Inhibiting notch activity in breast cancer stem cells by glucose functionalized nanoparticles carrying gamma-secretase inhibitors. Mol Ther. 2016;24(5):926–36.
67. Wang X, Zhao Y, Li X, Zhang Q, He J, Liu Y, Li M, Luo Z. Liposomal STAT3-degrading PROTAC prodrugs promote anti-hepatocellular carcinoma immunity via chemically reprogramming cancer stem cells. Nano Lett. 2024;24(16):4858–68.
68. Wang W, Zhang D, Jiang Z, Zhang X, Jiang Y, Luan Y. A nanodrug-enabled chemosensitization of cancer stem cells against tumor progression and metastasis. Chem Eng J. 2023;477:147121.
69. Yoldi G, Pellegrini P, Trinidad EM, Cordero A, Gomez-Miragaya J, Serra-Musach J, Dougall WC, Munoz P, Pujana MA, Planelles L, Gonzalez-Suarez E. RANK signaling blockade reduces breast cancer recurrence by inducing tumor cell differentiation. Cancer Res. 2016;76(19):5857–69.
70. Gupta PB, Fillmore CM, Jiang G, Shapira SD, Tao K, Kuperwasser C, Lander ES. Stochastic state transitions give rise to phenotypic equilibrium in populations of cancer cells. Cell. 2011;146(4):633–44.
71. Doustmihan A, Fathi M, Mazloomi M, Salemi A, Hamblin MR, Jahanban-Esfahlan R. Molecular targets, therapeutic agents and multitasking nanoparticles to deal with cancer stem cells: a narrative review. J Control Release. 2023;363:57–83.
72. de The H. Differentiation therapy revisited. Nat Rev Cancer. 2018;18(2):117–27.
73. Shen S, Xu X, Lin S, Zhang Y, Liu H, Zhang C, Mo R. A nanotherapeutic strategy to overcome chemotherapeutic resistance of cancer stem-like cells. Nat Nanotechnol. 2021;16(1):104–13.
74. Xie X, Jiang K, Li B, Hou S, Tang H, Shao B, Ping Y, Zhang Q. A small-molecule self-assembled nanodrug for combination therapy of photothermal-differentiation-chemotherapy of breast cancer stem cells. Biomaterials. 2022;286:121598.
75. Wu C, Zhang F, Li B, Li Z, Xie X, Huang Y, Yao Z, Chen Y, Ping Y, Pan W. A self-assembly Nano-prodrug for combination therapy in triple-negative breast cancer stem cells. Small. 2023;19(41):e2301600.
76. Li L, Ni R, Zheng D, Chen L. Eradicating the tumor "seeds": nanomedicines-based therapies against cancer stem cells. Daru. 2023;31(1):83–94.

77. Datta NR, Puric E, Klingbiel D, Gomez S, Bodis S. Hyperthermia and radiation therapy in locoregional recurrent breast cancers: a systematic review and meta-analysis. Int J Radiat Oncol Biol Phys. 2016;94(5):1073–87.
78. Yao Y, Zhao Z, He J, Ali B, Wang M, Liao F, Zhuang J, Zheng Y, Guo W, Zhang DY. Iridium nanozyme-mediated photoacoustic imaging-guided NIR-II photothermal therapy and tumor microenvironment regulation for targeted eradication of cancer stem cells. Acta Biomater. 2023;172:369–81.
79. Ning S, Zhang T, Lyu M, Lam JWY, Zhu D, Huang Q, Tang BZ. A type I AIE photosensitiser-loaded biomimetic nanosystem allowing precise depletion of cancer stem cells and prevention of cancer recurrence after radiotherapy. Biomaterials. 2023;295:122034.
80. Bai X, Ni J, Beretov J, Graham P, Li Y. Cancer stem cell in breast cancer therapeutic resistance. Cancer Treat Rev. 2018;69:152–63.
81. Hua Z, White J, Zhou J. Cancer stem cells in TNBC. Semin Cancer Biol. 2022;82:26–34.
82. Lei G, Zhuang L, Gan B. Targeting ferroptosis as a vulnerability in cancer. Nat Rev Cancer. 2022;22(7):381–96.
83. Wu K, Zhang W, Chen H, Wu J, Wang X, Yang X, Liang XJ, Zhang J, Liu D. An iron oxyhydroxide-based nanosystem sensitizes ferroptosis by a "three-pronged" strategy in breast cancer stem cells. Acta Biomater. 2023;160:281–96.
84. Feng Q, Fang W, Guo Y, Hu P, Shi J. Nebulized therapy of early orthotopic lung cancer by iron-based nanoparticles: macrophage-regulated ferroptosis of cancer stem cells. J Am Chem Soc. 2023;145(44):24153–65.
85. Li Z, Xiao C, Yong T, Li Z, Gan L, Yang X. Influence of nanomedicine mechanical properties on tumor targeting delivery. Chem Soc Rev. 2020;49(8):2273–90.

Nanotechnology in Traditional Chinese Medicine for Cancer Treatment

13

Chao Hong, Jinshuai Lan, Zhe Li, Ruifeng Zeng, Tong Zhang, and Yue Ding

Abstract

The rapid advancement of nanotechnology, when integrated with traditional Chinese medicine (TCM), is not only expanding the application scope of TCM but also driving its modernization. This chapter focuses on exploring the potential of combining nano-delivery systems with traditional Chinese herbal medicines and TCM theories for tumor treatment. First, it will elaborate on the antitumor properties of TCM and the underlying mechanisms. Second, it emphasizes the synergistic effects between nano-delivery systems and traditional herbal remedies, with a particular focus on their role in boosting therapeutic efficacy. Furthermore, this chapter discusses how such integration can facilitate the modernization of herbal medicines and enhance their global accessibility. Finally, it analyzes the current challenges and future research directions in this emerging interdisciplinary field. Overall, the application of nanotechnology in TCM offers

Chao Hong and Jinshuai Lan have equally contributed to this work.

C. Hong
School of Pharmacy, Shanghai University of Traditional Chinese Medicine, Shanghai, P. R. China

J. Lan · Z. Li · R. Zeng · T. Zhang
School of Pharmacy, Shanghai University of Traditional Chinese Medicine, Shanghai, P. R. China

State Key Laboratory of Integration and Innovation of Classic Formula and Modern Chinese Medicine, Shanghai University of Traditional Chinese Medicine, Shanghai, P. R. China

Y. Ding (✉)
School of Pharmacy, Shanghai University of Traditional Chinese Medicine, Shanghai, P. R. China

National Innovation Platform for Medical Industry-Education Integration, Shanghai University of Traditional Chinese Medicine, Shanghai, P. R. China

M. Chu (ed.), *Modern Nanomaterials and Nanotechnology for Cancer Therapy*,
https://doi.org/10.1007/978-981-95-6849-9_13

a promising strategy for advancing tumor therapy and reinforcing the relevance of TCM in contemporary medical practice.

Keywords

Nanotechnology · Traditional Chinese Medicine · Tumor · Drug Delivery

1 Introduction

Cancer constitutes a major global public health challenge, with an estimated 20 million new cases and 9.7 million deaths ascribed to this disease in 2022 [1]. The pathogenesis of cancer is complex, and its prevention and management are complicated by a broad range of risk factors. Currently, the primary treatment approach involves surgical resection of the tumor, which is often supplemented with chemotherapy and/or radiotherapy—given that surgical intervention alone frequently fails to eradicate all cancerous tissue [2, 3]. However, conventional chemotherapy agents may exert varying degrees of toxicity to healthy tissues, resulting in intolerable side effects in many patients [4].

Traditional Chinese medicine (TCM) serves as a promising alternative for the treatment of various cancers. Compared with synthetic anticancer agents, these plant-derived formulations exert multiple synergistic or additive mechanisms of action, induce fewer and milder side effects, and have a lower tendency to trigger drug resistance [5]. Oral administration of TCM has been shown to improve chemotherapy outcomes in patients with non-small cell lung cancer (NSCLC), particularly regarding quality of life, anemia, and neutropenia [6]. Specifically, PHY906—a formulation consisting of *Scutellaria baicalensis*, *Ziziphus jujuba*, *Paeonia lactiflora*, and *Glycyrrhiza glabra*—was reported to alleviate the adverse effects of capecitabine in a phase I/II clinical study involving patients with advanced hepatocellular carcinoma [7]. Moreover, recent progress in chemical separation and structural analysis has enabled the identification of several key antitumor components in TCM [8]. For instance, *PNAS* reported that epigallocatechin gallate (EGCG), which targets the laminin receptor (Lam 67R), exhibits promising efficacy in prostate cancer treatment [9]; *Signal transduction and targeted therapy* demonstrated that gambogic amide suppresses glioma by targeting WDR1-dependent cytoskeleton remodeling [10]; and *Journal of Ethnopharmacology* described that a compound formulation containing andrographolide, wogonin, and oroxylin A (derived from *Andrographis paniculata* (Burm.f.) Nees) inhibits hepatocellular carcinoma cell proliferation by regulating the cell cycle, AGE-RAGE, and leptin signaling pathways [11]. Notably, another key advantage of natural small molecules from Chinese medicinal herbs is their improved biological safety when used as primary drugs, excipients, or carriers for other anticancer agents—this is attributed to their lower inherent cytotoxicity and reduced tissue accumulation [12]. However, the clinical application of TCM and its bioactive compounds in cancer treatment is hindered by several challenges, including limited administration routes, poor solubility, instability, short biological half-lives, suboptimal targeting efficiency, high susceptibility to metabolic degradation, and rapid clearance from the body [13, 14].

The application of nanoscale delivery systems is a cutting-edge area in pharmaceutical science, enabling precise control over drug administration and therapeutic outcomes. These systems employ nanoparticles or nanocarriers to leverage their unique nanoscale properties for the encapsulation, delivery, and release of therapeutic agents with high precision. The multifunctional features of these nano-delivery platforms—such as enhanced drug solubility, prolonged circulation time, and targeted delivery—have made them a core focus of innovation in the pharmaceutical field [15]. Over the past decade, the use of these systems has advanced the development of Chinese herbal formulations and sparked interest in the synergistic potential between TCM and modern medicine [16]. By improving the solubility and stability of bioactive compounds, extending their blood circulation time, and enabling targeted delivery to specific tissues or organs, nano-delivery systems open new avenues for the integration of nanotechnology with TCM, laying the foundation for collaborative advancements in both fields [17–19].

This chapter focuses on the convergence of these two distinct yet complementary fields, highlighting the potential synergistic effects and applications arising from the integration of nano-delivery systems into TCM. It will first introduce the antitumor properties and associated mechanisms of TCM. Furthermore, we will discuss the potential of nano-delivery systems in enhancing the therapeutic effects of traditional herbal medicines and elaborate on the integration of fundamental TCM theories with nano-delivery systems. Finally, we will address the challenges and future directions in this integrated field. Overall, this chapter aims to provide valuable insights into the fusion of nanotechnology and TCM, offering innovative perspectives for developing therapeutic strategies that can improve patient outcomes and promote human health.

2 Traditional Chinese Medicine and Its Mechanisms of Antitumor Activity

TCM has a long history of tumor treatment. Guided by the "concept of holism and treatment based on syndrome differentiation," it adopts approaches such as clearing heat and detoxifying, strengthening the body's vital qi to expel pathogenic factors, promoting blood circulation to resolve blood stasis, eliminating dampness and phlegm, and using toxic agents for detoxification. These approaches play an irreplaceable role in the comprehensive treatment of tumors. TCM-based antitumor therapy offers several advantages: minimal toxic side effects, improved overall bodily functions, and a significant ability to enhance efficacy and reduce toxicity when combined with chemotherapeutic agents. Additionally, it improves patients' prognosis and quality of life and prolongs their survival. In recent years, the multi-link, multi-target, and multi-pathway molecular mechanisms underlying the antitumor effects of TCM have gradually gained recognition.

2.1 Development of Traditional Chinese Medicine in Tumor Treatment and Overview of Antitumor Traditional Chinese Medicine

2.1.1 Development of Traditional Chinese Medicine in Tumor Treatment

The development of TCM oncology spanned an early stage from the Yin and Zhou dynasties (1600 BC–221 BC) to the Sui and Tang dynasties (581 AD–907 AD), established a solid theoretical foundation during the Song and Yuan periods (960 AD–1368 AD), and gradually matured in the Ming and Qing dynasties (1368 AD–1911 AD). The term "tumor" as a disease entity was first recorded on Shang dynasty (1600 BC–1046 BC) oracle bone inscriptions, representing the earliest documented reference to tumors in TCM. Compiled over 2000 years ago, the Rites of Zhou referred to physicians specializing in tumor treatment as "ulcer doctors." The Huangdi Neijing also documented tumor-related conditions, such as skin tumors, muscle tumors, intestinal inversion, esophageal obstruction, and accumulations, summarizing their pathogenesis as "blockage of nutrient qi," "cold qi invading the intestines and conflicting with defensive qi," "deficiency of healthy qi," and "dominance of pathogenic qi." This classic work laid the groundwork for the formation and development of TCM oncology. Hua Tuo pioneered surgical anesthesia for treating internal "accumulations" (equated to tumors) and noted in Zhongzang Jing: "All abscesses, ulcers, and swellings arise from toxin retention in the five zang-organs and six fu-organs, not merely from obstruction of nutrient and defensive qi." He emphasized that tumors are local manifestations of systemic disorders, with internal factors playing a leading role in pathogenesis. Zhang Zhongjing, the "Sage of Medicine" in the late Eastern Han Dynasty (25 AD–220 AD), authored the Treatise on Cold Damage and Miscellaneous Diseases, establishing a syndrome differentiation system centered on zang-fu, meridian, and collateral theories—thus setting clinical diagnosis and treatment standards for TCM oncology. Synopsis of the Golden Chamber (a section of the above work) stated: "Accumulation is a zang-organ disorder with fixed localization; gathering is a fu-organ disorder with intermittent occurrence, shifting pain, and treatable outcomes." It stressed that tumor treatment should "examine pulses and symptoms to identify pathogenic targets, and intervene based on syndrome patterns." Subsequent generations of physicians advanced TCM oncology from theory to clinical practice. Chao Yuanfang of the Sui Dynasty (581 AD–618 AD) in General Treatise on the Etiology and Symptoms of Diseases systematically described tumor names, etiologies, pathologies, and symptoms. Sun Simiao of the Tang Dynasty (618 AD–907 AD) first recorded over 50 tumor-specific prescriptions in Qianjin Yaofang, highlighting the use of insect medicines, highly toxic agents, and drugs for resolving phlegm and stasis. Dou Hanqing of the Song Dynasty (960 AD–1279 AD) noted in Complete Book of Experience on Sores and Ulcers: "Breast cancer feels as hard as stone when pressed—hence its name. Early treatment saves life; late treatment leads to internal ulceration, exposure of the five zang-organs, and death." The four prominent physicians of the Jin and Yuan dynasties (1115 AD–1368 AD) made distinct contributions to tumor therapy: Liu Hejian focused on pathogenic fire and frequently used heat-clearing and detoxifying drugs; Li Dongyuan proposed "nourishing healthy qi to eliminate accumulations" based on the theory "internal injury to the spleen and stomach causes various diseases"; Zhu Danxi advocated "resolving phlegm to dissipate tumors" from the perspective that "most lumps in the body are caused by phlegm"; Zhang Zihe excelled in using sweating, vomiting, and purging to dispel pathogens and dissolve tumors—for example, purging could "remove stagnant accumulations, clear the gastrointestinal tract, eliminate tumors completely, and boost nutrient and defensive qi." Outstanding surgeons of the Ming and Qing dynasties (1368 AD–1911 AD) summarized previous experiences and detailed the etiology, pathogenesis, symptoms, internal treatment, and external therapy of tumors. From the late Qing Dynasty to the early Republic of China (1900 AD–1920 AD), Wang Qingren (in Corrections on the Errors of Medical Works), Zhang Xichun (in Integrating Chinese and Western Medicine), and Tang Rongchuan (in Treatise on Blood Syndromes) reached the peak of using stasis-resolving methods for cancer treatment. With the spread of Western learning to the East, missionaries established Western hospitals in coastal and major Chinese cities, prompting widespread exploration of tumor management from an integrated TCM-Western medicine perspective. The term "Ai" was translated as "cancer," and "cancer" was included in the Chinese Etymology Dictionary published during the Guangxu period (1875 AD–1908 AD). In modern times, TCM oncology has continued to develop and emerged as a distinct field from TCM internal medicine, surgery, gynecology, and miscellaneous diseases, becoming a dynamic clinical discipline within TCM. It has also fostered numerous contemporary TCM oncologists, including Yu Guiching, Yu Rencun, Zhang Daizhao, Pan Mingji, Zhou Daihan, Liu Jiaxiang, Li Peiwen, and Pan Minqiu [20]. In conclusion, TCM oncology has evolved from its embryonic stage in ancient times to an independent discipline today. Its historical development gave rise to the "holistic concept" and "syndrome differentiation-based treatment." In clinical practice, core therapeutic principles include "treating the root cause," "nourishing healthy qi to dispel pathogens," and "harmonizing yin and yang to restore balance." To address tumor pathogenesis—encompassing cancer toxin, deficient healthy qi, blood stasis, and phlegm coagulation—specific therapeutic strategies have been established, such as heat-clearing and detoxification, nourishing healthy qi to dispel pathogens, promoting blood circulation to resolve stasis, eliminating dampness and phlegm, and using toxic agents to detoxify. These strategies form the mainstay of TCM cancer therapy and play an irreplaceable role in the comprehensive clinical management of cancer. The key milestones in the history of TCM tumor treatment are illustrated in Fig. 13.1.

2.1.2 Overview of Antitumor Traditional Chinese Medicine

Exploring potent and low-toxicity agents in TCM and its compound formulations is a key focus in the research and development of clinical antitumor drugs. Numerous studies have shown that TCM and its compound formulations can play a unique role in the clinical management of tumors through diverse pathways and mechanisms. Herein, the TCM herbs and compound formulations with antitumor effects are introduced (Fig. 13.2).

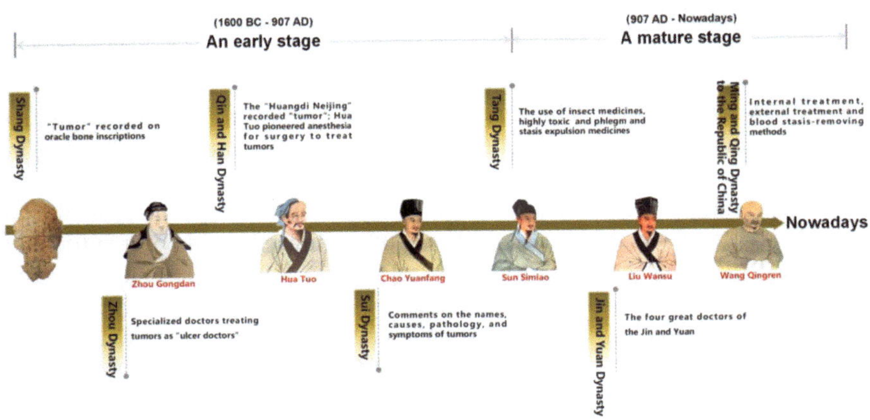

Fig. 13.1 The history of traditional Chinese medicine for tumor treatment

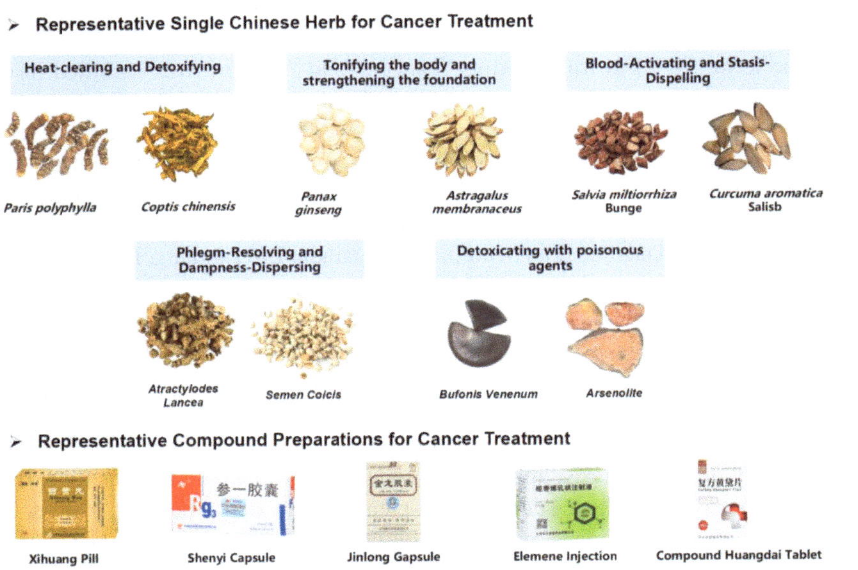

Fig. 13.2 Representative Chinese medicines and compound preparations with antitumor effects

Single Traditional Chinese Herbs for Cancer Treatment

1. *Heat-Clearing and Detoxifying Traditional Chinese Medicine for Cancer Treatment*

   The heat-clearing and detoxifying approach aims to clear internal heat, eliminate toxins, and alleviate the symptoms of heat-toxin accumulation in advanced tumors, which serves as a basic principle of TCM for treating malignant tumors. Currently, commonly used heat-clearing and detoxifying traditional Chinese medicines for cancer treatment include *Coptis chinensis*, *Hedyotis diffusa*, *Scutellaria baicalensis*, *Anemarrhena asphodeloides*, *Herba Rabdosiae*, *Lithospermum erythrorhizon*, *Artemisia carvifolia*, and *Paris polyphylla*.

2. *Tonic Traditional Chinese Medicine for Cancer Treatment*
   The "tonifying vital qi and fortifying the root" approach refers to the use of tonic TCM to preserve the body's vital qi, strengthen physical constitution, and dispel pathogens, with the aim of treating tumors. Any agent that can support and nourish the body's vital qi, reinforce the body's foundation, and address deficiencies is defined as a tonic medicine. Tumors are chronic consumptive diseases that deplete the body's vital qi over an extended period. Patients often present with "internal deficiency," and insufficient vital qi renders the body susceptible to pathogenic invasion. Tonic TCM enhances the body's immune function and strengthens its disease-resistant capacity; rather than directly inhibiting tumor cells, it achieves the goal of "dispelling pathogens" through the "tonifying vital qi and fortifying the root" strategy. The main tonic traditional Chinese medicines for cancer treatment include *Panax ginseng*, *Astragalus membranaceus*, *Glycyrrhiza uralensis* Fisch, *Aconitum carmichaeli* Debx. and *Epimedium brevicornu* Maxim.

3. *Blood-Activating and Stasis-Dispelling Traditional Chinese Medicine for Cancer Treatment*
   The formation and development of tumors are closely related to blood stasis. The lumps, stasis points, and stasis spots that appear in cancer patients can be treated with the method of activating blood and dispelling stasis. The traditional Chinese medicine with blood-activating and stasis-dispelling effects includes *Salvia miltiorrhiza* Bunge, *Curcuma aromatica* Salisb., *Corydalis yanhusuo*. *Boswellia carteri*, *Commiphora myrrha* Engl., *Curcuma zedoaria Roscoe*, and *Polygonum cuspidatum*.

4. *Phlegm-Resolving and Dampness-Dispersing Traditional Chinese Medicine for Cancer Treatment*
   The occurrence of tumors is also related to the coagulation of phlegm and dampness. If the accumulation lasts for a long time, tumors are likely to occur. Therefore, phlegm-resolving and dampness-dispersing medicines are also often added in the treatment of malignant tumors, such as the following: *Mangnolia officinalis*, *Atractylodes Lancea*, *Stephania tetrandra*, and *Semen Coicis*.

5. *Toxic Traditional Chinese Medicine for Cancer Treatment*
   The method of detoxicating with poisonous agents involves the use of highly toxic drugs to treat toxic stagnation, which can directly inhibit cancer cells. Commonly used drugs are mainly some toxic Chinese medicines, such as *Bufonis Venenum*, *Garcinia morella Desv.*, *Tripterygium wilfordii*, *Mylabris*, *Realgar*, *Arsenolite*, and *Fructus Bruceae*. It is well known that *Arsenolite* is a highly toxic drug, but it is also a good antitumor drug. The US Food and Drug Administration has officially approved the use of As_2O_3, the main component in *Arsenolite*, to treat acute promyelocytic leukemia [21].

Traditional Chinese Medicine Compound Prescriptions and Patent Medicines for Cancer Treatment

Chinese traditional compound prescriptions and Chinese patent medicines, which exhibit multi-link, multi-target, and multi-pathway properties, possess unique advantages in inhibiting tumor proliferation, reducing toxicity, enhancing efficacy, preventing recurrence, prolonging survival, and improving quality of life. Commonly used Chinese traditional compound prescriptions and Chinese patent drugs are proved effective in cancer treatment, such as Compound antitumor drink, Zeqi decoction, Fuzheng antitumor prescription, Jieduxiaozheng decoction, Ruyiping formula, Qizhu formula, Tenglongbuzhong decoction, Shenyi capsule, Shenqifuzheng injection, Huaier granule, Zhenqifuzheng capsule, Shengmai injection, Jinfukang oral liquid, Jinlong capsule, Xiaoaiping tablet, Huachansu tablet, Elemene injection, Xihuang pill, Antike capsule, compound Kushen injection, compound Huangdai Tablet, and Icaritin soft capsules.

2.2 Active Ingredients of Traditional Chinese Medicine and Their Antitumor Mechanisms

The advantage of using TCM lies in its ability to target multiple cancer-associated signaling pathways and molecular targets. TCM can induce tumor cell death, inhibit tumor cell metastasis, reduce the proportion of cancer stem cells, reverse multidrug resistance, and enhance the body's immune function, thereby exerting an antitumor effect. The specific antitumor mechanisms are shown in Fig. 13.3 and Table 13.1.

2.2.1 Antitumor Effects of Traditional Chinese Medicine via Promoting Tumor Cell Death

A plethora of studies have confirmed that numerous traditional Chinese medicinal compounds can modulate cancer-associated signaling pathways through multiple mechanisms, including apoptosis, autophagy, ferroptosis, and the induction of cell cycle arrest, thereby promoting tumor cell death.

Apoptosis is a programmed cell death that is the major mechanism induced by cytotoxic anticancer drugs [22]. Traditional Chinese medicine can induce apoptosis by regulating the expression of some members of the Bcl-2 family (i.e., Bcl-2 and Bax) and apoptotic-related proteins (i.e., caspase-3 and caspase-9) [23]. Zhang DM et al. demonstrated that arenobufagin had antineoplastic effects on hepatocellular carcinoma (HCC) cells both in vitro and in vivo. Arenobufagin-induced apoptosis was associated with a decreased mitochondrial membrane potential, increased Bax/Bcl-2 expression ratio, and Bax translocation from the cytosol to the mitochondria to inhibit the PI3K/Akt/mTOR pathway [24]. Tanshinone IIA, an extract of *Salvia miltiorrhiza*, caused the aggregation of p53, altered the expression of p53-dependent targets, and caused p53-mediated apoptosis by modulating Bcl-2, Bax, caspase-3, and PARP cleavage in human papillomavirus-positive CaSki cells [25]. Oridonin, extracted from *Herba Rabdosiae*, is a tetracyclic diterpenoid natural compound of the ent-kaurane type. Oridonin can modulate the expression of target protein genes,

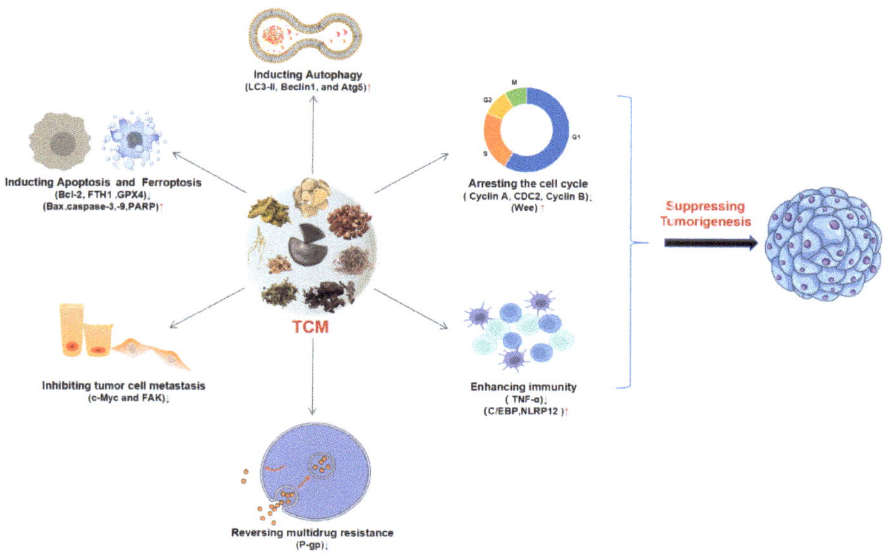

Fig. 13.3 Schematic diagram of the antitumor mechanism of traditional Chinese medicine

such as phosphoinositide 3-kinase regulatory subunit alpha (PIK3R1) and transforming growth factor-β2, activating the PI3K/Akt signaling pathway, thereby promoting apoptosis of tumor cells and inhibiting their proliferative capacity. Quercetin can upregulate the expression of several apoptosis-related genes, including P53, Bax, and Fas, blocking the activation of caspase proteases and inducing apoptosis in tumor cells. It has a strong inhibitory effect on non-small cell lung cancer, being able to induce caspase-independent cell death. Additionally, quercetin can also downregulate DNA methylation levels, suppressing the expression of oncogenes [26]. Triptolide, an extract from the Chinese herb *Tripterygium wilfordii*, inhibited NF-κB transcriptional activity via blocking the association of p65 subunit with CREB-binding protein (CBP)/p300 at the early stage and via decreasing p65 protein level at the late stage [27]. Zheng F et al. reported that berberine exerted apoptosis induction in human lung cancer cells through activation of p38 alpha MAPK signaling pathway [28, 29]. Icariin, a promising new star in liver cancer treatment, inhibited the proliferation of hepatocellular carcinoma HepG2 cells and promoted their apoptosis. The mechanism might be associated with the downregulation of Bcl-2 protein expression, the upregulation of Bax protein expression, and the enhancement of caspase-3 activity [30].

Autophagy is a cellular process in which damaged organelles or proteins are engulfed by autophagic vesicles and subsequently transported to lysosomes for degradation, ultimately leading to cell death. As a survival-preserving mechanism, autophagy enables tumor cells to proliferate rapidly by degrading long-lived intracellular proteins and cytoplasmic organelles, thereby overcoming nutrient deprivation and hypoxia. However, excessive autophagy activation can cause extensive

Table 13.1 A brief overview of the antitumor mechanism of the active components from traditional Chinese medicine

Mechanism		Component	Function	Reference
Promoting tumor cell death	Apoptosis	Arenobufagin	Decreased mitochondrial membrane potential, increased Bax/Bcl-2 expression	[24]
		Tanshinone IIA	Altering the expression of p53-dependent targets and causes p53-mediated apoptosis by modulating Bcl-2, Bax, caspase-3, and PARP cleavage	[25]
		Oridonin	Modulate the expression of target protein genes such as phosphoinositide 3-kinase regulatory subunit alpha (PIK3R1) and transforming growth factor-β2, activating the PI3K/Akt signaling pathway	[26]
		Quercetin	Upregulating the expression of several apoptosis-related genes, including P53, Bax, and Fas, and blocking the activation of caspase proteases	[26]
		Triptolide	Inhibiting NF-κB transcriptional activity via blocking the association of p65 subunit with CREB-binding protein (CBP)/p300 at the early stage and via decreasing p65 protein level at the late stage	[27]
		Icariin	Downregulation of Bcl-2 protein expression, the upregulation of Bax protein expression, and the enhancement of caspase-3 activity.	[30]
	Autophagy	Bufalin	Upregulating the expression of LC3-II, Beclin1, and Atg5 proteins	[31]
	Ferroptosis	Dihydroartemisinin	Modulating the mTOR/p70S6k signaling pathway and activating the autophagic degradation of FTH	[33]
		Gambogic acid	Inhibition of FTH1 and GPX4 expression	[34]
	Cell cycle arrest	Matrine	Elevating the cell cycle-related protein wee1 and reducing the expression levels of cyclin A, CDC2, and cyclin B	[35]
Inhibition of tumor cell metastasis		Cryptotanshinone	Degrading c-Myc and FAK and attenuating the FAK signaling pathway	[36]
Reversal of multidrug resistance		Polyphyllin VII	Impairing the Wnt/β-catenin pathway activity and leading to a reduction in the side population cells	[37]
		Timosaponin A-III	Regulation of the PI3K/Akt signaling pathway	[38]
		β-elemene	Decreasing mitochondrial membrane potential and P-gp expression activated the intracellular redox system	[39]
Regulation of immune function		Astragalin	Promoting the expression of recombinant human C/EBP homologous proteins	[40]
		Ginsenoside Rg1	Upregulating the expression of inflammatory body NLRP12, inhibiting the expression of interleukins and TNF-α	[41]

destruction of proteins and organelles, resulting in cell death. Bufalin is the primary component of the TCM *Bufonis Venenum*. It has been shown to induce autophagy in various liver cancer cells, including HepG2, SK-HEP-1, Huh7, Hep3B, and HA22T. Studies demonstrated that in human colon cancer cells (HT-29 and Caco-2), bufalin upregulated the expression of LC3-II, Beclin1, and Atg5, and stimulated reactive oxygen species (ROS) production. Inhibition of ROS with antioxidants, suppression of JNK using the inhibitor SP600125, or JNK2 knockdown attenuated bufalin-induced autophagy. These findings suggest that bufalin induces autophagic cell death in colon cancer through a mechanism involving ROS production and JNK activation [31].

In 2012, Dixon et al. induced a novel form of cell death distinct from apoptosis, autophagy, and necrosis using the synthetic chemical erastin; this cell death modality is characterized by iron-dependent reactive oxygen species (ROS) accumulation, independence from classical apoptotic pathways, and insensitivity to other cell death inhibitors, with morphological features, including mitochondrial condensation and increased intracellular double-membrane density, a process termed ferroptosis [32]. Dihydroartemisinin (DHA)—a classic ferroptosis inducer derived from *Artemisia carvifolia*—exerted dose-dependent cytotoxic effects on acute myeloid leukemia (AML) cells (including HL-60, KG1, and THP-1 cell lines) by modulating the mTOR/p70S6k signaling pathway and activating autophagic degradation of ferritin heavy chain (FTH) in AML cells, thereby triggering ferroptotic cell death, which could be mitigated by iron chelators, such as deferoxamine (DFO) and ferrostatin-1 (Fer-1) [33]. Gambogic acid (GA), an active component of the TCM *Garcinia morella Desv.*, can induce ferroptosis in renal clear cell carcinoma, and the underlying mechanism might involve the inhibition of FTH1 and GPX4 expression [34].

Dysregulation of the cell cycle constitutes one of the principal hallmarks of malignant neoplasms. Arresting the cell cycle represents an ideal therapeutic strategy in the treatment of cancer. Matrine, extracted from *Sophora flavescens* Alt., could exert antitumor activity when applied to HeLa cervical cancer cells in vitro; a significant alteration in the cell cycle distribution was observed. The expression level of the cell cycle-related protein Wee1 was elevated, while the expression levels of Cyclin A, CDC2, and cyclin B were reduced, resulting in the suppression of cell proliferation [35].

2.2.2 Inhibition of Tumor Cell Metastasis

Metastasis is a multistep process initiated by the dissemination of cancer cells from the primary tumor to other body sites; metastatic cells then invade vital organs, causing organ damage and progressive deterioration of patient health, and tumor metastasis is actually the leading cause of cancer-related mortality. Epithelial–mesenchymal transition (EMT) refers to a biological process wherein epithelial cells lose their epithelial traits and gain a mesenchymal phenotype to transform into mesenchymal cells; as the initial step in tumor metastasis (corrected for accuracy, as EMT primarily facilitates metastasis rather than tumorigenesis), EMT promotes tumor cell invasion and metastasis. Lipopolysaccharide (LPS) induces EMT-like phenotypic changes, which in turn enhance cell migration and invasion. Shikonin significantly reduces N-cadherin expression in MDA-MB-231 cells (standardized cell line nomenclature) and upregulates E-cadherin expression in MCF-7 cells. Cryptotanshinone (CT)—a natural compound isolated from *Salvia miltiorrhiza* Bunge—inhibits the growth and metastasis of primary ovarian tumors by degrading c-Myc and FAK, as well as attenuating the FAK signaling pathway [36].

2.2.3 Reversal of Multidrug Resistance

Chemotherapy is a primary therapeutic option for cancer treatment; however, incomplete chemotherapy often occurs due to dose-limiting toxicity associated with drug resistance. Multidrug resistance (MDR) is defined as cellular resistance to multiple drugs with distinct mechanisms of action and/or chemical structures, thereby limiting the efficacy of chemotherapy. To date, hundreds of agents are available to reverse MDR, but most are not widely used due to toxicity, high costs, or other non-negligible disadvantages. In contrast, TCM lacks these drawbacks and may thus serve as an effective MDR reversal agent. Polyphyllin VII, derived from *Paris polyphylla*, binds to moesin and induces its degradation via the ubiquitin-proteasome pathway, thereby impairing Wnt/β-catenin pathway activity and reducing side population cells to overcome bortezomib resistance [37]. Timosaponin A-III (TAIII), a saponin isolated from *Anemarrhena asphodeloides*, can reverse adriamycin (ADM) resistance in human chronic myeloid leukemia (CML) K562/ADM cells by regulating the PI3K/Akt signaling pathway [38]. β-elemene, a novel anticancer drug extracted from *Curcuma zedoaria* Roscoe (known as zedoary), enhances the sensitivity of A549/DDP cells to cisplatin and reverses their drug resistance through mechanisms involving decreased mitochondrial membrane potential, reduced P-gp expression, activated intracellular redox systems, and induced apoptosis [39].

2.2.4 Regulation of Immune Function

Immune dysfunction plays a critical role in tumor progression, recurrence, and metastasis, making it a key target for cancer therapy. TCM regulates immune function in a biphasic manner and restores the dynamic balance of the immune system. *Astragalus membranaceus* (also known as Huangqi), a TCM herb renowned for replenishing Qi, can enhance the lung recruitment capacity of natural killer (NK) cells by upregulating interleukin-17D (IL-17D) expression, thereby promoting anti-tumor immune responses in the lung. Astragalin—a flavonoid isolated from *Astragalus membranaceus*—can reduce the expression of IL-17 and vitamin D receptor-related orphan receptor γt (RORγt) in serum and tumor tissues; it also increases the spleen index and thymus index of tumor-bearing mice; downregulates the expression of X-box binding protein 1, inositol 1, and glucose-regulated protein 78; and upregulates the expression of recombinant human CCAAT-enhancer-binding protein (C/EBP) homologous proteins, thus regulating immune function [40]. Ginsenosides, the major active components isolated from *Panax ginseng*, have been extensively studied in cancer therapy; they can activate cytotoxic T lymphocytes and NK cells and inhibit tumor growth by regulating immune function.

Specifically, ginsenoside Rg1 can upregulate the expression of the NLRP12 inflammasome, inhibit the expression of interleukins and tumor necrosis factor α (TNF-α), and improve the inflammatory microenvironment [41].

3 Nano-delivery Systems for Antitumor Active Ingredients of Traditional Chinese Medicine

Currently, an increasing number of researchers have found that nanoscale particles (e.g., nanoparticles, micelles, and lipids) are commonly present in TCM decoctions. Self-assembly may occur during the decoction process, driven by a variety of intermolecular forces (e.g., electrostatic attraction, hydrogen bonding, hydrophobic association, and π-π stacking) between individual constituents, leading to the formation of naturally occurring nanoparticles. As reported in previous studies, colloid-like aggregates were observed in all decoctions of 60 medicinal herbs and 24 Chinese herbal formulae [42, 43]. Despite the therapeutic advantages of plant-derived natural nanoplatforms, several limitations remain. For nanoparticles derived from TCM decoctions, their formation mechanisms are not fully elucidated; in particular, the energy changes and structural characteristics of self-precipitation during decoction require further investigation. Although nanoparticles isolated from decoctions were discovered relatively early, research on plant-derived nanoparticles—especially in the biomedical field—remains insufficient, and most studies are still in the laboratory research stage. Therefore, it is necessary to develop novel drug delivery systems using modern nanotechnology.

3.1 Application of the Holistic Concept of Traditional Chinese Medicine in Nanomedicine Formulations

The core concept of TCM theory holds that the human body operates as an integrated whole, with interconnections and interactions between its internal and external components. TCM focuses not only on local pathological changes but also on restoring the body's overall balance and regulatory function. Meanwhile, the application of nanomedicine formulations in TCM is not merely a delivery system for active compounds. Through nanotechnology, multicomponent co-delivery of TCM ingredients can be achieved; alternatively, by integrating drugs and excipients—utilizing structurally unique active components of TCM to replace nanocarriers—nanomedicine formulations with enhanced functionality can be constructed; or via co-delivery of TCM components and chemotherapeutics, tumor growth can be inhibited while the tumor microenvironment is modulated. These strategies embody the wisdom of ancient Chinese people and represent valuable approaches in modern formulation technology that merit further exploration and adoption.

3.1.1 Multicomponent Nanomedicines Derived from Traditional Chinese Medicine

The diverse constituents of TCM formulations often exhibit complementary and synergistic pharmacological activities, which mutually enhance therapeutic efficacy and reduce adverse events. Nano-drug delivery systems enable the co-loading and compatible administration of multiple active phytochemical ingredients, thereby realizing the TCM principle of "multicomponent, multi-targeted" interventions through "synergistic interactions."

For instance, Zhu et al. [44] developed self-nanoemulsifying oil formulations (SNEFs) of naringenin (a poorly soluble TCM chemopreventive flavanone) using *Coix lacryma-jobi* seed oil, constructing the naringenin-Coix SNEF (NCNLC) platform. Compared with SNEFs formulated with Capryol 90 (naringenin-Capryol nanostructured lipid carrier; NDNLC) or oleic acid (naringenin-oleic nanostructured lipid carrier; NONLC), NCNLC exhibited enhanced tumor-targeted drug release. Furthermore, compared with conventional excipients, NCNLC improved the cytotoxicity, antiproliferative effects, and proapoptotic activity of naringenin in HepG2 cells. These findings highlight the favorable solubilization capacity and synergistic tumoricidal effects of the TCM-derived oil excipient, providing insights into the future development of phyto-oils in nano-pharmaceuticals.

This underscores the potential of advancing multi-phytochemical nanosystems toward clinical translation, offering guidance for future research on TCM-based nano-pharmaceuticals.

3.1.2 Co-formulation Nanomedicines Integrating Traditional Chinese Medicine Drugs and Excipients

The harmonious integration of active medicinal ingredients and auxiliary excipients has long been a core tenet guiding the development of TCM products, distinguishing such formulations from conventional chemical drug preparations. The therapeutic efficacy of TCM agents depends not only on the bioactivity and content of pharmacodynamic constituents but also on the precise selection of excipients with pharmacological properties: using these excipients to directly construct delivery carriers can significantly enhance treatment outcomes. By minimizing unnecessary additives, such "drug-excipient fusion" systems offer advantages of material efficiency and simplified procedures. With the advancement of these co-formulation technologies, pharmaceutical excipients are increasingly recognized as biologically functional components rather than inert carriers, generating substantial research interest in exploiting and maximizing their auxiliary therapeutic roles.

For instance, Yang et al. [45] prepared cholesterol-free liposomes loaded with *Bufonis Venenum* (BV) extracts—primarily composed of bufadienolide components—by replacing endogenous cholesterol with bufosteroidal compounds. This integrated delivery platform was designed to enhance the antineoplastic efficacy of BV components while reducing their toxicity and adverse effects. Results confirmed that cholesterol-free BV extract liposomes (NON-CHOL-BVE-LP) exhibited stronger tumor-suppressive effects than conventional BV injections and prolonged the

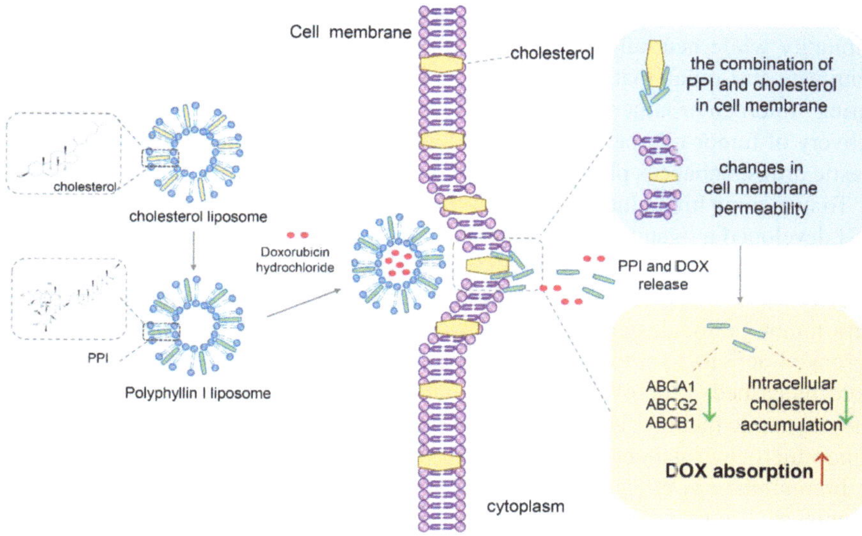

Fig. 13.4 A membrane polyphylin I component of liposomes with glycyrrhetinic acid for co-delivery with DOX. (Copyright © 2022 Elsevier Ltd. Reprinted with permission from ref. [46])

in vivo exposure of bioactive bufogenins, including bufotenine, desacetylbufotalin, arenobufagin, bufalin, and cinobufagin.

Additionally, as shown in Fig. 13.4, Jiang et al. [46] replaced the membrane cholesterol component of liposomes with glycyrrhetinic acid (a triterpenoid saponin that modulates cholesterol metabolism) for co-delivery with the chemotherapeutic agent doxorubicin (DOX). The resulting drug-loaded liposomes (DLPs) improved drug solubility, reduced toxicity, and exerted synergistic antineoplastic effects. Meanwhile, DLPs depleted cellular cholesterol reserves through the regulatory effect of glycyrrhetinic acid on metabolic pathways, reversing the high-cholesterol microenvironment that contributes to chemoresistance. Finally, membrane modification with glycyrrhetinic acid enhanced the cellular uptake and intracellular accumulation of DOX, increasing the sensitivity of tumor cells to chemotherapy. This study provides a mechanism-based strategy for the combined application of TCM triterpenoid saponins and first-line cytotoxic agents.

3.1.3 Modulation of the Tumor Microenvironment by Traditional Chinese Medicine-Based Nanomedicines

The tumor microenvironment denotes the intricate intercellular and acellular networks enveloping malignant cells, encompassing neighboring immune constituents, vascular endothelium, fibroblasts and extracellular matrix components. This specialized milieu exerts profound impacts on oncogenesis, dissemination, and therapeutic refractory phenotypes. Herbal medicines possess distinctive competencies to redress aberrant tumor microenvironment signaling. First, multicomponent botanical preparations simultaneously engage diverse molecular targets in neoplastic and stromal compartments. Additionally, immunomodulatory herbs strengthen host immunity while neutralizing tumor evasion tactics. Chinese materia medica also manifests anti-inflammatory, anti-angiogenic, and anti-oxidative properties, mitigating microenvironmental perturbations. Therefore, nanotechnology-assisted delivery of tumor microenvironment-modulating herbal agents, alongside antineoplastic drugs, enhances pharmacological outcomes.

To overcome high glutathione levels impeding photodynamic therapy, Lan et al. [47] developed a ligand-free nanomedicine (GA-Ce6-FA) by integrating the herbal component gambogic acid with the photosensitizer chlorin e6 and folic acid (Fig. 13.5). GA-Ce6-FA exploited folic acid-mediated active targeting to increase intra-tumor gambogic acid accumulation. Once internalized, gambogic acid's multipronged actions as a tumor microenvironment regulator and apoptosis inductor were unleashed. Moreover, gambogic acid depleted glutathione, thus enhancing photodynamic therapy. This represents a prototype engrafting herbs to optimize the tumor microenvironment for potentiated phototherapy.

Immunotherapy efficacy is hampered by the immunosuppressive tumor microenvironment, limiting immune infiltrates and antitumor immunity. To address this limitation, Lan et al. [48] developed a biomimetic drug delivery system called Tm@PDA-GA comprising gambogic acid, the polydopamine polymer, and tumor cell membrane components (Fig. 13.6). Within Tm@PDA-GA, gambogic acid normalized the aberrant tumor vasculature through anti-angiogenic activity, enhancing immune infiltration. Additionally, polydopamine generated photothermal effects upon 808 nm laser irradiation, eliciting immunogenic tumor cell death and elevating antitumor immune responses. The concerted actions of gambogic acid and

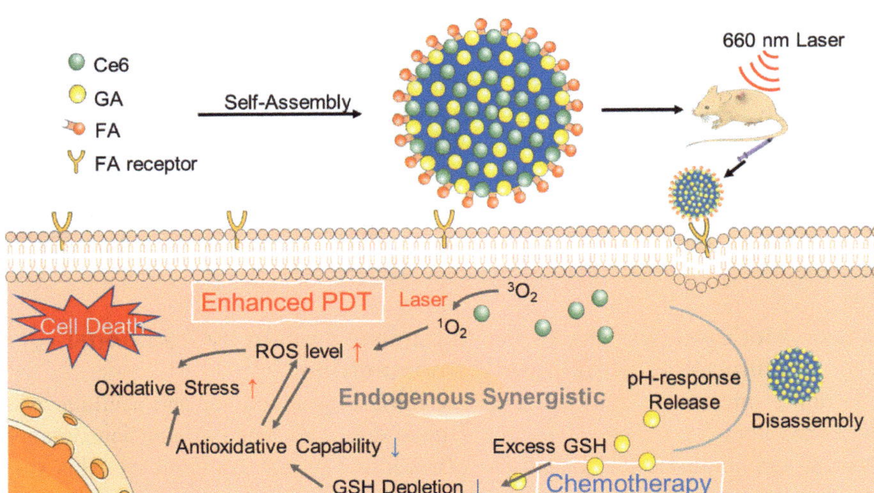

Fig. 13.5 A novel carrier-free GA-Ce6-FA NP with GSH depletion was developed to achieve a remarkable chemotherapy-PDT synergistic antitumor effect. (Copyright © 2022 Elsevier Ltd. Reprinted with permission from ref. [47]).

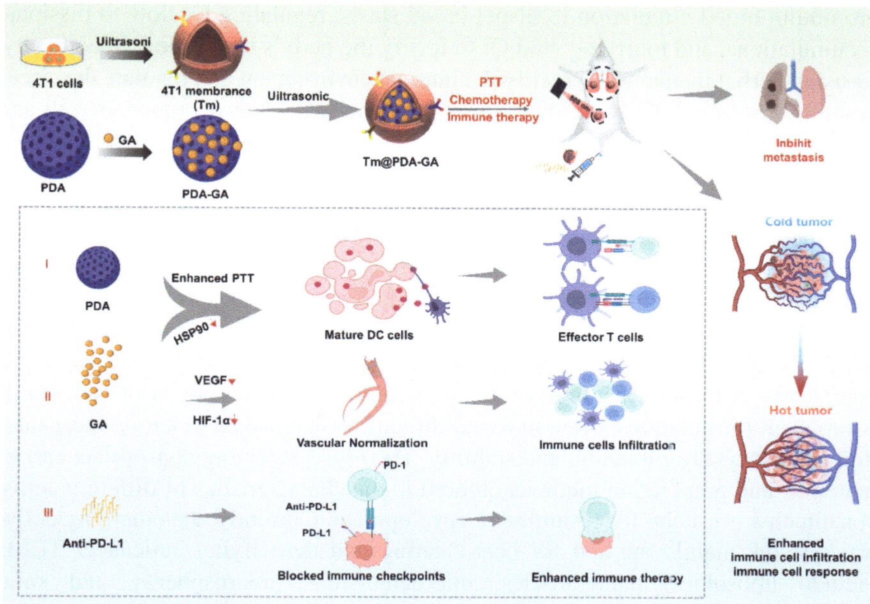

Fig. 13.6 A biomimetic drug delivery system Tm@PDA-GA could remodel the immunosuppressive microenvironment and strengthen cancer immunotherapy. (Copyright © 2024 Elsevier Ltd. Reprinted with permission from ref. [48])

photothermal therapy transformed poorly infiltrated, poorly immunogenic "cold" tumors into highly infiltrated, highly immunogenic "hot" tumors. This heightened tumor immunogenicity and immune surveillance bolstered checkpoint blockade efficacy using anti-PD-L1 antibodies. Moreover, sustained posttreatment tumor suppression and antimetastatic potency against breast cancer cells were achieved. Tm@PDA-GA highlights the feasibility of employing nano-formulated herbal ingredients to remodel the immunosuppressive microenvironment and strengthen cancer immunotherapy, holding promise for clinical translation.

3.2 Application of Traditional Chinese Medicine Theory in Nanomedicine Formulations

Nanomedicines incorporating traditional Chinese herbal ingredients have been extensively investigated for the development of anticancer therapeutics. Taking cancer as an example, to explore how TCM theory guides the construction of herbal nanomedicine formulations, one must first understand the TCM approaches to cancer treatment.

In TCM, tumor formation is attributed to Yin-Yang imbalance, internal accumulation of heat-toxin, and blood stasis. Thus, TCM therapeutic strategies for cancer include clearing heat and detoxifying, softening hard masses and resolving phlegm, promoting blood circulation to dispel blood stasis, regulating Qi flow to dissipate accumulations, and tonifying vital Qi to fortify the body's foundation. Specifically, by using herbal medicines to rectify the internal environment and regulate the circulation of Qi, blood, body fluids, and essence, the inhibition of tumor growth and elimination of cancer cells can be achieved.

3.2.1 Heat-Clearing and Detoxifying

Most herbal medicines and their constituents with heat-clearing, detoxifying, and anticancer activities—including alkaloids (e.g., berberine, matrine), flavonoids (e.g., baicalin, wogonin), terpenoids (especially diterpenes such as triptolide, oridonin), and fatty acids (e.g., castor oil)—face limitations such as poor aqueous solubility, low oral bioavailability, and insufficient stability. Additionally, the structural complexity of these active ingredients, particularly the development of new multi-component formulations, often makes it difficult to achieve satisfactory encapsulation efficiency, drug loading, and stability. Therefore, selecting appropriate carrier materials and preparation methods tailored to the characteristics of different active constituents is crucial for formulation development. Currently, the emerging delivery systems mainly applied for heat-clearing and detoxifying anticancer TCMs include liposomes, nanoparticles, microemulsions, microspheres, and solid dispersions.

For instance, berberine is a quaternary ammonium alkaloid isolated from the TCM herb *Coptis chinensis* and serves as its major active antibacterial component [49]. Owing to its isoquinoline alkaloid structure, berberine also exerts broad therapeutic effects, such as antiviral, antimicrobial, antidiarrheal, anti-inflammatory, and antitumor activities. Li et al. [50] constructed folate receptor-targeted, thermosensitive liposomes co-encapsulating berberine (BBR), and indocyanine green (ICG), designated as BI-FA-LP (Fig. 13.7). Via folate receptor-mediated targeting, BI-FA-LP can accumulate specifically in tumors. This combined chemo-photothermal treatment strategy utilizes BBR to inhibit the expression of heat shock proteins induced by photothermal therapy (PTT); meanwhile, BBR suppresses PTT-induced inflammation and metastasis and mitigates tissue damage.

Zhang et al. [51] selected Timosaponin AIII (TAIII), a steroid saponin extracted from *Anemarrhena asphodeloides* Bunge with heat-clearing properties, which possesses a similar structure to cholesterol and exhibits a wide range of antitumor activities. This makes it possible to develop a TAIII-based liposome, where TAIII could potentially stabilize the phospholipid bilayer as a substitute for cholesterol while also acting as a chemotherapeutic agent. Additionally, TAIII was shown to enhance the cellular uptake of doxorubicin hydrochloride (DOX) in human hepatocellular carcinoma (HCC) cells and demonstrate a synergistic effect. Thus, they designed a novel thermally sensitive multifunctional liposomal system composed of TAIII and lipids to deliver DOX for enhanced HCC treatment. The innovative dual application of TAIII as both a bilayer stabilizer and chemotherapeutic facilitates a stable multifunctional liposomal delivery system for synergistic therapy against HCC, aligned with the traditional Chinese medicine theory of "heat-clearing and detoxification."

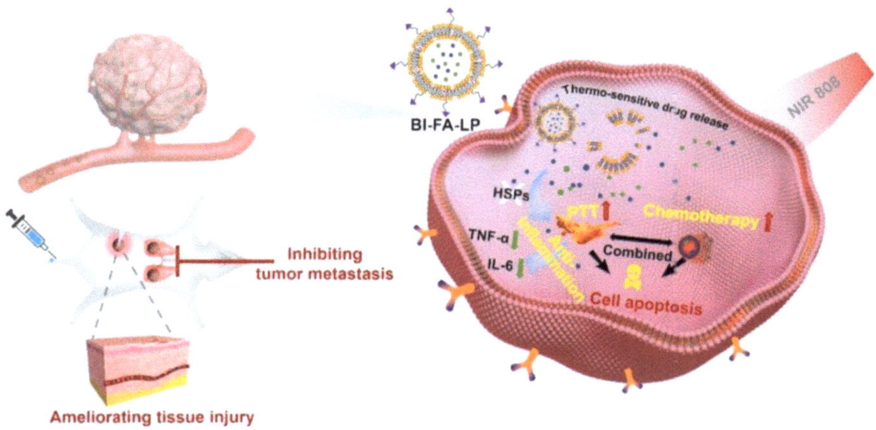

Fig. 13.7 A versatile thermosensitive liposome (BI-FA-LP) was developed and could reduce the expression of HSPs and overcome PPT-exacerbated inflammation, thus achieving synergetic chemo-PTT to inhibit tumor metastasis and ameliorate tissue injury in breast cancer therapy. (Copyright © 2024 Elsevier Ltd. Reprinted with permission from ref. [50])

This use of TAIII may be referred to for the development of other types of saponins with similar properties.

3.2.2 Tonifying Healthy Energy

The major components of tonifying traditional Chinese medicines include polysaccharides, saponins, essential oils, organic acids, steroids, amino acids, etc. Since long-term administration is frequently necessary and these medicines have complex total extracts with many components, large dosages are required in the clinic to ensure efficacy. This can lead to poor patient compliance and undermine the therapeutic effects. Hence, researching and developing new formulations of their effective components will aid in improving the clinical applicability of this class of medicines.

For instance, ginsenoside Rg3 is an active ingredient isolated from *Panax ginseng*, which can inhibit tumor angiogenesis, induce apoptosis of tumor cells, and selectively inhibit tumor metastasis while improving immune function [52]. As Fig. 13.8 shown, Lan et al. [53] utilized Rg3 as a liposome component to substitute for cholesterol, forming Rg3 liposomes to improve tumor-selective accumulation via interactions between the glycosyl moieties of Rg3 and overexpressed GLUT transporters on 4 T1 tumor cells. Additionally, they innovatively coated 4 T1 tumor cell membranes onto the Rg3-liposomes prior to co-loading them with gambogic acid (GA) to construct multifunctional biomimetic liposomes (GA/Rg3-Mlipo). As an adjuvant agent, Rg3 also enhanced the antitumor efficacy of GA through synergistic effects—killing tumor cells, inhibiting migration and invasion, achieving anti-angiogenesis, and improving antitumor immune cell infiltration to transform "cold" tumors into "hot" tumors.

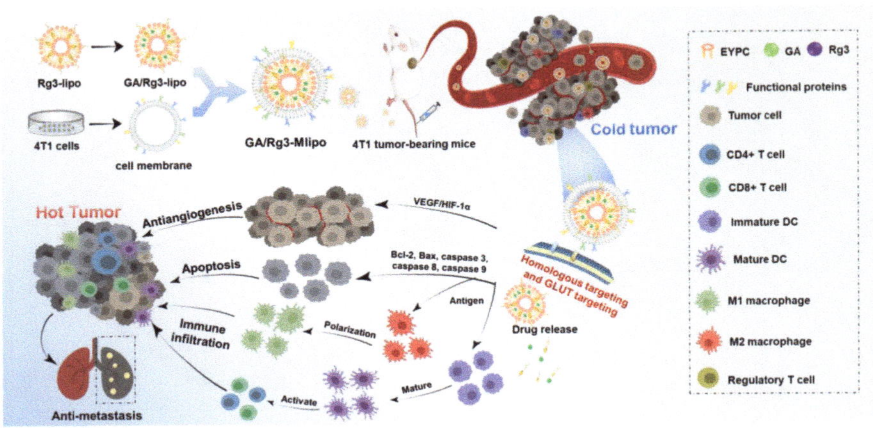

Fig. 13.8 Schematic illustration of GA/Rg3-Mlipo for the treatment of TNBC. GA/Rg3-Mlipo exerted a remodeling effect on the TME through the induction of tumor cell apoptosis, inhibition of angiogenesis, and promotion of immune cell infiltration. (Copyright © 2022 Elsevier Ltd. Reprinted with permission from ref. [53])

Astragalus membranaceus is a traditional Chinese medicine qi invigorating drug, and Astragalus polysaccharide is a water-soluble heteropolysaccharide made from the dried root of *Astragalus membranaceus* through extraction, concentration, and purification, which has been confirmed to significantly improve the function of human leukocytes to induce interferon, regulate body immunity, and promote the formation of antibodies. By regulating and inducing interferon, the virus in the body is destroyed to achieve a therapeutic effect. Xu et al. [54]. utilizes pH-responsive PLGA to form nanocarriers of Astragalus polysaccharides (APSPs) to enhance their bioavailability and accumulation at tumor sites. APSPs' immunomodulatory effects are evidenced by the inhibited growth of tumors in mice treated with pH-responsive APSPs and an increased ratio of CD4+/CD8+ T cells in the spleen lymphocytes. The results demonstrate the significant potential of traditional Chinese medicinal polysaccharides in boosting the immune system of cancer patients.

3.2.3 Blood-Activating and Stasis-Dispelling

Blood-activating and stasis-resolving herbs exert significant antitumor effects through multiple mechanisms, such as modifying blood rheology and improving microcirculation, making them a widely used class of anticancer agents in clinical practice. However, these herbs typically exhibit irritant properties and contain high levels of volatile oils and other lipophilic components with poor bioavailability, which severely limit their clinical application. Thus, utilizing novel delivery systems to enhance bioavailability and targeting represents a key research direction for this class of herbs, which has been explored by numerous researchers over the years.

The active constituents of blood-activating and stasis-resolving herbs—including volatile oils, tanshinone IIA, and resveratrol—often exhibit poor stability and aqueous solubility, leading to fluctuating blood concentrations (peak and trough effects) and low bioavailability. Volatile oils, also known as essential oils, are aromatic, oil-like liquids that evaporate at room temperature. They are abundant in commonly used blood-activating herbs, such as *Ligusticum chuanxiong*, *Curcuma aromatica*, *Curcuma kwangsiensis*, and *Boswellia carterii*, and are recognized as their major active components. Composed primarily of monoterpenes and sesquiterpenes, volatile oils have extremely low aqueous solubility and are highly volatile, with marked instability under light, heat, and air exposure. Additionally, many blood-activating herbs contain other lipophilic antitumor constituents—such as ferulic acid in *Ligusticum chuanxiong* and resveratrol in *Polygonum cuspidatum*—which also suffer from poor aqueous solubility and related limitations. Therefore, developing novel drug delivery systems to maintain the stability of volatile oils, improve solubility, and enhance the bioavailability of lipophilic antitumor constituents is of great importance.

Tanshinone is a fat-soluble phenanthraquinone compound extracted from the roots of the traditional Chinese medicine *Salvia miltiorrhiza*, with known antibacterial activity. It exhibits antibacterial, anti-inflammatory, blood-activating, and stasis-resolving effects, promotes wound healing, and has no obvious side effects. Liu et al. [55] successfully developed cationic bovine serum albumin-conjugated tanshinone IIA PEGylated nanoparticles. These nanoparticles significantly reduced infarct volume, improved neurological deficits, and alleviated histopathological severity, exerting remarkable neuroprotective effects in ischemic stroke by modulating inflammatory cascades and neuronal signaling pathways involved in cerebral ischemia.

3.2.4 Diuretic and Dampness-Dispersing

The main constituents of phlegm-resolving herbs include triterpenoid saponins (from *Panax ginseng*, *Poria cocos*, and *Polyporus umbellatus*), Lycium polysaccharides, salvianolic acids (from *Salvia miltiorrhiza*), and glucomannan (from *Pueraria lobata* roots). Although these herbs exhibit potent dampness-drying and phlegm-resolving effects, they are also associated with limitations, such as poor water solubility, low oral bioavailability, and insufficient stability—factors that restrict their clinical application in cancer treatment. Thus, researching and developing nanoformulations of phlegm-resolving herbs can help improve the clinical applicability of this class of herbs. These novel formulations can enhance efficacy, reduce dosage, improve patient compliance, and optimize drug delivery, thereby providing better therapeutic outcomes and quality of life for patients.

Wang et al. [56] developed a novel lung-targeted nanocarrier loaded with baicalin and curcumin for the treatment of non-small cell lung cancer. In vitro and in vivo experiments demonstrated that this nanoformulation exerted significant efficacy against non-small cell lung cancer and could effectively inhibit tumor growth. Baicalin clears heat and dries dampness, while curcumin activates blood circulation, dispels stasis, and relieves pain. These two components act synergistically—one focusing on "clearing" (heat/dampness) and the other on "activating" (blood circulation)—primarily by targeting pathogenic factors, which enables effective tumor regression. Guided by the TCM principle of "dispelling dampness and resolving phlegm," heat-clearing and dampness-resolving herbs are combined with qi-regulating and blood-activating herbs to achieve synergistic therapeutic effects.

3.3 Application of Meridian Tropism Theory in Nanomedicine Formulations

Enhancing tumor-targeted drug delivery can improve the antitumor efficacy of therapeutic agents while reducing their side effects, and this has become a major focus of research in anticancer nanomedicine. Strategies involving liposomes, polymeric micelles, folic acid, monoclonal antibodies, RGD peptides, and other targeting moieties all highlight the importance of solubility and targeting in drug development. However, current nanoparticle drug delivery systems still face substantial challenges, including the lack of highly selective and specific materials, the inability to penetrate biological barriers (e.g., the blood-brain barrier), and the failure to evade immune system detection and clearance. According to traditional Chinese medicine theory, herbs with different properties are thought to exhibit distinct in vivo distribution patterns. Furthermore, some TCM herbs can modulate the targeting ability and in vivo distribution of other components—an effect referred to as the "guijing" (meridian tropism) action in TCM theory.

For example, borneol is a traditional Chinese herb with a long history in China for opening orifices and alleviating pain. It is often used as an adjuvant in formulas. Guo et al. [57] developed a delivery system featuring borneol-conjugated distearoyl phosphoethanolamine-polyethylene glycol (DSPE-PEG-Bor) to facilitate blood-brain barrier (BBB) penetration. The system also includes Pep-1 modified DSPE-PEG (DSPE-PEG-Pep-1) to specifically target glioma cells, utilizing DSPE-PEG and dioleoyl phosphoethanolamine (DOPE) as the carrier matrix. This innovative system not only successfully breaches the BBB but also actively homes in on glioma cells, thereby enhancing the concentration of CMS within the tumor tissue. This is a significant stride forward in overcoming the challenges associated with CMS, such as its brief half-life, limited retention, and systemic toxicity. Moreover, the inclusion of Pep-1/Bor/CMS-M allows for the rapid release of the medication in acidic conditions, thanks to the pH-sensitive nature of DOPE. This introduces a novel approach for the targeted delivery of therapeutics to glioma-affected regions.

Diril et al. [58] synthesized a liver-targeted conjugate by attaching glycyrrhetinic acid (Gly), a liver-targeting ligand, to polyethylene glycol (PEG) using the Steglich reaction. This Gly-PEG conjugate was then integrated into liposomes, which were subsequently loaded with Dox-HCl utilizing a pH gradient technique. The most effective formulation, named Gly-Peg-Dox-ProLP-F6, exhibited a high Dox-HCl encapsulation efficiency of 90.0% ± 1.85% and a small particle size of 120 ± 3.2 nm. In vitro, this formulation significantly outperformed the standard Dox-HCl product in terms of cytotoxicity against hepatocellular carcinoma (HCC) cells, showing increased toxicity by factors of 1.14, 1.5, and 1.24 against HepG2, Mahlavu, and Huh-7 cells, respectively. Conversely, Gly-Peg-Dox-ProLP-F6 was found to be 1.86 times less toxic than the noncancerous AML-12 cell line. Additionally, it doubled the permeability from the apical to the basolateral side of cellular membranes. In a mouse model of liver cancer, Gly-Peg-Dox-ProLP-F6 proved to be significantly more effective at shrinking tumors, as measured by in vivo imaging systems (IVIS). After administering the Gly-Peg-Dox-ProLP-F6 formulation, isolated mouse liver tissue contained 2.48 times more Dox compared to when Dox-HCl was used, indicating a more targeted delivery. Moreover, there was significantly less accumulation of the drug in the heart tissue. These results suggest that glycyrrhetinic acid can significantly enhance the targeting capability of nanocarrier drug delivery systems to HCC.

In summary, the long history of TCM provides abundant examples and classical applications that can inform and inspire antitumor research in the field of nanomedicine. Although no TCM-derived nanomedicine for cancer treatment has yet entered clinical readiness, the unique properties of TCM-based nanomedicines have demonstrated potential in enhancing the body's immune response, inducing tumor cell apoptosis, improving the tumor microenvironment, and enhancing drug targeting. Therefore, leveraging the insights and advantages of TCM practices can offer new opportunities and perspectives for modern nanomedicine in combating cancer.

4 Challenges Faced by Nanomedicine in TCM and Future Directions

4.1 Challenges in the Implementation of Nanomedicine in TCM

Nanotechnology has emerged as a promising field in the application of TCM, offering numerous advantages and potential breakthroughs [59]. Nevertheless, nano-TCM remains in its infancy. Currently, nano-TCM is characterized by insufficient guidance from TCM theory and weak basic research, which imposes certain limitations and challenges that must be addressed for its successful integration (into clinical practice/modern medicine).

4.1.1 Complexity of TCM Formulations

The clinical application of TCM is relatively complex, as TCM typically exerts its effects through prescriptions guided by TCM theory. However, in the modern development of nanomedicine, most TCM-based nanomedicines are developed based on specific active ingredients of TCM, rather than complex TCM formulations [60]. Since the pharmacodynamic mechanisms of most TCM prescriptions have not been fully elucidated, current TCM nanoformulations mostly exhibit partial advantages of a single ingredient or a group of bioactive compounds derived from TCM prescriptions. Additionally, nanotechnology has regulated or fundamentally altered certain physicochemical interactions between TCM ingredients, leading to uncertainties regarding the effective components of TCM—this poses a challenge to the application of TCM theory.

4.1.2 Encapsulation of TCM in Nanocarriers

The application of nanotechnology to TCM research has undoubtedly opened a new avenue for TCM modernization. Nonetheless, many fundamental and exploratory studies need to be strengthened. Lagging behind nanotechnology preparation techniques, and the mismatch between the requirements for encapsulation of TCM pharmacodynamic components and nanoparticles are the main problems that urgently need to be solved. Moreover, the compounds in TCM formulations mostly display different physicochemical properties, making it challenging to encapsulate them efficiently using one strategy of nanocarriers; Solubility issues: Many Chinese medicine compounds have poor solubility in water or organic solvents, which can further hinder their efficient encapsulation into nanocarriers. This can result in low drug loading capacity and affect the overall effectiveness of the drug delivery system. Stability concerns: Chinese medicine compounds can be sensitive to environmental factors such as temperature, pH, and light. The encapsulation process within nanocarriers can potentially impact the stability of these compounds, leading to degradation and reduced drug loading. Size compatibility: Nanocarriers used in drug delivery systems have specific size limitations. Chinese medicine compounds, particularly herbal extracts, can have large molecular sizes that may not be easily accommodated within nanocarriers, leading to low drug loading efficiency.

4.1.3 Clinic Effectiveness of Nanodrugs

Despite the emergence of numerous nanodrugs with increasingly complex designs over the past three decades, only a limited number of them, namely ten, have been granted FDA approval [61]. Moreover, the clinical efficacy of these approved nanodrugs remains relatively low, with only 14% demonstrating superior outcomes. The suboptimal effectiveness of nanodrugs can be attributed to several factors. Firstly, the poor accumulation of nanoparticles at specific lesion sites poses a significant challenge. Additionally, the pharmacokinetics of nanodrugs may be compromised, further impacting their therapeutic efficacy. It is worth noting that various parameters of nanoparticles, including their material composition, geometry, surface chemistry, charge, and mechanical properties, can influence their interaction with biological barriers. Consequently, these parameters may also impact the extravasation process, which plays a critical role in determining the nanoparticles' efficacy.

4.1.4 Safety Concerns Associated with Nanoparticles

While the safety of lipids and polymers has been extensively studied, there is still a lack of sufficient exploration regarding the in vivo biodistribution and toxicity of other synthetic components, such as ligands and coatings, which are frequently used in the construction of multifunctional nanocarriers [62]. The physical characteristics of nanoparticles, particularly their size and shape, significantly impact their biological behavior. Therefore, it is essential to fully understand their absorption, distribution, metabolism, and excretion properties. Additionally, the unintended biological interactions of nanocarriers, the chronic exposure to nonbiodegradable materials, and their enhanced penetration through physiological barriers raise further safety concerns. Most safety investigations of nanocarriers are conducted using cell lines and animal models. However, it is important to recognize that the biological responses in humans may differ, limiting the relevance of safety assessments based solely on animal studies. Therefore, thoroughly addressing the safety evaluation of nanocarriers is an urgent and necessary task.

4.1.5 Regulatory Challenges

The regulatory frameworks for TCM and nanocarriers vary across different countries and regions, which can pose challenges in terms of approval and commercialization. The safety and efficacy of nanocarrier-based TCM formulations need to be thoroughly evaluated and validated to meet regulatory requirements.

4.2 Strategies to Address the Issue of Nanocarrier Application in TCM

To address these challenges, more research is needed to understand the mechanisms of action of TCM formulas and to develop safe and effective nanotechnology-based drugs.

4.2.1 Understanding the Components of TCM and Loading of Active Components

It is essential to have a deep understanding of the various components of traditional Chinese medicine, their interactions, potential synergies, and active components. This understanding will help in selecting the appropriate components to be loaded into the nano drug delivery system, ensuring the safety of nanocarriers in clinical applications and providing guidance for the rational design of nanocarriers.

For the Active Monomer Components of TCM with Clear Efficacy, the Nanotechnologies Can Help Them Achieve Clinical Application or Enhance Clinical Efficacy

Since 1980, Kingkong elemene liposome has been widely studied and applied as an anticancer drug worldwide. Its discovery can be traced back to the last century when scientists in the field of traditional Chinese medicine innovatively proposed the principle of "eliminating tumor and resolving masses to support the healthy qi" for the treatment of malignant tumors. This principle is different from the traditional "attacking with toxins" and "clearing heat and detoxifying" principles [63]. Based on the records of traditional Chinese herbal medicine and years of clinical practice, it was found that nontoxic Chinese herbal medicine, such as *Curcuma* zedoary, radix curcumae, and other *Zingiberaceae* plants, is very effective in the treatment of liver cancer, lung cancer, and gynecological tumors. Finally, effective and nontoxic small molecular anticancer active components—elemene—were isolated from these plants. It was found that elemene has significant inhibitory effects on the growth and proliferation of cancer cells; significantly activates the body's antitumor immune response; reverses tumor cell resistance; synergistically enhances the effect of chemotherapy/radiotherapy; reflects the network pharmacological effect of multiple targets, channels, and ways to play an anticancer role [64]. However, because elemene is not soluble in water, if it is made into a common water injection, it needs to be added with castor oil, Tween-80, and other solubilizers. These excipients have strong irritation, are prone to hemolysis and allergy, and are toxic to the organs like the liver, kidney, and heart. In order to improve the bioavailability and efficacy of elemene, researchers successfully developed Kingkong elemene liposome. In subsequent studies, scientists confirmed the efficacy and safety of Kingkong elemene liposome in the treatment of cancer, which showed that this formulation has significant efficacy in the treatment of various cancers, especially for non-small cell lung cancer, liver cancer, gastric cancer, and other malignant tumors [65]. The discovery history and process of Kingkong elemene liposome are a long and ongoing process, representing scientists' in-depth research and exploration of effective components of traditional Chinese medicine as well as innovative integration with modern nano-pharmaceutical technology.

Currently, numerous active compounds have been isolated from traditional Chinese medicine (TCM) and are considered the main pharmacological components responsible for the medicinal properties of TCM [66]. Some of these compounds have been successfully developed into products and are now serving in clinical practice. For example, the Horny Goat Weed Extract Capsule (icaritin), which was approved for market in 2022, is primarily composed of icariin extracted and purified from the traditional Chinese herb Epimedium [67]. Other TCM active ingredient products similar to icariin include *Panax notoginseng* capsule, olanzapine oral suspension, and artemisinin, etc. [68]. The use of nanoparticle formulations can not only address their defects in physical and chemical properties, inherent toxic side effects, and enhance their efficacy but also facilitate their combination with other drugs, maximizing their synergistic effects.

Explore Multifunctional Active Ingredients in TCM, and Develop New Nano-drug Delivery Systems Based on "Unifications of Medicines and Excipients (UME) [69]"

UME, an essential principle in the pharmacy of TMC, means that parts of the drug in the prescription not only exert their therapeutic benefits but also serve as auxiliary materials. The modern study [70] of the basic pharmacodynamic substances of Chinese medicine and the in vivo processes of drugs have deepened the scientific understanding of the UME, which is that there are extra functions in some medicines, such as altering the solubility and absorption rate of other drugs and reducing the toxicity to create synergies. The diversity of active component structures in TCM also expands the application of active ingredients as materials. Especially for molecules with amphiphilic structures, this feature endows them with a strong ability to self-assemble to form drug delivery systems like micelles, nanoparticles, etc. For example, saponins have an amphiphilic structure, which promotes their application in natural surfactants [71]. They can form micelles in solution to encapsulate poorly water-soluble drugs and enhance their water solubility and bioavailability. Meanwhile, saponins have a cholesterol-like structure [72], which also prompts them to replace cholesterol as one of the materials for synthesizing liposomes to reduce the health problems caused by excessive cholesterol intake. The unique rheological properties of polysaccharides make them potential carrier materials for mucoadhesive drug delivery systems [73]. At the same time, the hydrophilicity and structural characteristics of polysaccharides make them have great potential to combine with hydrophobic drugs to form amphiphilic molecules, which will help to synthesize novel drug delivery systems, such as micelles and nanoparticles. Essential oils [74] and flavonoids [75] have also been excavated for excipients.

There are multiple forms of the carrier in the nano-drug delivery systems, like liposomes, microspheres, gels, microemulsions, microcapsules, nanoparticles, etc. [76]. The components include phospholipids, surfactants, oils and fats, amphiphilic polymers, etc. According to the previous studies, many active ingredients in TCM also exhibit the property of acting as an excipient. They can be applied in different carrier forms to design a drug delivery system with the function of UME on the grounds of their diverse structure, physicochemical properties, and biological activities [77]. Therefore, the excavation of the property of "unification of medicines and excipients" from TCM may help us to search for more active compounds that can be used as materials in drug delivery systems purposefully, prepare more novel carriers, and understand the formation of nano-carries further in the future.

Identify the Active Components of TCM That Have Clear Auxiliary Therapeutic Effects on Existing Clinical Drugs, and Co-deliver the TCM Active Components with Clinical Drugs by Using Nano-carriers to Achieve the Synergistic Effect and Reduce Toxicity in Clinical Treatment

In recent years, the combined use of TCM active ingredients with clinical drugs has gained significant attention in the field of drug discovery and development [78].

Enhanced Therapeutic Efficacy Combining TCM active ingredients with clinical drugs has the potential to enhance therapeutic efficacy through various mechanisms [51, 79]. Synergistic interactions between these substances can lead to improved drug absorption, distribution, metabolism, and excretion, ultimately enhancing the overall pharmacological response. Moreover, the combination therapy approach may target multiple disease pathways simultaneously, offering a comprehensive treatment strategy.

Reduced Toxicity One of the major challenges in drug therapy is the occurrence of adverse effects associated with high drug doses [80, 81]. The combination of TCM active ingredients and clinical drugs has been shown to reduce drug toxicity by enabling dose reduction without compromising efficacy. This reduction in toxicity can be attributed to the complementary actions of TCM active ingredients, which may protect normal cells and tissues from the harmful effects of clinical drugs, thereby improving patient tolerance and quality of life.

Reversal of Drug Resistance The emergence of drug resistance poses a significant obstacle to successful therapeutic outcomes [82, 83]. Combination therapy involving TCM active ingredients has shown promise in circumventing drug resistance mechanisms. These active ingredients can modulate drug transporters, inhibit drug-metabolizing enzymes, and restore cellular signaling pathways, thereby sensitizing drug-resistant cells to the effects of clinical drugs. The inclusion of TCM active ingredients in combination regimens holds great potential for overcoming drug resistance and improving treatment outcomes in patients.

To maximize the synergistic effects of TCM active ingredients and clinical drugs, it is crucial to ensure their efficient delivery to target sites. Nanoparticle drug delivery systems offer a promising solution by providing controlled release, improved stability, enhanced cellular uptake, and targeted delivery. These systems can encapsulate both TCM active ingredients and clinical drugs, allowing for their co-delivery to specific tissues or cells. By precisely controlling the release kinetics and localization of these substances, nanoparticle drug delivery systems can optimize the synergistic interactions, thereby achieving superior therapeutic outcomes. Further research and development in this field will pave the way for the integration of TCM principles into modern medicine, ultimately benefiting patients worldwide.

4.2.2 Preparation of Nanoparticles

The selection of appropriate nanomaterials is crucial. Based on the intended purpose and desired properties, suitable nano-carriers can be prepared using various methods such as nano-emulsion, nano-spray drying, nano-molding, etc. [84]. In addition, explore novel nanocarrier materials and engineering to develop novel nanocarriers with high drug loading capacity, enhanced stability, and improved compatibility with various drugs can also help.

4.2.3 Cross-Communication

It is crucial for the modernization of TCM and the harmonization of TCM and nanomedicine to have a thorough understanding of both disciplines. This includes being familiar with the different concepts and terminologies used in each field. It is worth noting that there is a notable difference between TCM and Western medical doctors in this regard. TCM doctors are often trained in both TCM and Western medicine, while this is not as common in Western countries. Many instances where TCM has been considered unscientific can be attributed to a lack of understanding among non-trained scientists and doctors. Therefore, incorporating a significant portion of TCM training into core medical education can help bridge the cultural and language gaps. In turn, TCM can gradually introduce modern language terminology to describe the scientific basis and analysis of its knowledge. The advancement of communication technology nowadays facilitates cultural exchanges, which should be approached with the goal of mutual understanding.

4.2.4 Clinical Translation of Nano-TCM Products

Clinical translation of nano-TCM products faces similar challenges to the clinical translation of nanomedicines and TCM products separately. Recent reviews have highlighted the critical determinants that need to be carefully addressed in order to translate nano-TCM products into clinical practice. These determinants include toxicity, safety, long-term fate, and pharmacokinetics, which require the development of ADME models to understand their precise biodistribution and evolution in the human body. Additionally, nano-TCM products must be produced under good manufacturing practice (GMP) conditions. However, this is challenging because GMP reagents for nanomaterial preparation are not typically part of the drug discovery pipeline of pharmaceutical companies and chemical producers, and there is high variability in proposed TCM products. Therefore, the development of pharmacological products will require a strict definition of the characteristics of nano-TCM products, including the dosage amount of the active principle, size, purity, colloidal stability, and the presence of excipients and potential by-products.

5 Conclusion and Future Perspectives

The emergence of nano-delivery systems represents a notable convergence of modern science and technology with traditional medicine, particularly revitalizing the field of traditional Chinese medicine (TCM) with innovative possibilities. Bibliometric analyses indicate a steady rise in global interest in the application of nano-delivery systems within TCM. Significant contributions are being made by researchers from countries, including China, the United States, and India, highlighting the worldwide importance of this field. The adoption of nano-delivery systems offers numerous advantages that enhance both research and practical applications in TCM. Notably, these systems significantly improve the bioavailability and efficacy of active compounds found in traditional Chinese herbs. By encapsulating these bioactive ingredients within nanocarriers, drugs can be delivered more reliably and precisely to targeted tissues or cells, thereby optimizing therapeutic outcomes. Moreover, nano-delivery systems enable the formulation of intricate combinations of TCM ingredients and provide a technological foundation for developing novel herbal medicines. This methodology supports a deeper understanding of the molecular compatibility theory in TCM, promoting the integration of traditional concepts with contemporary scientific principles to enhance predictability and personalize treatment options. Additionally, these systems offer modern technological support for traditional therapeutic practices, such as acupuncture. By integrating drug carriers with acupuncture techniques, we can enhance treatment precision, thereby increasing efficacy and expanding therapeutic options. As nanotechnology continues to progress, the potential for uniting other traditional Chinese practices, such as cupping, moxibustion, and massage, with nanotechnology injects new vitality into the field, creating broader application scenarios and accelerating the modernization of TCM.

Despite significant progress in the research of nano-delivery systems within TCM, several challenges persist. First, the preparation techniques need ongoing improvement to enhance the stability, drug-loading capacity, and production efficiency of nano-delivery systems. Second, the prioritization of drug safety and toxicity assessment, especially in extensive and prolonged utilization, is of utmost importance. In addition, there is a need for additional refinement in the optimization of storage conditions and transportation issues for nano-delivery systems to guarantee the enduring efficacy and safety of drugs. Nevertheless, in terms of prospects, a diverse array of applications for nano-delivery systems within the field of TCM is anticipated. The utilization of nano-delivery systems holds significant promise in facilitating the modernization and internationalization of TCM, thereby bolstering its global recognition and application.

References

1. Bray F, Laversanne M, Sung H, Ferlay J, Siegel R, Soerjomataram I, Jemal A. Global cancer statistics 2022: GLOBOCAN estimates of incidence and mortality worldwide for 36 cancers in 185 countries. CA Cancer J Clin. 2024;74:229–63.
2. Xu MJ, Hu KL, Liu YP, Huang YK, Liu SS, Chen Y, Wang DY, Zhou SL, Zhang Q, Mei N, Lu HP, Li FG, Gao XL, Chen J. Systemic metastasis-targeted nanotherapeutic reinforces tumor surgical resection and chemotherapy. Nat Commun. 2021;12:3187.
3. Anand U, Dey A, Singh Chandel AK, Sanyal R, Mishra A, Kumar Pandey D, De Falco V, Upadhyay A, Kandimalla R, Chaudhary A, Dhanjal JK, Dewanjee S, Vallamkondu J, de la Lastra JMP. Cancer chemotherapy and beyond: current status, drug candidates, associated risks and progress in targeted therapeutics. Genes Dis. 2023;10:1367–401.
4. Juthani R, Punatar S, Mittra I. New light on chemotherapy toxicity and its prevention. BJC Rep. 2024;2:1–6.
5. Liu YQ, Wang XL, He DH, Cheng YX. Protection against chemotherapy- and radiotherapy-induced side effects: a review based on the mechanisms and therapeutic opportunities of phytochemicals. Phytomedicine. 2021;80:153402.
6. Chen SS, Flower A, Ritchie A, Liu JP, Molassiotis A, Yu H, Lewith G. Oral Chinese herbal medicine (CHM) as an adjuvant treatment during chemotherapy for non-small cell lung cancer: a systematic review. Lung Cancer. 2010;68:137–45.
7. Yen Y, So S, Rose M, Saif MW, Chu E, Liu S, Foo A, Jiang Z, Su T, Cheng Y. Phase i/ii study of phy906/capecitabine in advanced hepatocellular carcinoma. Anticancer Res. 2009;29:4083–92.
8. Xue HK, Zhang PQ, Zhang C, Gao YC, Tan JQ. Research progress in the preparation, structural characterization, and biological activities of polysaccharides from traditional Chinese medicine. Int J Biol Macromol. 2024;262:129923.
9. Shukla R, Chanda N, Zambre A, Upendran A, Katti K, Kulkarni RR, Nune SK, Casteel SW, Smith CJ, Vimal J, Boote E, Robertson JD, Kan P, Engelbrecht H, Watkinson LD, Carmack TL, Lever JR, Cutler CS, Caldwell C, Kannan R, Katti KV. Laminin receptor specific therapeutic gold nanoparticles (198AuNP-EGCG) show efficacy in treating prostate cancer. Proc Natl Acad Sci USA. 2012;109:12426–31.
10. Qu JR, Qiu B, Zhang YX, Hu Y, Wang ZX, Guan ZA, Qin YM, Sui TT, Wu F, Li BY, Han W, Peng XZ. The tumor-enriched small molecule gambogic amide suppresses glioma by targeting WDR1-dependent cytoskeleton remodeling. Signal Transduct Target Ther. 2023;8:1–15.
11. Gao SY, Tan HX, Gang J. Inhibition of hepatocellular carcinoma cell proliferation through regulation of the cell cycle, AGE-RAGE, and leptin signaling pathways by a compound formulation comprised of andrographolide, wogonin, and oroxylin a derived from Andrographis Paniculata(Burm.f.) Nees. J Ethnopharmacol. 2024;329:118001.
12. Tan W, Lu JJ, Huang MQ, Li YB, Chen MW, Wu GS, Gong J, Zhong ZF, Xu ZT, Dang YY, Guo JJ, Chen XP, Wang YT. Anti-cancer natural products isolated from Chinese medicinal herbs. Chin Med. 2011;6:27.
13. Esmeeta A, Adhikary S, Dharshnaa V, Swarnamughi P, Ummul Maqsummiya Z, Banerjee A, Pathak S, Duttaroy AK. Plant-derived bioactive compounds in colon cancer treatment: an updated review. Biomed Pharmacother. 2022;153:113384.

14. Zou B, Long Y, Gao R, Liu QZ, Tian XF, Liu B, Zhou Q. Nanodelivery system of traditional Chinese medicine bioactive compounds: application in the treatment of prostate cancer. Phytomedicine. 2024;135:155554.
15. Anwar DM, El-sayed M, Reda A, Fang JY, Khattab SN, Elzoghby AO. Recent advances in herbal combination nanomedicine for cancer: delivery technology and therapeutic outcomes. Expert Opin Drug Deliv. 2021;18:1609–25.
16. Zhou JH, Li M, Liu ZW, Luo W, Han SQ, Xiao F, Tao W, Wu QB, Xie T, Kong N. Unleashing the potential: integrating nano-delivery systems with traditional Chinese medicine. Nanoscale. 2024;16:8791–806.
17. Liu YF, Liang YS, Yuhong J, Xin P, Han JL, Du Y, Yu XR, Zhu RH, Zhang MX, Chen W, Ma YJ. Advances in nanotechnology for enhancing the solubility and bioavailability of poorly soluble drugs. Drug Des Devel Ther. 2024;18:1469–95.
18. Tie SS, Xue MM, Chen YN, Wu Y, Gu SB, Tan MQ. Enhancing stability, bioavailability and nutritional intervention effect of procyanidins using bio-based delivery systems: a review. Int J Biol Macromol. 2025;287:138517.
19. Jayanta JK, Das G, Fraceto LF, Ramos Campos EV, Rodriguez-Torres MDP, Acosta-Torres LS, Diaz-Torres LA, Grillo R, Swamy MK, Sharma S, Habtemariam S, Shin HS. Nano based drug delivery systems: recent developments and future prospects. J Nanobiotechnol. 2018;16:71.
20. Zhou DH. Reflection and prospect of traditional Chinese medicine oncology development. Onco Chin Med. 2019;1:1–5+9.
21. Long YL, Long XM. Clinical classification of anti-tumor traditional Chinese medicine under the guidance of therapeutic principle of TCM. Chin Med Mod Dis Edu Chi. 2011;9:88–9.
22. Cotter TG. Apoptosis and cancer: the genesis of a research field. Nat Rev Cancer. 2009;9:501–7.
23. Lin JM, Wei LH, Chen YQ, Liu XX, Hong ZF, Sferra TJ, Peng J. Pien Tze Huang induced apoptosis in human colon cancer ht-29 cells is associated with regulation of the bcl-2 family and activation of caspase 3. Chin J Integr Med. 2011;17:685–90.
24. Chou SM, Lai WJ, Hong T, Tsai SH, Chen YH, Kao CH, Chu R, Shen TL, Li TK. Involvement of p38 MAPK in the anticancer activity of cultivated cordyceps militaris. Am J Chin Med. 2015;43:1043–57.
25. Tian XW, Li YL, Shen YY, Li QQ, Wang QL, Feng LS. Apoptosis and inhibition of proliferation of cancer cells induced by cordycepin. Oncol Lett. 2015;10:595–9.
26. Li XJ, Wu MX, Kong FM, Yang PY, Zhu JL, Jia YJ. Review of anti-tumor activity and mechanism of Chinese materia medica active ingredients. Chin Tradit Herb Drug. 2020;51:2587–92.
27. Luo X, Peng JM, Su LD, Wang DY, Yu YJ. Fangchinoline inhibits the proliferation of spc-a-1 lung cancer cells by blocking cell cycle progression. Exp Ther Med. 2016;11:613–8.
28. Greening DW, Gopal SSK, Mathias RA, Liu L, Sheng JY, Zhu HJ, Simpson RJ. Emerging roles of exosomes during epithelial–mesenchymal transition and cancer progression. Semin Cell Dev Biol. 2015;40:60–71.
29. Aling S, Lin W, Chen YQ, Liu LY, Chen HW, Zhuang QC, Lin JM, Sferra TJ, Peng J. Pien Tze Huang inhibits metastasis of human colorectal carcinoma cells via modulation of TGF-β1/ZEB/mir-200 signaling network. Int J Oncol. 2015;46:685–90.
30. Wang ZM, Song N, Ren YL. Anti-proliferative and cytoskeleton-disruptive effects of icariin on hepg2 cells. Mol Med Rep. 2015;12:6815–20.
31. Xie CM, Chan WY, Yu SD, Zhao J, Cheng CHK. Bufalin induces autophagy-mediated cell death in human colon cancer cells through reactive oxygen species generation and JNK activation. Free Radic Biol Med. 2011;51:1365–75.
32. Zhang Y, Song L, Cai L, Wei RX, Hu H, Jin W. Effects of baicalein on apoptosis, cell cycle arrest, migration and invasion of osteosarcoma cells. Food Chem Toxicol. 2013;53:325–33.
33. Du J, Wang TT, Li YC, Zhou Y, Wang X, Yu XX, Ren XY, A YH, Wu Y, Sun WD, Fan WM, Zhu QJ, Wang Y, Tong XG. DHA inhibits proliferation and induces ferroptosis of leukemia cells through autophagy dependent degradation of ferritin. Free Radic Biol Med. 2019;131:356–69.
34. Wang M, Li SS, Wang YL, Cheng H, Su JJ, Li QL. Gambogenic acid induces ferroptosis in melanoma cells undergoing epithelial-to-mesenchymal transition. Toxicol Appl Pharmacol. 2020;401:115110.

35. Cheng Y, Guo QY, Yu SY, Huang N, Qin K, Zhao J, Xiong HH. Inhibitory effect of matrine on proliferation of human cervical cancer hela cells cultured in vitro. Herald Med. 2019;38:1255–9.
36. Guo HJ, Zhang WJ, Wang JX, Zhao GN, Wang YH, Zhu BM, Dong PX, Watari H, Wang BJ, Li W, tigyi G, Yue JM. Cryptotanshinone inhibits ovarian tumor growth and metastasis by degrading C-Myc and attenuating the FAK signaling pathway. Front Cell Dev Biol. 2022;10:959518.
37. Wang HQ, Xiao XJ, Li ZZ, Luo SQ, Hu L, Yi H, Xiang RH, Zhu Y, Wang YP, Zhu L, Xiao L, Dai CW, Aziz A, Yuan LL, Cui YJ, Li RJ, Gong FJ, Liu XF, Liang L, Peng HL, Zhou H, Liu J. Polyphyllin VII, A novel moesin inhibitor, suppresses cell growth and overcomes bortezomib resistance in multiple myeloma. Cancer Lett. 2022;537:215647.
38. Chen JR, Jia XH, Wang H, Yi YJ, Wang JY, Li YJ. Timosaponin A-III reverses multi-drug resistance in human chronic myelogenous leukemia K562/ADM cells via downregulation of MDR1 and MRP1 expression by inhibiting PI3K/Akt signaling pathway. Int J Oncol. 2016;48:2063–70.
39. Yao CC, Jiang J, Tu YR, Ye SF, Du HX, Zhang Y. β-elemene reverses the drug resistance of A549/DDP lung cancer cells by activating intracellular redox system, decreasing mitochondrial membrane potential and P-glycoprotein expression, and inducing apoptosis. Thorac Cancer. 2014;5:304–12.
40. Yang B, Yu GH, Li MY, Gu HM, Chen YP, Feng L, Jia XB. Mechanism of flavonoid components in astragali radix in inhibiting tumor growth and immunoregulation in C57BL/6 tumor bearing mice based on "invigorating Qi for consolidation of exterior.". China J Chin Mater Med. 2019;44:5184–90.
41. Zhu G, Wang HN, Wang TC, Shi FS. Ginsenoside Rg1 attenuates the inflammatory response in DSS-induced mice colitis. Int Immunopharmacol. 2017;50:1–5.
42. Zhuang Y, Yan JJ, Zhu W, Chen L, Liang DH, Xu XJ. Can the aggregation be a new approach for understanding the mechanism of traditional Chinese medicine? J Ethnopharmacol. 2008;117:378–84.
43. Zhang YL, Wang YL, Y K, Li HY, Zhang XY, Essola JM, Ding CC, Chang KX, Qing GC, Zhang FX, Tan Y, Peng TT, Wang X, Jiang M, Liang XJ, Hua Q. Traditional Chinese medicine formulae QY305 reducing cutaneous adverse reaction and diarrhea by its nanostructure. Adv Sci. 2024;11:2306140.
44. Zhu JY, Huang YL, Zhang JQ, Feng Y, Shen L. Formulation, preparation and evaluation of nanostructured lipid carrier containing naringin and coix seed oil for anti-tumor application based on "unification of medicines and excipients". Drug Des Devel Ther. 2020;11:1481–91.
45. Yang SQ, Lan JS, Li Z, Li M, Wu Y, Sun LY, Zhang T, Ding Y. Bufonis venenum extract loaded novel cholesterol-free liposome for the treatment of hepatocellular carcinoma. Front Pharmacol. 2024;25:1486742.
46. Jiang M, Chen LX, Hong C, Zhang LJ, Lu L, Zhao WJ, Ding Y, Zhang T. A novel polyphyllin I-based liposome delivery system sensitizes hepatic carcinoma to doxorubicin via cholesterol modulation. J Drug Deliv Sci Technol. 2022;78:103925.
47. Lan JS, Liu L, Zeng RF, Qin YH, Hou JW, Xie SS, Yue S, Yang J, Ho RY, Ding Y, Zhang T. Tumor-specific carrier-free nanodrugs with GSH depletion and enhanced ROS generation for endogenous synergistic anti-tumor by a chemotherapy-photodynamic therapy. Chem Eng J. 2021;407:127212.
48. Lan JS, Zeng Rf, Li Z, Yang XG, Liu L, Chen LX, Sun LY, Shen Y, Zhang T, Ding Y. Biomimetic nanomodulators with synergism of photothermal therapy and vessel normalization for boosting potent anticancer immunity. Adv Mater. 2024;36:2408511.
49. Pandey S, Mewada A, Thakur M, Shau R, oza G, Sharon M. Biogenic gold nanoparticles as fotillas to fire berberine hydrochloride using folic acid as molecular road map. Mater Sci Eng C Mater Biol Appl. 2013;33:3716–22.
50. Li Z, Lan JS, Liu L, Wang Y, Chen LX, Zeng RF, Gu DH, Hu RL, Zhang T, Ding Y. Versatile thermo-sensitive liposomes with HSP inhibition and anti-inflammation for synergistic chemo-photothermal to inhibit breast cancer metastasis. Int J Pharm. 2024;25:124583.

51. Zhang LJ, Zhang SG, Jiang M, Lu L, Ding Y, Ma NH, Zhao Y, Xuchen SH, Zhang NL. Novel timosaponin AIII-based multifunctional liposomal delivery system for synergistic therapy against hepatocellular carcinoma cancer. Int J Nanomedicine. 2021;16:5531–50.
52. Li J, Wang XH, Guo YM, Zhang YJ, Zhu A, Zeng W, Di LQ, Wang RN. Ginsenoside Rg3-engineered exosomes as effective delivery platform for potentiated chemotherapy and photo-immunotherapy of glioblastoma. Chem Eng J. 2023;471:144692.
53. Lan JS, Chen LX, Li Z, Liu L, Zeng RF, He YT, Shen Y, Zhang T, Ding Y. Multifunctional biomimetic liposomes with improved tumor-targeting for TNBC treatment by combination of chemotherapy, antiangiogenesis and immunotherapy. Adv Healthc Mater. 2024;13:2400046.
54. Xu SW, Wusiman A, Liu ZG, Gu PF, Ni HY, Zhang Y, Hu YL Liu JG, Wu Y, Wang DY. PH-responsive astragalus polysaccharides-loaded poly(lactic-co-glycolic acid) nanoparticles and their in vitro immunogenicity. Int J Biol Macromol. 2019;125:865–75.
55. Liu X, An CY, Jin P, Liu XS, Wang LH. Protective effects of cationic bovine serum albumin-conjugated PEGylated tanshinone IIA nanoparticles on cerebral ischemia. Biomaterials. 2013;34:817–30.
56. Wang BJ, Zhang W, Zhou XD, Liu MN, Hou XY, Cheng ZT, Chen DQ. Development of dual-targeted nano-dandelion based on an oligomeric hyaluronic acid polymer targeting tumor-associated macrophages for combination therapy of non-small cell lung cancer. Drug Deliv. 2019;26:1265–79.
57. Guo XY, Wu GJ, Wang H, Chen LK. Pep-1&borneol–Bifunctionalized carmustine-loaded micelles enhance anti-glioma efficacy through tumor-targeting and BBB-penetrating. J Pharm Sci. 2019;108:1726–35.
58. Diril M, Özdokur KV, Yıldırım Y, Karasulu HY. In vitro evaluation and in vivo efficacy studies of a liposomal doxorubicin-loaded glycyrretinic acid formulation for the treatment of hepatocellular carcinoma. Pharm Dev Technol. 2023;28:915–27.
59. Zheng YH, Wang Y, Xia MY, Gao Y, Zhang L, Song YN, Zhang C. The combination of nanotechnology and traditional Chinese medicine (TCM) inspires the modernization of TCM: review on nanotechnology in TCM-based drug delivery systems. Drug Deliv Transl Res. 2022;12:1306–25.
60. Zeng ML, Guo DD, Guillermo Fernández-Varo G, Zhang X, Fu SQ, Ju SJ, Yang HL, Yang XL, Liu XF, Wang YC, Zeng Y, Casals G, Casals E. The integration of nanomedicine with traditional Chinese medicine: drug delivery of natural products and other opportunities. Mol Pharm. 2022;20:886–904.
61. Halwani AA. Development of pharmaceutical nanomedicines: from the bench to the market. Pharmaceutics. 2022;14:106.
62. Liu XS, Tang I, Wainberg ZA, Meng H. Safety considerations of cancer nanomedicine—a key step toward translation. Small. 2020;16:2000673.
63. Xie T, Xu K, Thoughts and practices on integrating translational medicine with traditional Chinese medicine—take the industrialization and application of jin gang elemene liposomes targeted antitumor natural medicine as an example. In: The fourth Chinese traditional medicine scientific and technological achievements forum. 2014; 17–21.
64. Zhai BT, Zhang NN, Han XM, Li QJ, Zhang MM, Chen XY, Li GH, Zhang RN, Chen P, Wang WG, Li CX, Xiang Y, Liu SP, Duan T, Lou JS, Xie T, Sui XB. Molecular targets of β-elemene, a herbal extract used in traditional Chinese medicine, and its potential role in cancer therapy: a review. Biomed Pharmacother. 2019;114:108812.
65. Xie T, Li CL, Wang SL, Zeng ZW, Chen HB, Chai KQ, Xu K, Peng XQ, Peng ZZ, Liu Y. Kingkong elemene liposome: an original drug with network pharmacological anti-cancer effects. Basic Clin Pharmacol Toxicol. 2014;115:9–9.
66. Gu XJ, Hao DC, Xiao PG. Research progress of Chinese herbal medicine compounds and their bioactivities: fruitful 2020. Chin Herb Med. 2022;14:171–86.
67. Zhang CJ, Sui XX, Jiang YD, Wang XY, Wang SM. Antitumor effects of icaritin and the molecular mechanisms. Discov Med. 2020;29:5–16.

68. Luo H, Tengvong C, Chen HB, Gao Y, Lyu P, Qiu L, Zhao MM, Liu Q, Cheng ZH, Yao JF, Gao CF, Wei JC, Ung CO, Wang SP, Zhong ZF, Wang YT. Naturally occurring anti-cancer compounds: shining from Chinese herbal medicine. Chin. Med. (London, U. K.). 2019;14:48.
69. Feng MF, Dai XX, Yang CT, Zhang YY, Tian YT, Qu QS, Sheng M, Li ZX, Peng XH, Cen S, Shi XY. Unification of medicines and excipients: the roles of natural excipients for promoting drug delivery. Expert Opin Drug Deliv. 2023;20:597–620.
70. Hu HQ, Cheng GW, Wu YF, Zhang YT, Wu Y, Yang YQ, Yang M, Wu ZF. Conversion of traditional Chinese medicine (TCM) into nanomedicine:application of theory of unification of medicines and excipients. China J Chin Mater Med. 2023;48:1800–7.
71. Liao YY, Li ZX, Zhou Q, Sheng MK, Qu QS, Shi YS, Yang JQ, Lv LJ, Dai XX, Shi XY. Saponin surfactants used in drug delivery systems: a new application for natural medicine components. Int J Pharm. 2021;603:120709.
72. Hong C, Liang LM, Xia JX, Zhu Y, Guo YZ, Wang A, Lu CY, Ren HW, Chen C, Li SY, Wang D, Zhan HX, Wang JX. One stone four birds: a novel liposomal delivery system multi-functionalized with ginsenoside Rh2 for tumor targeting therapy. Nano Micro Lett. 2020;12:129.
73. Luo W, Bai LY, Zhang J, Li ZW, Liu YN, Tang XY, Xia PP, Xu MX, Shi A, Liu X, Zhang DJ, Yu P. Polysaccharides-based nanocarriers enhance the anti-inflammatory effect of curcumin. Carbohydr Polym. 2023;311:12078.
74. Thakur D, Kaur G, Puri A, Nanda R. Therapeutic potential of essential oil-based microemulsions: reviewing state-of-the-art. Curr Drug Deliv. 2021;18:1218–33.
75. Petrisor G, Motelica L, Cracium LN, Operea OC, Ficai D, Ficai A. Melissa officinalis: composition, pharmacological effects and derived release systems—a review. Int J Mol Sci. 2022;23:3591.
76. Bhatia R, Sharma A, Narang RK, Rawal RK. Recent nanocarrier approaches for targeted drug delivery in cancer therapy. Curr Mol Pharmacol. 2021;14:350–66.
77. Li QD, Chen XR, Lin W, Guo XF, Ma Y. Application of a novel multicomponent nanoemulsion to tumor therapy based on the theory of "unification of drugs and excipients". Pharm Dev Technol. 2023;28:351–62.
78. Li B, Shao H, Gao L, et al. Nano-drug co-delivery system of natural active ingredients and chemotherapy drugs for cancer treatment: a review. Drug Deliv. 2022;29:2130–61.
79. Shen CS, Yang FZ, Xiang YY, Chen GY, Ji JJ, Zhang ZG. Study on the synergistic effect and mechanism of traditional Chinese medicine active compound galangin and gefitinib on non-small cell lung cancer. J Nanjing Univ Tradit Chin Med. 2021;37:72–6.
80. Zhang XY, Qiu H, Li CS, Cai PP, Qi FH. The positive role of traditional Chinese medicine as an adjunctive therapy for cancer. Biosci Trends. 2021;15:283–98.
81. Sun YK, Zhang YF, Xie L, Rong F, Zhu XY, Xie J, Zhou H, Xu T. Progress in the treatment of drug-induced liver injury with natural products. Pharmacol Res. 2022;183:106361.
82. Wei GQ, Wang Y, Yang G, Wang Y, Ju R. Recent progress in nanomedicine for enhanced cancer chemotherapy. Theranostics. 2021;11:6370–92.
83. Sang SJ, Jiang JJ, Gong YB. Molecular mechanism and research progress of traditional Chinese medicine combined with EGFR-TKIs in the treatment of NSCLC drug resistance. Jiangxi J Tradit Chin Med. 2023;54:80–4.
84. Zhang YY, Ai LS, Gong YS, Jin YX. Preparation and usage of nanomaterials in biomedicine. Biotechnol Bioeng. 2023;120:2777–92.

FSC
www.fsc.org
MIX
Papier aus verantwortungsvollen Quellen
Paper from responsible sources
FSC® C105338

If you have any concerns about our products,
you can contact us on
ProductSafety@springernature.com

In case Publisher is established outside the EU,
the EU authorized representative is:
Springer Nature Customer Service Center GmbH
Europaplatz 3, 69115 Heidelberg, Germany

Printed by Libri Plureos GmbH
in Hamburg, Germany